Porous Membranes

Scrivener Publishing
100 Cummings Center, Suite 541J
Beverly, MA 01915-6106

Publishers at Scrivener
Martin Scrivener (martin@scrivenerpublishing.com)
Phillip Carmical (pcarmical@scrivenerpublishing.com)

Porous Membranes

Breakthroughs in Manufacturing and Applications

Edited by

Annarosa Gugliuzza

Research Institute on Membrane Technology, National Research Council of Italy

and

Wanqin Jin

Department of Chemical Engineering, Nanjing Tech, Jiangsu, University, China

Scrivener Publishing

WILEY

This edition first published 2025 by John Wiley & Sons, Inc., 111 River Street, Hoboken, NJ 07030, USA and Scrivener Publishing LLC, 100 Cummings Center, Suite 541J, Beverly, MA 01915, USA

For more information about Scrivener publications please visit www.scrivenerpublishing.com.

Wiley Global Headquarters
111 River Street, Hoboken, NJ 07030, USA

For details of our global editorial offices, customer services, and more information about Wiley products visit us at www.wiley.com.

Library of Congress Cataloging-in-Publication Data

ISBN 978-1-394-30345-8

Front cover images courtesy of Annarosa Gugliuzza
Cover design by Russell Richardson

Set in size of 11pt and Minion Pro by Manila Typesetting Company, Makati, Philippines

Printed in the USA

10 9 8 7 6 5 4 3 2 1

To my beloved parents

Contents

Preface

Membrane technology is envisioned as a practical pathway for implementing sustainable solutions to environmental and energy challenges, natural resource management, and forward-looking tools for social and economic policy planning. These solutions are not only suitable for community and industry development but also contribute to environmental health, thereby benefiting end-users. Membrane processes are recognized as pivotal in achieving specific objectives across various fields, engaging companies, research institutions, academic organizations, citizens, and public administrations. Implementing a membrane process can serve as the first step toward innovative problem-solving, which is increasingly vital in areas facing emergencies.

A new vision for completing successful projects using membrane technology must, however, include a clear understanding of how to design an archetype of membranes. Morphological and chemical features are essential for creating sophisticated architectures, as they play a critical role in controlling mechanisms and host-guest interfacial forces that facilitate or hinder the separation of molecular species.

One of the major characteristics of membranes is their porosity, including the concept of pore size. Membrane pores are typically regarded as micro- to sub-nanometer channels that allow free pathways for molecular species in liquid or gas states. Parameters such as size, density, distribution, distance, length, and tortuosity must be carefully considered to associate a discriminative ability with the pores. It is crucial to differentiate between geometrical and effective pores, larger and mean pore sizes, open and dead-end channels, and various pore shapes such as round, conical, and funnel-shaped. Precisely controlled pore morphology can enhance desired separation mechanisms, amplify process productivity and efficiency, and address critical events effectively both at the surface and within the bulk of membranes.

Reasonably, pores become key elements in reducing resistance to mass transfer, increasing heat flow, supporting suspended nanofilms, influencing the mechanical resistance of polymeric films, spreading or preventing wetting, assisting adsorption, or countering foulant adhesion. However, pores are not merely passive morphological features; they can be transformed into dynamic and intelligent gates in polymeric films.

This book uses accessible language to guide readers through the world of porous membranes, offering a narrative exploration of innovative strategies for fabricating and characterizing free pathways in interfaces with high levels of organization, structure, and dynamics. These pathways exhibit properties related to wetting, separation, adsorption, transport, sensing, and biocatalysis.

The book reveals how porous membranes with tailored architectures and specific properties can be derived from traditional materials and flexibly applied in fields such as sustainable natural resource management, environmental remediation, food processing, waste treatment, advanced fuel cells, and the recovery of valuable products. It also explores the potential of novel membrane types for cleaner purification processes through numerous case studies, including water desalination, ion filtration, and the removal of pollutants such as pesticides, bacteria, pharmaceuticals, and heavy metals from aqueous streams, as well as hydrogen purification and CO_2 capture.

This book covers a choice of topics and split into three sections, each dealing with aspects related to the role of pores in membranes and related fabrication, characterization and application.

Part I consists of two introductive chapters. Chapter 1 (Gugliuzza) is an introduction to basic pore concepts and transport mechanisms needed to drive a wide set of membrane operations. Chapter 2 (Gordano) examines from more traditional to advanced methods to estimate membrane porosity within a range of a few micron to nanometer values.

Part II comprises four chapters focused on achieving greener and more sustainable approaches for fabricating membranes with highly defined and dynamic pores. Chapter 3 (Liu) presents advanced grafting-based procedures to create responsive gates in membranes. The pore size can be adjusted in response to external stimuli, which subsequently affects selectivity. Size sieving and affinity effects are highlighted as powerful tools for smart pollutant removal. Case studies are examined alongside process-oriented design, stability, and scalable membrane production. Chapter 4 (Díaz) explores the use of click chemistry as a viable method for preparing anion exchange membranes with customizable functionalities for fuel cell applications. A straightforward and broadly applicable strategy for all functional groups is proposed, emphasizing the cost-effectiveness and scalability of the synthesis. Chapter 5 (Cardea) discusses the application of supercritical CO2 phase separation to generate porous membranes in a solvent-free manner, aligning with green and circular strategies. The effects of process parameters on membrane porosity are analyzed, and case studies are presented in specific application areas, including filtration, biomedical, pharmaceutical, and electronics. Chapter 6 (Gugliuzza) focuses on the fabrication of highly defined porous membranes for membrane contactor processes. A range of techniques is reviewed, from traditional phase inversion methods to innovative procedures capable of achieving lithographic precision in polymeric architectures. The chapter delves into the technological fundamentals of advanced membrane science, such as membrane contactors, while examining top-down and bottom-up fabrication approaches. The advantages and limitations of each process are analyzed, and structure-transport relationships are discussed, detailing the influence of individual morphological parameters on final membrane properties and transport performance.

Part III consists of four chapters dedicated to advances in separation membranes based on porous materials with a special focus on bio-purification, environmental remediation, recovery of value-added products, water desalination and gas separation. Chapter 7 (Cui) explores membrane bioreactor technology, which combines biological degradation with the complete physical retention of bacterial flocs and suspended particles. An overview of membrane fabrication with pore sizes ranging from 0.002 to 0.05 μm is provided, along with a classification of membrane bioreactors and their commercialization at scale. Chapter 8 (Donato) focuses on porous imprinted membranes for the recovery of target compounds and environmental remediation. After introducing the fundamental concepts and mechanisms of molecular imprinting for selective separations, the chapter examines the potential applications of molecularly imprinted membranes in fields such as agro-food, bioactive molecule recovery, water treatment, hazardous compound removal, and ion filtration. Chapter 9 (Gugliuzza) investigates the role of few-layered materials in porous membranes for water desalination. The discussion covers ion filtration and membrane distillation, emphasizing the creation of interlayered nanochannels capable of distinguishing between mono- and divalent ions. Additionally, the confinement of defective few-layer materials within porous membranes is analyzed for their ability to enhance transport properties. The chapter highlights successful water desalination processes for freshwater production and the recovery of high-quality salts through sorption-desorption mechanisms. Chapter 10 (Jin) discusses how gas flux is maximized through sub-nanometer channels in two-dimensional material membranes due to their atom-level thickness. In-plane nanopores and sub-nanometer interlayer channels in monolayer, laminar, and nanosheet-based mixed-matrix membranes are examined within the context of gas separation. The chapter covers fabrication, testing with various gas pairs for hydrogen purification, CO2 capture, and the purification of gaseous streams, all based on molecular sieving mechanisms. Future perspectives are explored, with a particular focus on graphene, its derivatives, and graphene oxide.

We are pleased to have edited this book, which documents advances in breakthrough manufacturing techniques for the fabrication of solvent-free, high-definition porous membranes and explores their potential applications in the efficient and sustainable processing of waste, water, gases, energy sources, food, and bio-products.

We believe this book can foster an interdisciplinary exchange of ideas and actions, contributing to the innovative design of high-performing porous membranes that provide rapid and practical solutions to critical global challenges. We hope you enjoy reading this work and find it valuable for your research endeavors.

Annarosa Gugliuzza
Wanqin Jin
March 2025

Acknowledgements

We acknowledge financial grant from "The Italian Ministry of Foreign Affairs and International Cooperation" within the framework of the Great Relevance International Project Italy (MAECI)-China (NSFC) 2018-2020 - New Materials, with particular reference to Two-dimensional Systems and Graphene (2DMEMPUR), MAE00691702020-06-26.

A special memory and thanks to Prof. Enrico Drioli who was a pioneer in the development of membrane processes for strategic technology areas, where he has left a permanent mark with his groundbreaking research.

PART I

BASIC CONCEPTS ON POROUS MEMBRANES

1

Porous Membranes: A Brief Introduction to Basics Concepts and Fields of Applications

Annarosa Gugliuzza

Research Institute on Membrane Technology-National Research Council (CNR-ITM), Rende, Italy

Abstract

Pores can be regarded as channels and sub-nanometer free pathways, in which transport can take place through. Porous membranes can be classified by pore size and subsequent capability to discriminate among molecular species according to various mechanisms. Transport mechanisms through porous media can be described according to mathematical equations, while structural elements can be regarded as critical issue for membrane productivity and efficiency. A classification is herein proposed for membranes used in pressure-driven membrane processes, including microfiltration, ultrafiltration, nanofiltration, and reverse osmosis, with a brief outline of dialysis and electrodialysis. Structure-property relationships are critically discussed for other advanced membrane operations, including membrane distillation, membrane crystallization, membrane condensation, and membrane emulsification. Finally, a consideration on free volume compared to free paths on an Å length scale is also provided for membranes worked in gas separation and pervaporation.

Keywords: Porous membranes, transport mechanism, microfiltration, ultrafiltration, nanofiltration, reverse osmosis, membrane contactors, gas separation and pervaporation

Email: a.gugliuzza@itm.cnr.it

Annarosa Gugliuzza and Wanqin Jin (eds.) Porous Membranes: Breakthroughs in Manufacturing and Applications, (3–26) © 2025 Scrivener Publishing LLC

1.1 Introduction

Traditionally, a membrane is regarded as a perm-selective interface enabling mass, energy, charge, and signals transport under predefined driving forces [1]. More recently, the membrane has been regarded as an interactive and/or a dynamic and adaptable film; its permeability and selectivity can be controlled and readdressed under external triggers, if necessary [2]. Undoubtedly, morphological features decide transport and discriminating property of membranes. Pores, defined as free gaps through which molecular diffusion takes place, represent a decisive structural element for the occurrence of specific transport mechanisms. Of course, pore size, shape, and tortuosity generate the morphological environment wherein particles and various molecular species can be separated according to their size, geometry, molecular weight, as well as affinity and condensability [3]. The size and geometry of pores are the result of thermodynamic and kinetic events that control the formation of a membrane during *i*) phase separation or *ii*) imprinting action of pore formers (Figures 1.1a, b). During phase separation, regions poorer in polymer form pores in membranes in a more random way, leading to porous structures with low control of pore geometry and density distribution. When geometrically regular templates are assembled in predefined volumetric spaces, well-shaped and sized pore can be obtained throughout the membrane surface once removed. Sometimes, pores can be simply free gaps generated by the rearrangement of segment polymer chains, or they can consist of the intrinsic porosity of nanofillers, or more simply, tracks and voids shaped at organic/inorganic interfaces (Figure 1.1c). Further, the switching on/off of pores can be generated by self-assembly or stimulated arrangements of materials, enabling changes in size and shape (Figure 1.1d) [4].

A significant morphological aspect is the propagation of the pore length through membranes. Frequently, the length of the diffusion path is higher than the length of a straight capillary. In this case, a tortuosity factor is generated, which can compromise or affect molecular diffusion due to local restrictions in the pathways [5]. The active pore could be not placed on the membrane surface but rather confined within the bulk of the membranes. This stresses how the membrane morphology, as a whole, can affect its efficiency in regions that are part of the skin or sublayer of the film. Different morphological situations can usually occur in a membrane: a) pores connected to the surface and working as free channels; c) pores open on the surface but closed at the other end (dead-end); d) pores closed at both ends; and e) pores with restrictions in the middle path. So, a careful evaluation

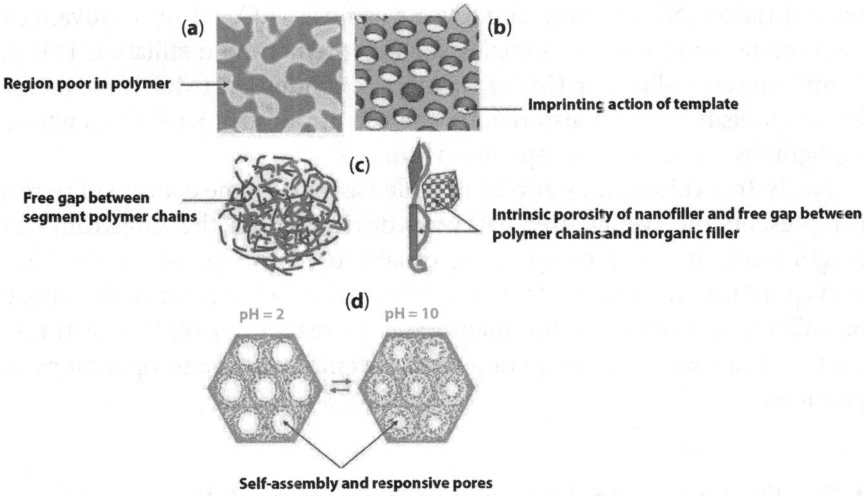

Figure 1.1 Representation of mechanisms for pore generation in membrane: (a) regions poorer in polymer during phase separation; (b) imprinting action of organic/inorganic template; (c) free gaps formed at the organic/organic and organic/inorganic interfaces; and (d) self-assembly of materials with responsive properties.

of the pore size, tortuosity, and free access throughout the membrane is always needed [6]. The number and density of pores in a membrane determine its overall porosity, expressed as the ratio of the volume of air embedded in the matrix to the volume of the solid film. This other morphological parameter is fundamental because it controls the degree of resistance to molecular transport. High porosity is strongly desired to increase the productivity of a membrane process. Highly porous films are, in fact, used as sublayers for tiny selective layers to provide good mechanical resistance without adding resistance to transport. On the other hand, too large a pore size can affect seriously the selectivity of the process.

Another important issue is that pore size, shape, and distribution can affect the final surface properties of membranes, producing singular textures able to amplify rugosity factors and subsequently enhance the repellence properties of the film [7]. Also, pores could be preferable sites for the adhesion of molecular compounds as a first step towards undesired clogging, polarization concentration, and fouling, which are responsible for flux decline and loss of performance.

Pore size resolves which kind of membrane operation can be implemented for each membrane-type. Usually, porous films are indicated for pressure-driven separation processes [8], including microfiltration (MF, 100 nm to 2 μm), ultrafiltration (UF, 2–100 nm), dialysis (2–5 nm), nanofiltration (NF 2–1 nm), and reverse osmosis (RO, < 1 nm). Advanced membrane contactor technologies such as membrane distillation (MD), membrane crystallization (MCr), membrane condensers (MCe), and membrane emulsifiers (ME) also require the use of porous membranes whose topography can decide the operation's success.

Lastly, free volume may also be regarded as an extreme concept of pores; it represents the average free pathways distributed on the Ångström (Å) length scale in dense membranes, usually used for gas separation and pervaporation. It could be frozen or fluctuant as a function of the glassy or rubbery properties of the membrane. Hereafter, an outline of transport mechanisms and membranes for specific membrane operations is provided.

1.2 Overview on Pore Size Concept and Transport Mechanisms

Pore size is the key structural element that determines the type of transport across a membrane through mechanisms that discriminate among molecular size, affinity, and condensability. Passive, active, and assisted transport can be distinguished depending on morphology, chemistry of membranes, and nature of penetrants as well as working conditions. The most common transport mechanisms through porous media include Poiseuille flow, Knudsen diffusion, surface diffusion, capillary condensation, molecular sieving, and solution-diffusion (Figure 1.2).

Based on the discriminant ability of porous media, the mechanisms could be classified into the following categories: a) bulk flow; b) restricted diffusion; c) selective absorption; and d) solution-diffusion. A short mathematical description of each single transport mechanism is herein given.

1.2.1 Poiseuille Flow

Poiseuille flow is a viscous, unselective flux that takes place in macropores with sizes wider than 50 nm. The pore diameter is usually larger than the mean free path of the molecular penetrant. This kind of transport is well described by *eq. 1.1*:

$$J = \frac{\epsilon \eta r^2}{8 \mu R T} p_{av} \tag{1.1}$$

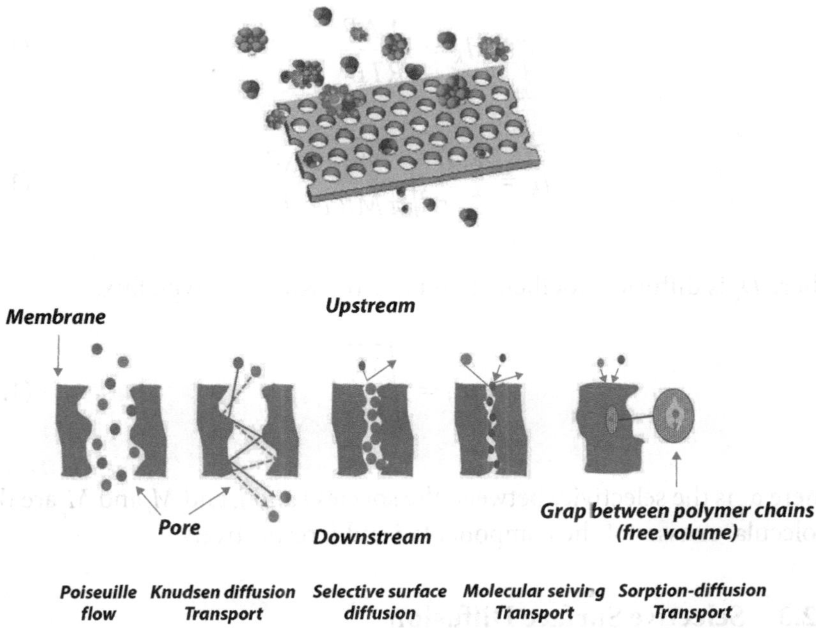

Figure 1.2 Schematic representation of classic transport mechanisms through porous membranes.

where ε is the porosity; μ is the viscosity of the penetrant; η is the shape factor (assumed to be equal to the reciprocal tortuosity of the medium), r is the pore radius; p_{av} is the mean pressure.

When the membrane has a densified skin, the occurrence of Poiseuille flow yields a clear indication of the presence of defects, cracks, or pinholes through the active layer of the film.

1.2.2 Knudsen Diffusion

Microporous membranes with pore sizes in the range of 2–50 nm make Knudsen diffusion predominant due to the collision of the molecules with the pore wall [9]. In this case, the mean free path of the molecules is greater than the pore size, while the selectivity is proportional to the ratio of the inverse square root of the molecular weights as hereafter described:

$$D_k = \frac{d_p}{3}\sqrt{\frac{8RT}{\pi M}} \tag{1.2}$$

$$J_k = \frac{D_k \Delta P}{RTL} \tag{1.3}$$

$$J_k = \frac{d_p}{3}\frac{\theta}{\tau}\sqrt{\frac{8}{\pi MRT}}\frac{\Delta P}{L} \tag{1.4}$$

where D_k is diffusion coefficient and J_k is the Knudsen-type flux.

$$\alpha_{ij} = \sqrt{\frac{M_j}{M_i}} \tag{1.5}$$

where α_{ij} is the selectivity between the species *i* and *j*, and M_j and M_i are the molecular masses of the components *j* and *i*, respectively.

1.2.3 Selective Surface Diffusion

Surface diffusion is a separation mechanism that exploits the different amounts of absorption of molecules—generally at the state of gas/vapor—along the pore walls forming the membrane. Non-absorbable or weakly absorbable penetrants can be separated from absorbable compounds. Generally, more condensable components in pores can exclude the passage of the smaller ones. This kind of diffusion in microporous films is well described by the Dubinin-Radushkevich isotherm:

$$W = W_0 exp\left[-B\left(\frac{T}{\beta}\right)^2 log^2\left(\frac{P_0}{P}\right)\right] \tag{1.6}$$

where *W* is the volume of the species adsorbed in the micropores, W_0 is the maximum of adsorbent per adsorbed mass, *B is* the structural parameter of the microporous membrane, and β is the affinity coefficient of the typical curve [10].

1.2.4 Molecular Sieving

Molecular sieving takes place in pores of 3.0–5.2 Å [11]. Bigger molecules are stopped, while the diffusion of the smallest penetrants is allowed.

The activation energy (E_a) plays a determinant role in the separation process and consists of the amount of energy necessary to promote the diffusion of one molecule through the narrow pathways. In this case, the flux can be expressed as:

$$J_{MS} = \frac{\Delta P}{RTL} D_i(T) exp\left[-\frac{E_{a,MS}}{RT}\right] \tag{1.7}$$

where J_{MS} is the molecular sieving flux and $E_{a,MS}$ is the molecular sieving activation energy.

1.2.5 Solution-Diffusion Transport

This mechanism takes place in densified films where transient free gaps of the order of Å—free volume—allow penetrants to diffuse [12]. This mechanism can be diffusion-dependent and/or solution-dependent according to *eq. 1.8:*

$$P = DS \tag{1.8}$$

where P is the permeability coefficient, D is the diffusion coefficient, and S is the solubility coefficient.

For single gases, the selectivity is given by:

$$\alpha_{ij} = \frac{P_i}{P_j} = \frac{D_i}{D_j}\frac{S_j}{S_i} \tag{1.9}$$

If diffusion is the discriminating factor, the separation depends on the molecular size of the penetrants and subsequently on their diffusivity; in the case of solubility predominance, the species with higher affinity to the matrix will pass faster [13]. This kind of mechanism controls the transport through membranes equipping gas separation (GS), pervaporation (PV), and reverse osmosis (RO) plants.

In these kinds of processes, molecules are adsorbed onto the membrane face coming in contact with the upstream, diffused through the membrane, and are then desorbed on the downstream face of the membrane [13]. In the case of RO operation [14, 15], *eq. 1.8* can be rewritten as follows:

$$J_{Aw} = L(\Delta p - \Delta \pi) \tag{1.10}$$

where J_{Aw} is the water flux, Δp is the transmembrane pressure, $\Delta\pi$ is the difference in osmotic pressure between feed and permeate, and L is a constant related to membrane physical parameters described as:

$$L = \frac{DSV}{RTl} \tag{1.11}$$

where D is the water diffusion coefficient, S is the water solubility coefficient, V is the molar volume of water, R is the ideal gas constant, T is the room temperature, and l is the membrane thickness.

The osmotic pressure plays a key role in the overall RO process because it is necessary to cause water to leave the saline solution and move towards pure water (permeate) through the membrane. In the case of complete dissociation of salt ions, the osmotic pressure is termed as:

$$\pi = CRT \tag{1.12}$$

where π is the osmotic pressure, C is the concentration of salt ions expressed as the ratio between number of ions per gram of water and specific volume of water, and T is the solution temperature. While the diffusion of water through the membrane depends on the pressure, the flux of salt is concentration dependent:

$$J_s = B\left(C_f - C_p\right) \tag{1.13}$$

where J_s is the flux of salt through the membrane, B is the permeability constant of salt, and C_f and C_p are the concentration of salt at feed and permeate side, respectively. The constant B can be described as:

$$B = \frac{D_s K_s}{l} \tag{1.14}$$

where D_s is the diffusion coefficient of salt, K_s is the partition coefficient of salt, and l is the membrane thickness.

The salt rejection can be calculated from:

$$R = \left(1 - \frac{C_{permeate}}{C_{feed}}\right) \cdot 100\% \tag{1.15}$$

It is relevant to observe the relationship between water and salt fluxes

$$\frac{J_{Aw}}{J_s} = \frac{C_{wp}}{C_{sp}} \tag{1.16}$$

where C_{wp} and C_{sp} are the concentration of water and salt in the permeate.

Rearranging *eqs. 1.10* and *1.13*, the rejection can be described as follows:

$$R = \left(\frac{\frac{L}{B}(\Delta P - \Delta\pi)}{1 + \frac{L}{B}(\Delta P - \Delta\pi)} \right) \cdot 100\% \tag{1.17}$$

In this way, it becomes evident how the salt selectivity can be predicted from the evaluation of experimental conditions and membrane properties.

1.2.6 Mixed Transport Mechanisms

Membranes exhibit somewhat complex structures compared to simplified theoretical models, so multiple transport mechanisms frequently occur during a specific separation. Wider and sub-nanometer channels can coexist within the same matrix causing the occurrence of different molecular diffusion types [16]. In the case of gases, molecular sieving and adsorption diffusion could characterize the molecular separation through the same porous membrane at low temperatures. This mixed transport can be described by *eq. 1.18*:

$$J_{\frac{AD}{MS}} = \frac{\Delta P}{RTL} D_i(T) exp - \left[\frac{E_{aMS} - E_{a,ADS}}{RT} \right] \tag{1.18}$$

where $J_{AD/MS}$ is the ratio of the adsorption diffusion and molecular sieving fluxes, while $E_{a,MS}$ and $E_{a,ADS}$ are the activated molecular sieving and adsorption diffusion energies.

Mixed molecular sieving and Knudsen diffusion may occur at higher temperatures. In this case, the transport can be described according to *eq. 1.19*:

$$J_{\frac{MS}{K}} = k_0 \frac{d_p}{3} \frac{\theta}{\tau} \sqrt{\frac{8}{\pi MRT}} \frac{\Delta P}{L} exp\left[\frac{E_{a,MS}}{RT}\right] \tag{1.19}$$

An increase in temperature could cause a loss of selectivity shifting the transport toward a Knudsen-type diffusion:

$$\lim_{T\to\infty} \frac{E_{a,MS}}{RT} = 0 \tag{1.20}$$

$$\lim_{T\to\infty} D_i(T) = D_k \tag{1.21}$$

$$J = \frac{D_k \Delta P}{RTL} \tag{1.22}$$

Summarizing, the main mechanisms of transport through porous membranes include the concept of bulk flow, selective absorption, restricted diffusion, and solution-diffusion.

1.2.7 Active and Assisted Transport

In a more modern vision, the membrane is designed to be an interactive interface allowing one to direct, accelerate, or yield on-demand mass, energy, charge, or signal transport according to cooperative mechanisms. In recent years, there has been a larger interest in composite and responsive materials and manufacturing techniques, which allow materials with complementary functions to be allocated in predefined volumetric spaces. Like a piece of a jigsaw, each compound/moiety plays a role thereby leading to new functions and performance not otherwise achievable working individually. This prerogative amplifies the pore action in membranes, increasing the productivity, but at the same time enhancing the discernment ability of the membrane making the separation much more selective.

Today, there is great attention towards the integration of organic and inorganic nanofillers and macromolecules with the intent to assist the transport of mass [17–19], electrical [20, 21], ionic charge [22, 23], as well as to stimulate release [24, 25], local rearrangement [26, 27] self-healing [28, 29], and self-cleaning events [30, 31].

1.3 Porous Membranes for Membrane Processes

Processes that use porous membranes include microfiltration (MF), ultrafiltration (UF), nanofiltration (NF), reverse osmosis (RO), membrane distillation (MD), membrane crystallization (MCr), membrane condensation (MCe), and membrane emulsification (ME). For all these processes, the structural elements of the membrane become key issues [15]. Currently, the market does not provide membranes whose structural features are close to those considered ideal for some processes such as MD, MCr, MCe, and ME, which are part of advanced and sustainable membrane contactor technology [32]. The performance of each process strongly depends on structure-property relationships, so controlled pore size and pore distribution, along with the degree of porosity and thickness, become a cause for high-quality separation. Well-patterned surfaces are envisaged to amplify point by point desired events with great benefits for the productivity and/or selectivity of separation, as well as high pore density is expected to reduce the resistance to transport [33]. Low thicknesses increase productivity but affect the mechanical stability [33] and efficiency of the film [34]. Controlled textures can enhance anti-wetting/anti-adhesion properties so that fouling and/or scaling can be limited [35].

The design of new nanostructured membranes becomes necessary to meet the requirement of developing sustainable membrane technologies

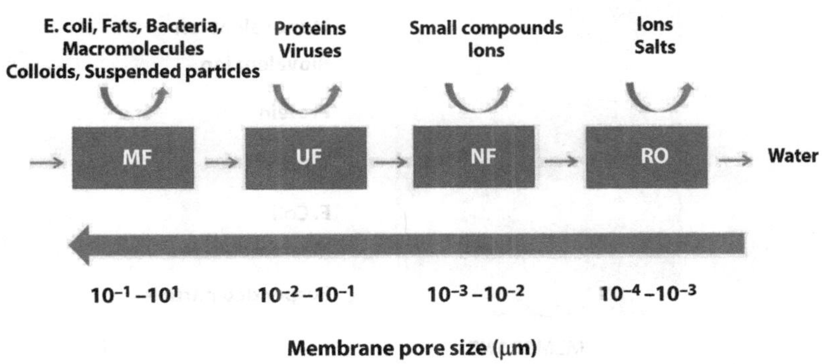

Figure 1.3 Scheme of membrane processes working with porous membranes for selective separations.

on a larger scale. Based on mechanism types, each membrane process requires membranes with specific pore sizes to promote desired molecular separation (Figure 1.3).

MF uses membranes with pore sizes ranging from 10^{-1}–10^{1} μm; UF membranes exhibit pore sizes of 10^{-2}–10^{-1} μm, while RO equips membranes with pores within the range of 10^{-4}–10^{-3} μm. NF membranes fall in the range between UF and RO (10^{-3}–10^{-2} μm). Gas separation (GS) and pervaporation (PV) are in the region of Å where the free volume concept becomes expression of free ways through which the penetrant diffuses [15].

1.3.1 Microfiltration Membranes

MF operations find large applications in the clarification and stabilization of the beverage industry, filtration of fluids, sterilization of food and pharmaceutical industrial chains, as well as the removal of suspended solids (Figure 1.4) [36].

When applying low pressures, macromolecules, suspended particles, and fats can be retained easily from waste. Hydrophilic membranes are usually preferred for the filtration of aqueous solutions, while hydrophobic membranes are proposed for applications (MD/MCr) where water is separated as a vapor rather than a liquid. Also, microporous membranes can be used as support for a liquid membrane solution, which usually consists of an organic phase containing carriers and/or modifiers for the stabilization

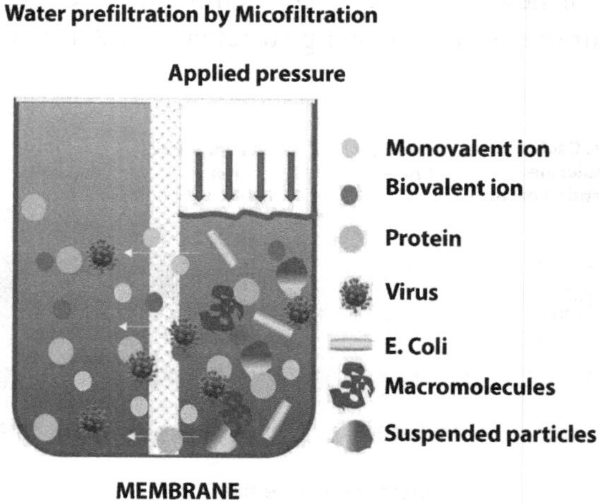

Figure 1.4 Representative scheme of pressure-driven MF process: prefiltration of wastewater.

of the phases coming in contact or the facilitated transport of selected molecular species [37].

The fabrication techniques for MF membranes include traditional phase inversion, stretching, track-etching, and electrospinning. Depending on the selected manufacturing procedures, different materials can be used from traditional polysulfone (PS), poly(ether sulfone) (PES), polyacrylonitrile (PAN), cellulose acetate (CA) to polyethylene (PE), polypropylene (PP), polytetrafluoroethylene (PTFE, Teflon), and poly (vinylidene fluoride) (PVDF).

1.3.2 Ultrafiltration Membranes

UF processes are preferred for the concentration or pretreatment of various chemical streams, food and beverage clarification, emulsion separation, pretreatment of water in integrated membrane processes, etc. [38]. UF is a low-pressure-driven process, which bases the separation on size-exclusion mechanisms, retaining molecular species from 2000 to 500000 Da. However, UF offers poor decontamination for potentially toxic elements making the process more highly susceptible to clogging and fouling (Figure 1.5).

The prerogatives of these membranes are mechanical, thermal, and chemical stability. For this reason, the materials used for membrane fabrication are traditional poly (ether ether ketone) (PEEK), polyacrylonitrile (PAN), polysulfone (PS), poly(ether sulfone), cellulose acetate (CA), polyimide (PI), poly(etherimide) (PEI), poly(vinylidenefluoride) (PVDF). In this case, phase

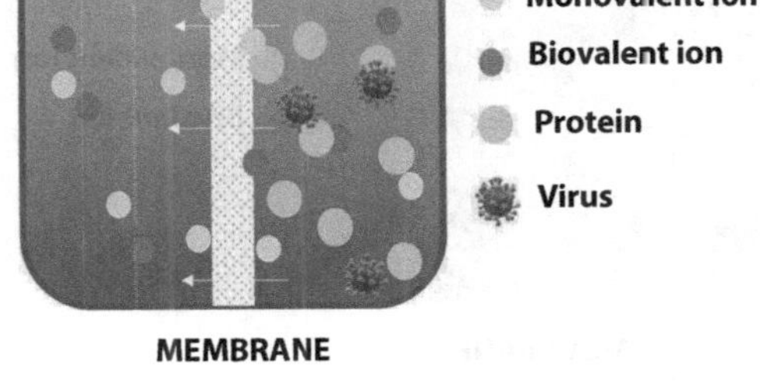

Figure 1.5 Representative scheme of pressure-driven UF: removal of proteins and viruses from waste.

inversion is the preferred manufacturing technique. Recently, zeolites, metal–organic frameworks (MOFs), and graphene oxide (GO), carbon-based, and other composite-based nanomaterials have also been incorporated into polymers to remove harmful elements from water through better controlled fouling phenomena.

Selective diffusion is also the mechanism that controls the separation of smaller molecules from larger compounds or dissolved substances from colloidal particles. This kind of process is well known as dialysis and takes place under a gradient of concentration. Usually, the molecules are neutral, but in the case of electrolytes or charged membranes, Donnan effects take place as a subsequence of the unequal distribution of ions. The Donnan membrane principle inspires various engineered processes and materials to achieve better sustainability [39].

1.3.3 Nanofiltration Membranes

NF membranes find applications in water softening, water concentration, product recovery, and bivalent ion removal (Figure 1.6). This membrane process allows for the removal of impurities, sediments, and hazardous substances like arsenic and bivalent ions [40]. UF membranes usually exhibit an ionic charge, which is generated during the fabrication of thin layers on porous supports by interfacial polycondensation. So, their preferred use is for bivalent ion removal. Typical materials selected for NF

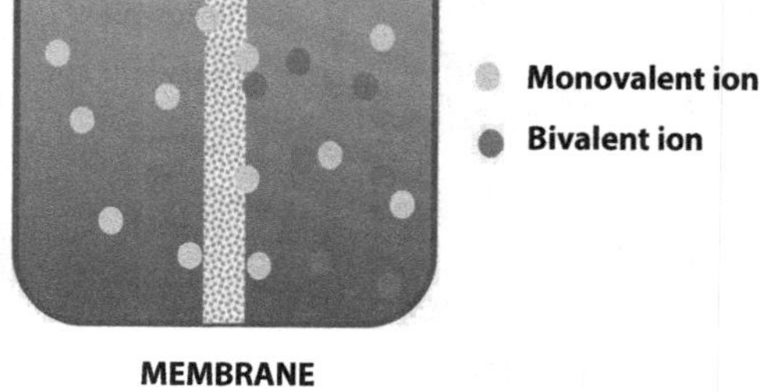

Figure 1.6 Representative scheme of pressure-driven NF: selective removal of bivalent ions.

membranes are polymers with carboxylic and sulfonic acids, which dissociate to become negatively charged.

A particular family of positively or negatively charged membranes is classified as membranes for electrodialysis (ED), dedicated to selectively removing ionic species. Anion exchange membranes exhibit a positive charge and are used to remove cations only, while cationic exchange membranes are negatively charged and allow the diffusion of cations only. Generally, functional polymers with quaternary ammonium salts are used for anionic membranes, while dissociated sulfonic or carboxylic acid moieties are used for cationic membranes. This class of membranes is used for the desalination of brackish water, wastewater treatment, concentration of RO brines, and the treatment of chemical, drug, and beverage streams. Currently, major limitations for the scale up of this technology are represented by fouling, which affects the durability of NF purification, selectivity, and cost-benefit ratio for the environment. New types of highly defined and patterned membranes have demonstrated extreme promise for enhancing productivity-efficiency trade-offs [41].

1.3.4 Reverse Osmosis Membranes

RO plays a major role in water desalination and waste water treatment processes, as well as in the recovery of by-products from various chemical streams (Figure 1.7) [15, 42].

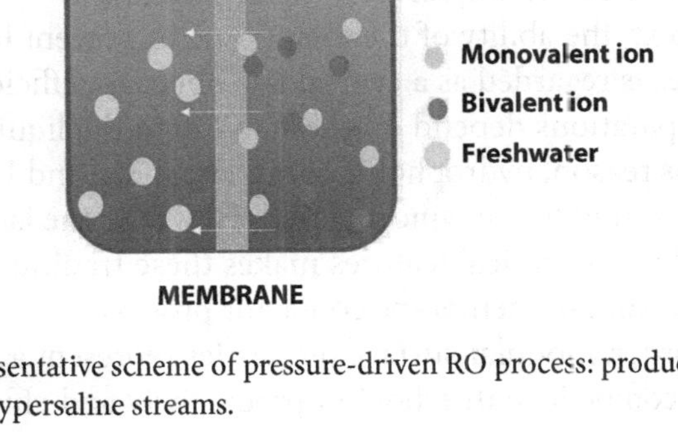

Figure 1.7 Representative scheme of pressure-driven RO process: production of freshwater from hypersaline streams.

In this kind of process, membranes with sub-nanometer pore sizes are chosen for molecular separation according to mechanisms that also depend on the intrinsic structure of the membrane. RO membranes are generally prepared by phase inversion or by interfacial *in situ* polymerization. They can be formed from unique or composite materials and can have asymmetric or layered structures. Among the various polymers, cellulose acetate is the most used, together with aromatic polyamides. In both cases, (bio) fouling and scaling represent a great limitation, causing flux decline over time. The use of disinfectants, anti-scaling agents, and other pre-treatment steps are proposed to contrast the adhesion of pollutants on the membrane surface. Despite that, effectiveness and durability of the process appears to be somewhat compromised with time. So, a good chance could be to optimize materials and manufacturing technologies as a practical route for giving longer stability and efficiency to the membrane process. However, RO membranes continue to dominate the membrane market for water desalination mainly, with a further increase of 8–10% by 2026.

1.3.5 Membrane Contactors Processes

This branch of membrane technology encompasses advanced and promising ecofriendly processes such as MD, MCr, MCe, and ME (Figure 1.8) [43]. Currently, there are no available membranes with structural and chemical features suitable for implementing these membrane operations on an industrial size. So, traditional PVDF, PP, and PTFE membranes are adapted according to the basics of the specific technology.

MD and MCr processes are targeted to recover mainly freshwater and minerals from hypersaline streams in the field of water management [19, 44]. These technologies exploit the ability of water or other volatile compounds to change from a liquid to a vapor state, diffuse through membrane pores as vapors, and then move again to a liquid state on the permeate side (Figure 1.8a). A difference in temperature applied across the membrane is one of the most used driving forces in these processes.

In this context, the ability of the membrane to prevent liquid intrusion inside the pores is regarded as a critical issue because efficiency and durability of the separations depend on the stability of the liquid-vapor-liquid phases. For this reason, hydrophobic commercial MF and UF membranes, i.e., PVDF, PP, and PET, are typically used. However, the lack of particular morphological and chemical features makes these traditional membranes unideal for this kind of membrane contactor process.

Structural order, modular surface properties represent a real bottleneck in achieving a competitive threshold in productivity and efficiency [18, 45].

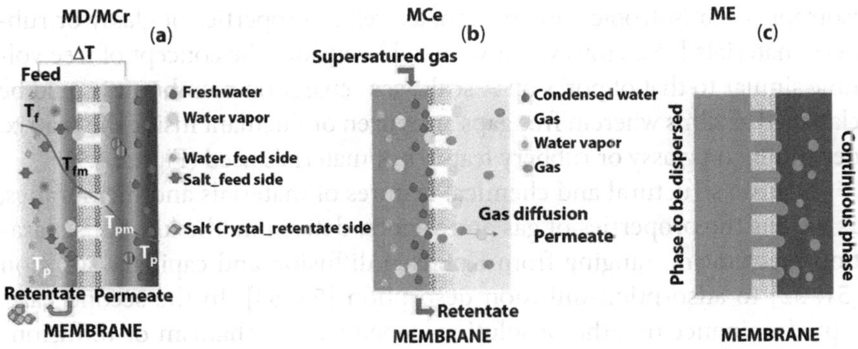

Figure 1.8 Schemes of (a) MD/MCr; (b) MCe; and (c) ME.

High interfacial area, anti-wetting properties, and low thermal conductivity are other prerogatives for making the processes competitive on an industrial size. Omniphobic membranes, as well as double hydrophobic/hydrophilic membranes, have also been proposed to yield faster productivity [46, 47]; however, the design of targetable membranes continues to be an important challenge for the development of membrane MC/MCr on an industrial size.

Hydrophobic membranes with specific structural features are in high demand for MCe [48]. This technology is usually proposed for the dehydration of flue gas (Figure 1.8b). In this case, water and other condensable gases are retained, while non-condensable gases are diffused through membranes and collected on the permeate side.

Without neglecting the role of interfacial forces and wall shear stress, membranes for ME processes also need to be equipped with pores having suitable geometry, size, and distance for the production of stable and uniform micro and nanoemulsions (Figure 1.8c). Tendentially, droplet sizes range between 2 and 10 times the nominal membrane pore diameter [49]. As is known, the choice of materials and surface texture also determines the final hydrophilic or hydrophobic properties of the membrane surfaces, which are crucial for the uniform dispersion of droplets in a continuous phase. Actually, a major limitation to the scale up of this technology is due to the low levels of production of the dispersed phase through membranes for submicron droplets.

1.3.6 Gas Separation and Pervaporation Membranes

Gas separation and pervaporation membranes exhibit dense skin layers, which can be supported by symmetric or asymmetric porous sublayers, isotropic or anisotropic composition, as well as properties of glassy or rubbery materials [15]. However, if we would consider the concept of free volume similar to that of pores on Å scale, gas separation membranes could be classified as films wherein free gaps are frozen or fluctuant inside the matrix, depending on glassy or rubbery features of materials used (Figure 1.9).

Intrinsic structural and chemical features of materials and membranes, as well as the properties of gas and vapors, determine the kind of separation mechanism, ranging from Knudsen diffusion and capillary sorption [51, 52] to adsorption-diffusion-desorption [53, 54]. In the second case, a predominance of either a solution-dependent mechanism or diffusion-dependent mechanism can be prevalent, depending on the combined properties of the membrane and molecular species to be separated (Figures 1.9b, c).

Without going into the discussion of more complex aspects of the processes, we like to highlight the use of these membranes in gas separation

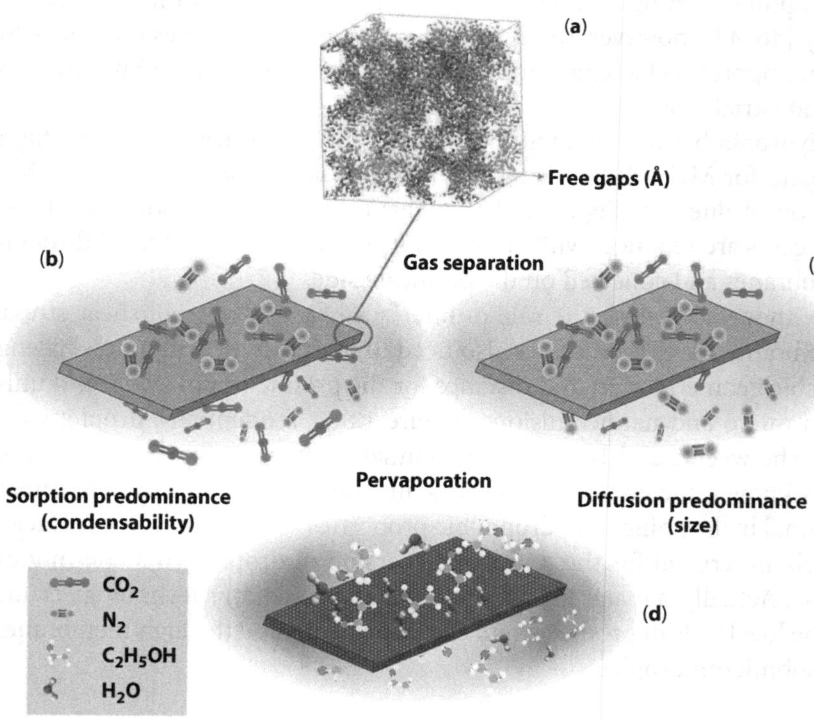

Figure 1.9 Representative schemes of sorption-diffusion-desorption gas and vapor separation: (a) free gaps on Å scale; (b) controlled sorption mechanism in GS; (c) controlled diffusion mechanism in GS; (d) separation of azeotropes by PV. Figure 1.9a is adapted from and reprinted from [50] (2023) with permission from MDPI.

processes of condensable from incondensable gases, air purification, CO_2 capture, biogas processing, dehydration of gaseous and liquid streams, hydrogen recovery, as well as organic/organic separation, azeotrope separation, and the removal of organics from water through a process well-known as pervaporation (PV). These processes exploit different condensability and diffusivity of molecules through dense matrices (Figure 1.9d) [55, 56]. Various manufacturing approaches can be used to realize membranes for gas and vapor separation, including phase separation, multilayered deposition, polycondensation, etc. In each case, the most important aspect is to prevent the formation of holes and defects in the active skin layer to prevent selective properties of membranes from deterioration. On the other hand, tiny defect-free films should be desirable to reduce the resistance to transport, so that highly productive separation can be implemented. With this purpose, macroporous membranes are frequently used as sublayers to yield mechanical resistance without affecting transport resistance. A permeable gutter layer is placed between the sublayer and the selective thin film to make it compatible and defect-free.

Conclusions

This chapter offers an outline of pore concept in membranes, exploring their role in transport, surface properties, and interfacial phenomena across different length scales. Morphological aspects of membranes are discussed concerning transport mechanisms and effects on permeability-selectivity trade-off. Mathematical descriptions of fluxes through membranes are provided along with an examination of the most suitable membranes for specific pressure-driven processes, membrane contactor operations, and gas and vapor separation. Structure-property relationships are assessed concerning with productivity and efficiency targets.

Acknowledgment

We acknowledge the financial grant from 'the Italian Ministry of Foreign Affairs and International Cooperation' within the framework of the Great Relevance International Project Italy (MAECI)- China (NSFC) 2018-2020 - New Materials, with particular reference to Two-dimensional systems and Graphene (2DMEMPUR), MAE00694722021-05-20.

References

1. Strathmann, H., *Introduction to Membrane Science and Technology*, pp. 1–544, Wiley-VCH Verlag GmbH, Weinheim, Germany, 2011.
2. Gugliuzza, A., *Smart Membranes and Sensors*, pp. 1–427, Wiley-Scrivener, Beverly MA, USA, 2014.
3. Bungay, M.C., Transport Principles - Porous Membranes, in: *Synthetic Membranes: Science, Engineering and Applications, NATO ASI Series*, vol. 181, M.C. Bungay, H.K. Lonsdale, M.N. de Pinho (Eds.), pp. 109–153, Springer, Portugal, 1986.
4. Pan, Y., Liu, Y., Yang, S., Zhang, C., Ullah, Z., Recent research progress on the stimuli-responsive smart membrane: A review. *Nanotechnol. Rev.*, 12, 20220538, 2023.
5. Ledesma-Durán, A., Hernández, S.I., Santamaría-Holek, I., Relation between the porosity and tortuosity of a membrane formed by disconnected irregular pores and the spatial diffusion coefficient of the Fick-Jacobs model. *Phys. Rev. E*, 95, 052804, 2017.
6. Roberge, E., Moreau, P., Couallier, E., Abellan., P., Determination of the key structural factors affecting permeability and selectivity of PAN and PES polymeric filtration membranes using 3D FIB/SEM. *J. Membr. Sci.*, 653, 120530, 2022.
7. Wassmann, T., Kreis, S., Behr, M., *et al.*, The influence of surface texture and wettability on initial bacterial adhesion on titanium and zirconium oxide dental implants. *Int. J. Implant Dent.*, 3, 32, 2017.
8. Seader, J.D. and Henley, E.J., *Separation process principles*, John Wiley & Sons, Inc., USA, 1998.
9. Gilron, J. and Soffer, A., Knudsen diffusion in microporous carbon membranes with molecular sieving character. *J. Memb. Sci.*, 209, 339–352, 2002.
10. Shelekhin, A.B., Dixon, A.G., Ma, Y.H., Theory of gas diffusion and permeation in inorganic molecular sieve membranes. *AlChE J.*, 41, 58–67, 1995.
11. Liu, L., Tan, S.J., Horikawa, T., Do, D.D., Nicholson, D., Liu, K., Water adsorption on carbon - A review. *Adv. Coll. Interf. Sci.*, 250, 64–78, 2017.
12. Gugliuzza, A. and Drioli, E., Evaluation of CO_2 permeation through functional assembled mono-layers: Relationships between structure and transport. *Polymer*, 46, 9994–10003, 2005.
13. Baker, R.W. and Wijmans, J.G., The solution-diffusion model: A review. *J. Membr. Sci.*, 107, 1–21, 1995.
14. Bird, R.B., Stewart, W.E., Lightfoot, E.N., *Transport Phenomena*, 2nd, John Wiley & Sons, Inc, New York, USA, 2002.
15. Baker, R.W., *Membrane Technology and Applications*, 2nd, John Wiley & Sons, Ltd, Chichester, USA, 2004.
16. Kanezashi, M. and Tsuru, T., *Gas Permeation Properties of Helium, Hydrogen, and Polar Molecules Through Microporous Silica Membranes at High*

Temperatures. Correlation with Silica Network Structure, Elsevier B.V, The Netherlands, 2011.
17. Gontarek, E., Macedonio, F., Militano, F., Giorno, L., Lieder, M., Politano, A., Drioli, E., Gugliuzza, A., Adsorption-assisted transport of water vapour in super-hydrophobic membranes filled with multilayer graphene platelets. *Nanoscale*, 11, 11521–11529, 2019.
18. Gugliuzza, A., Aceto, M.C., Drioli, E., Interactive functional poly (vinylidene fluoride) membranes with modulated lysozyme affinity: a promising class of new interfaces for contactor crystallizer. *Polym. Int.*, 58, 1452–1464, 2009.
19. Perrotta, M.L., Macedonio, E.F., Tocci, E., Giorno, L., Gugliuzza, A., Graphene stimulates the nucleation and growth rate of NaCl crystals from hypersaline solution via membrane crystallization. *Envir. Sci. Water Res. Tech.*, 6, 1723–1736, 2020.
20. Gugliuzza, A., Pingitore, V., Miriello, D., Drioli, E., Functional carbon nanotubes for high-quality charge transfer and moisture regulation through membranes structure-functions insights. *Phys. Chem. Chemi. Phys.*, 17, 12919, 2015.
21. Pingitore, V., Miriello, D., Drioli, E., Gugliuzza, A., Integrated carboxylic carbon nanotubes pathway with membranes for voltage-activated humidity detection and microclimate regulation. *Soft Matter*, 11, 4461–4468, 2015.
22. Xu, L., Liu, S., Yu, L., Li, K., Zhang, Y., Wang, J., Wang, J., Tuneable ion transport by electrically responsive membranes under electrical assistance. *J. Membr. Sci.*, 663, 121046, 2022.
23. Zhang, M., Guan, K., Ji, Y., Liu, G., Jin, W., Xu, N., Controllable ion transport by surface-charged graphene oxide membrane. *Nat. Commun.*, 10, 1253, 2019.
24. Yi, Y.T., Sun, J.Y., Lu, Y.W., Liao, Y.C., Programmable and on-demand drug release using electrical stimulation. *Biomicrofluidics*, 9, 022401, 2015.
25. Aytar Çelik, P., Erdogan-Gover, K., Barut, D., Enuh, B.M., Amasya, G., Sengel-Türk, C.T., Derkus, B., Çabuk, A., Bacterial Membrane Vesicles as Smart Drug Delivery and Carrier Systems: A New Nanosystems Tool for Current Anticancer and Antimicrobial Therapy. *Pharmaceutics*, 15, 1052, 2023.
26. Lafarge, E.J., Muller, P., Schroder,.E.P., Marques, C.M., Activation energy for pore opening in lipid membranes under an electric field. *PNAS*, 120, e2213112120, 2023.
27. Kabelka, I. and Vácha, I., Optimal conditions for opening of membrane pore by amphiphilic peptides. *J. Chem. Phys.*, 143, 243115, 2015.
28. Huang, C.-H. and Liu, Y.-L., Self-healing polymeric materials for membrane separation: an example of a polybenzimidazole-based membrane for pervaporation dehydration on isopropanol aqueous solution. *RSC Adv.*, 7, 38360–38366, 2017.

29. Wong, E.C., Lim, Y.H., Siew, M.P., Chong, W.C., Ong, Y.H., Pang, Y.L., Chong, K.C., A review of self-healing composite films and its development in membrane for water filtration. *J. Water Process Eng.*, 55, 104123, 2023.
30. Han, Y., Giorno, L., Gugliuzza, A., Photoactive Gel for Assisted Cleaning during Olive Mill Wastewater Membrane Microfiltration. *Membranes*, 66, 15, 2017.
31. Alebrahim, E. and Moreau, C.A., Comparative Study of the Self-Cleaning and Filtration Performance of Suspension Plasma-Sprayed TiO_2 Ultrafiltration and Microfiltration Membranes. *Membr. (Basel)*, 13, 750, 2023.
32. Gugliuzza, A. and Basile, A., Membrane contactors: fundamentals, membrane materials and key operations, in: *Handbook of membrane reactors: Fundamental materials science, design and optimization*, A. Basile (Ed.), Woodhead Publishing Limited, UK, 2013.
33. Speranza, V., Trotta, F., Drioli, E., Gugliuzza, A., High-definition polymeric membranes: Construction of 3d lithographed channel arrays through control of natural building blocks dynamics. *ACS Appl. Mater. Interf.*, 2, 459–466, 2010.
34. Hwang, S.T. and Kammermeyer, K., Effect of Thickness on Permeability, in: *Permeability of Plastic Films and Coatings. Polymer Science and Technology*, H.B. Hopfenberg, (Ed.), Springer, Boston, MA, 1974.
35. Lu, X., Peng, Y., Qiu, H., Liu, X., Ge, L., Anti-fouling membranes by manipulating surface wettability and their anti-fouling mechanism. *Desalination*, 413, 127–135, 2017.
36. Scott, K., Microfiltration, in: *Handbook of Industrial Membranes*, K. Scott (Ed.), pp. 373–429, Elsevier Science, UK, 1995.
37. Pont, N., Salvadó, V., Fontàs, C., Applicability of a Supported Liquid Membrane in the Enrichment and Determination of Cadmium from Complex Aqueous Samples. *Membranes*, 8, 21, 2018.
38. Shoshaa, R., Ashfaq, M.Y., Al-Ghouti, M.A., Recent developments in ultrafiltration membrane technology for the removal of potentially toxic elements, and enhanced antifouling performance: A review. *Environ. Technol. Innovation*, 31, 103162, 2023.
39. Sarkar, S., Gupt, A.K.S., Prakash, P., The Donnan Membrane Principle: Opportunities for Sustainable Engineered Processes and Materials. *Environ. Sci. Technol.*, 44, 4, 1161–1166, 2010.
40. Maroufi, N. and Hajilary, N., Nanofiltration membranes types and application in water treatment: a review. *Sustain. Water Resour. Manag.*, 9, 142, 2023.
41. Lang, C., Kumar, M., Hickey, R.J., Current Status and Future Directions of Self-Assembled Block Copolymer Membranes for Molecular Separations. *Soft Matter*, 17, 10405–10415, 2021.
42. Perry, R.H. and Green, D.W. (Eds.), *Perry's Chemical Engineers' Handbook*, 7th, McGraw-Hill, New York, 1997.

43. Rezakazemi, M. and Younas, M., *Membrane Contactor Technology: Water Treatment, Food Processing, Gas Separation, and Carbon Capture*, Wiley-VCH GmbH, Germany, 2022.
44. Perrotta, M.L., Saielli, G., Casella, G., Macedonio, F., Giorno, L., Drioli, E., Gugliuzza, A., An ultrathin suspended hydrophobic porous membrane for high-efficiency water desalination. *Appl. Mater. Today*, 9, 1–9, 2017.
45. Frappa, M., Del Rio Castillo, A.E., Macedonio, F., Di Luca, G., Drioli, E., A. Gugliuzza, A., Exfoliated Bi2Te3-enabled Membranes for New Concept Water Desalination: Freshwater Production Meets New Routes. *Water Res.*, 203, 117503, 2021.
46. Jaiswar, G., Dabas, N., Chaudhary, S., Jain, V.P., *et al.*, Progress in absorption of environmental carbon dioxide using nanoparticles and membrane technology. *Int. J. Environ. Sci. Technol.*, 20, 10385–10404, 2023.
47. Khayet, M., Mengual, J.I., Matsuura, T., Porous hydrophobic/hydrophilic composite membranes: Application in desalination using direct contact membrane distillation. *J. Membr. Sci.*, 252, 101–113, 2005.
48. Drioli, E. and Macedonio, F., Membrane-Assisted Condenser. *Clean Technol.*, 1, 1–7, 2018.
49. Joscelyne, S.M. and Trägårdh, G., Membrane emulsification — a literature review. *J. Membr. Sci.*, 169, 107–117, 2000.
50. Cheng, Z., Li, S., Tocci, E., Saielli, G., Gugliuzza, A., Wang, Y., Pathway for Water Transport through Breathable Nanocomposite Membranes of PEBAX with Ionic Liquid $[C_{12}C_{1im}]Cl$. *Membranes*, 13, 749, 2023.
51. Eickmann, U. and Werner, U., Porous Membranes in Gas Separation Technology, in: *Membranes and Membrane Processes*, E. Drioli, and M. Nakagaki, (Eds.), Springer, Boston, MA, 1986.
52. Isobe, T., Nishimura, M., Takada, Y., Matsushita, S., Honda, S., Iwamoto, Y., Nakajima, A., Gas separation using Knudsen and surface diffusion II: Effects of surface modification of epoxy/porous SiO2 composite. *J. Asian Ceram. Soc.*, 2, 190–194, 2014.
53. Gugliuzza, A. and Drioli, E., Role of additives in the water vapor transport through block co-poly(amide/ether) membranes: Effects on surface and bulk polymer properties. *Eur. Polym. J.*, 40, 2381–2389, 2004.
54. Rahman, M.M., Material Design Concepts and Gas Separation Mechanism of CO2 Selective Polyether-Based Multiblock Copolymers. *Macromol. Rapid Commun.*, 44, 2300114, 2023.
55. Pandey, P. and Chauhan, R.S., Membranes for gas separation. *Prog. Polym. Sci.*, 26, 853–893, 2001.
56. Vane, L.M., Review of Pervaporation and Vapor Permeation Process Factors Affecting the Removal of Water from Industrial Solvents. *J. Chem. Technol. Biotechnol.*, 95, 495–512, 2020.

2

Approaches to Characterize Pores in Membranes

Amalia Gordano

Research Institute on Membrane Technology, ITM-CNR, Rende, Cosenza, Italy

Abstract
This chapter focuses on some techniques currently used to characterize the porosity of membranes in indirect or indirect ways. The concept of porosity includes mean pore size, pore distribution, total void volume fraction in membranes, and geometrical and effective pores. Basic concepts are introduced together with mathematical equations and description of methods to estimate membrane porosity, which includes capillary balance, permeation of solutes, bubble pressure, liquid-liquid porosimetry, permeability, gas adsorption/desorption, mercury intrusion porosimetry, thermometry, perporometry, positron annihilation duration spectroscopy, scatter radiation, and atomic force microscopy.

Keywords: Pore size, pore distribution, porosity, porosimetry

2.1 Introduction to Porous Membranes

What is a scientist after all? It is a curious man looking through a keyhole, the keyhole of nature, trying to know what's going on (Jaques Yves Costeau).

And what is a door? It is like a keyhole that allows us to look beyond. The pores of a membrane? Well, they are doors! These doors allow passage, block it, or remain ajar to the mystery of science, in continuous discovery and always thirsty for a new journey.

We have all heard of membranes, and it is easy to imagine them as a sieve that act as a barrier for some substances and let others pass (Figure 2.1).

Email: a.gordano@itm.cnr.it

Annarosa Gugliuzza and Wanqin Jin (eds.) Porous Membranes: Breakthroughs in Manufacturing and Applications, (27–46) © 2025 Scrivener Publishing LLC

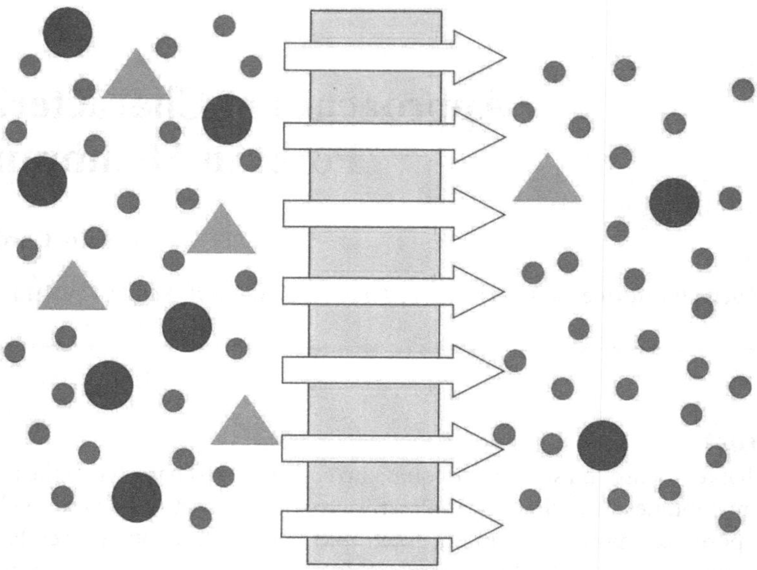

Figure 2.1 Schematic functioning of a membrane as a molecular sieve.

However, as with biological membranes, the process develops inside them. Similarly, in artificial membranes, the factors that influence a membrane separation are manifold. Membranes are not simple films with different pore sizes. The properties of the material constituting the membrane significantly contribute to the success of the separation, and in any case, if the pores represent the "passage point", it is necessary to evaluate the driving force that promotes the passage and the "path" through which it occurs. Pores are not simply holes; they are paths. As such, they can be more or less linear, more or less regular, more or less tortuous, more or less uphill, and more or less smooth.

Membrane separation processes operate without administering heat; therefore, there is less energy consumption, compared to traditional thermal separation processes. Although membrane separation techniques remained virtually unused until the 1970s, today they are widely used in various fields, including food technologies, biotechnology, and pharmaceutical industries [1, 2]. Even in the treatment of waste water [3] from various sources, membrane technology is gaining momentum due to its numerous advantages of use, as well as in fuel cells [4, 5] and osmotic power plants.

A simplification of the purely physical separation process involves the treatment of a compound of interest on a membrane and the subsequent formation of two fractions: the permeate (what passes through the membrane) and the retentate (what is rejected). For mass transfer across the membrane, two basic models are distinguished: the solution diffusion model and the hydrodynamic model.

In the solution diffusion model, transport takes place by diffusion. The component to be transported must first be dissolved inside the membrane. The general approach of this model is the existence of a chemical potential equilibrium at the interface of the adjacent membrane surface. This model is more important for dense membranes without natural pores, such as those for reverse osmosis and in fuel cells.

The hydrodynamic process involves passage by convection, and therefore, the size of the pores plays a fundamental role.

Membrane separation processes differ according to the separation mechanisms and the size of the separated particles They include microfiltration (MF), ultrafiltration (UF), nanofiltration (NF), reverse osmosis (RO), electrolysis, dialysis, electrodialysis, gas separation, pervaporation (PV), membrane distillation (MD), and membrane exchangers. All processes, with the exception of pervaporation, do not involve a phase change. In all processes, except electrodialysis, pressure is the driving force. MF and UF are widely used in food and beverage processing, biotechnology applications, the pharmaceutical industry, sewage and wastewater treatment, microelectronics industry, etc. NF and RO are mainly used for water purification. Dense membranes are used in gas separations and sometimes in MD. Figure 2.2 shows a schematic of the membrane processes with relative pore sizes.

What distinguishes the most common pressure-driven membrane processes, i.e., MF, UF, NF, and RO, is the application of hydraulic pressure to allow the transport process through the membrane. However, it is the very nature of the membrane that controls which components can permeate and which ones are retained.

In its ideal definition, RO retains all components except the solvent (for example, water), while UF retains only macromolecules or particles with dimensions greater than about 0.0001–0.02 μm. MF, on the other hand, is used to retain particles of the order of the micrometer, that is, suspended particles with dimensions between 0.1 and 5 μm. Hence, in a broad sense, RO is essentially considered a concentration technique, while UF is seen as a method for purifying and concentrating macromolecules or fine colloidal suspensions. MF is basically used as a clarification technique to separate suspended particles from dissolved substances, once these particles are of the size required by this type of membrane.

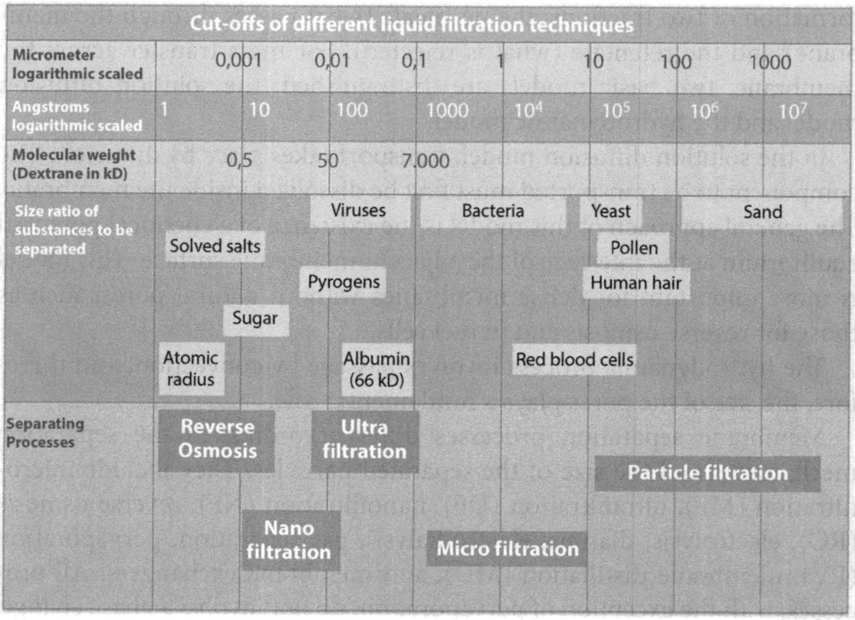

Figure 2.2 Membrane processes and pore size schema.

2.2 Porosity

Membranes are generally classified according to their molecular weight cut-off (MWCO) rather than the size of the particles they are capable of separating. This is because, at the dawn of this technology, UF membranes were characterized by studying their permeability to proteins and polyethylene glycols, which are typically distinguished based on their molecular weight. Although it is now well known that the molecular weight alone does not determine the size of proteins, this terminology has remained in common use and indicates the molecular weight of those species that have a rejection of more than 90%. Thus, if the MWCO is higher than 10000 Da, we speak of UF; if lower than 500 Da, of RO. On the other hand, another classification, suggested by some authors, is based on the identification of the transport of matter as a discriminating element: if the solvent flow is mainly convective, we speak of UF; if, on the contrary, the diffusive flow is decisive, we talk about RO. There is also the presence of membranes for which the types of solvent transport are both convective and diffusive, without a clear prevalence of one mechanism over the other. There is a cut-off interval, between 500 and 10000 Da, within which neither the one nor the other processes fall. Under these conditions is the NF process, in which charged membranes are used with pores larger than those of RO, but too small to allow the permeation of many organic compounds, such as sugars. These membranes allow for disparate applications where UF becomes scarcely selective and a complete RO barrier is not required. In fact, thanks to its selectivity characteristics, NF allows, with the same driving force, higher flows or, with the same flow, lower driving forces. Furthermore, these membranes allow for flows measured with water much higher than those recorded through the typical RO membranes and have better resistance to fouling, in the case in which oils, proteins, and other organics are treated.

Porous membranes can be classified as microporous or asymmetrical. Microporous membranes are then further divided into isotropic, which have uniform pores throughout the membrane, or anisotropic, in which, instead, the size of the pores changes passing from one surface to another MF membranes belong to this category and are generally designed to retain all particles above a certain size. For example, a 0.45 μm membrane does not allow the passage of particles larger than 0.45 μm. However, this does not imply that the pore size is 0.45 μm or less. Indeed, it has been observed that pores are usually larger than the particles they hold and that there is a certain distribution of pore size along the surface of the membrane, so particles that are approximately the same size as the pores can partially penetrate and obstruct them. The largest pores, therefore, which initially have the largest flows inside them, are the first to become blocked, and this justifies the classic rapid decrease in flows in the first minutes of operation, which is observed with MF membranes. Asymmetrical membranes, on the other hand, are characterized by a thin skin, which constitutes the active layer, on the surface of the membrane, while the layers below are characterized by high porosity and have the purpose of supporting the active layer. Separation occurs once again on the surface, but due to this particular structure, retained particles or macromolecules above a certain MWCO cannot penetrate the main body of the membrane. Most membranes from UF, NF, and RO have this structure. An important difference between microporous and asymmetrical membranes lies in the definition of the rejection limits. For microporous membranes, there is an absolute limit in terms of the maximum equivalent diameter of the pore, and, that is, they retain all particles with dimensions above this value, up to specific concentration limits in the feed. For asymmetrical membranes, on the other hand, nominal values are provided, which refer to the size of the molecules or to the molecular weight above which a certain percentage of solute in the feed solution is retained by the membrane, under specific and controlled operating conditions.

In all types of membrane filtration, the ratio between the size of the solutes and those of the pores of the membrane is the first factor to consider in the separation. When the pores are smaller than the solute, the molecules cannot cross the membrane. In NF and RO, typically, membranes are characterized in terms of pore radius, and salts and small organic molecules such as glucose are used to determine their size. Obviously, the characterization of a membrane is not limited to the choice of the cut-off but also takes into consideration the various aspects related to the reorganization of the membrane in response to the environment in which it is immersed. The main parameters to be characterized are the size of the pores, the hydrophilicity of the material, and the charge that originates on the surface (often responsible for fouling phenomena).

Operating on a molecular scale, the main characteristic of a membrane is constituted by the pores and their diameter, which helps to classify the membrane in terms of MWCO. The cut-off, therefore, represents the most distinctive feature of the membrane in determining the characteristics of the permeate. Although each type of membrane is classified according to its own cut-off, the latter is basically an ideal measure of reference, given that the membrane production techniques do not allow for obtaining of repeatable and accurate diameters for all pores. The cut-off value is therefore indicative, as it is average, to the point that, with the same nominal cut-off, the comparative performance between the membranes can be different. In industrial practice, the availability of various types of membranes, each characterized by specific operating characteristics (density, pore size, etc.), allows for extremely diversified separation or fractionation effects to be obtained for the same product.

The porosity of a material is defined as the ratio between the sum of the void volumes, Vp, and the total volume, Vm, of the material under consideration. The effective porosity of a membrane represents its holding capacity and can be determined through porometric techniques. Using these techniques we try to establish, in statistical terms, a characteristic size of the average pores of the membrane, and then go back to the porosity through the knowledge of the pore density.

$$\varnothing = Vp/Vm. \tag{2.1}$$

The particular mechanism responsible for the diffusion of matter and the characteristics and phenomena associated with it depend on the size of the pores. For example, in the case of mesopores, the Knudsen diffusion mechanism is particularly relevant and the phenomenon of capillary condensation can occur. When studying the porosity of a membrane, however, it is necessary to take into account a series of variables that intervene in the morphology of the matrix, due in part to the characteristics of the materials and in part to the preparation techniques. Not all pores, for example, are passers-by; it would be advisable to know the geometry and the tortuosity of the "path", precisely because the pores, as we said at the beginning, are not simple "holes". As always happens in scientific research, but also a little in life, to achieve a result you have to look at things from different perspectives, put together more information, and draw more or less complete conclusions. This is because, beyond the conformation of the pores, the thickness of the membrane determines the length of the passage path through it and, therefore, also the ease.

If the empty space of a porous membrane is saturated with a liquid until a constant weight is obtained and the excess liquid is removed, fractional porosity can be defined from the weight change:

$$Porosity = \frac{W_W - W_d}{V \, x \, \rho} \tag{2.2}$$

where W_W and W_d are the weights of the wet and dry membrane, respectively, ρ is the density of the liquid and V is the volume of the wet membrane.

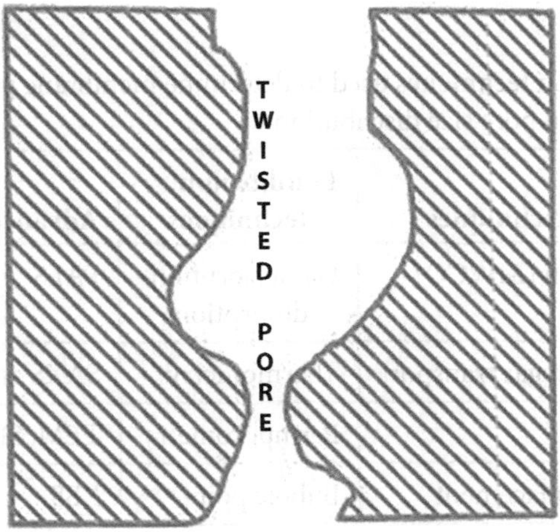

Figure 2.3 Schematic non-regular pore.

Within a membrane, the pores are rarely uniform (Figure 2.3). For this reason, their distribution is studied. Many indirect methods are used, some even dated, but simple and precise, and today, executable on automated tools.

The pores are generally divided into: macropores (r > 50 μm), mesopores (2 μm ≤ r ≤ 50 μm), and micropores (r < 2μm) [6]. However, there are also pores that are interrupted, therefore, they are present only on one surface of the membrane. These contribute to the porosity but not to the transport, and to differentiate them one has to resort to several measurement techniques.

Tortuosity is defined as the factor by which the observed transport speed would increase if the pores were perfectly straight, with the same porosity and pore size. The tortuosity of a membrane falls in the range between 1.5 and 5 [7]. Generally, porosity and tortuosity are independent parameters, but there would be a correlation according to the percolation theory [8], valid for many randomly anisotropic porous media.

2.3 Methods to Estimate Porosity

Table 2.1 shows the common methods used to estimate the porosity of a membrane, the mean pore diameter, and the distribution.

Indirect methods, excluding solute permeation, use the capillary balance of two fluids in a pore or pore flow. Balance methods count in dead-end pores, while pore flow does not.

Table 2.1 Common techniques used to determine membrane porosity, mean pores diameter, and pore size distribution.

Transport/flow techniques	Equilibrium techniques	Advanced techniques
Solute permeation (MWCO)	Gas adsorption-desorption	PALS
Bubble point with gas transport	Thermometry	SAXS and SANS
Gas permeability	Permporometry	TEM, SEM, and AFM
Liquid-liquid displacement	Bubble point	RBS
Mercury intrusion porosimetry		Nanoparticles

2.3.1 Methods of Capillary Balance

The methods of capillary balance are based on the relationship between the capillary pressure (Laplace) and the radius of the pores:

$$r_P = \frac{2\gamma cos\theta}{P} \tag{2.3}$$

where γ is the surface tension coefficient of the liquid, θ the contact angle of the liquid on the pore walls and P the critical pressure required to open or fill the pore. Pore flow methods measure the transmembrane flow of a fluid, which can be analyzed according to the Hagen-Poiseuille equation:

$$\Delta Q = \frac{\pi r_P^4 N \Delta P}{8\mu L} \tag{2.4}$$

where ΔP is the change in volume flow following a change in the transmembrane pressure along the pores ΔP, N the number of pores per unit area, μ the viscosity of the fluid and L is the length of the pores. Assumptions must be made, which risk oversimplifying the pore structure. In particular, it is assumed that all the pores are cylindrical, without tortuosity, parallel to each other and perpendicular to the surface of the membrane, and that the thickness of the membrane corresponds to the length of the pores. It follows that the results make sense if used as a comparison between membranes to study their differences. However, to account for a particular pore geometry, correction factors would be needed in the equations. Figure 2.4 reports an example of different porous structures in cross section of two PEEK-WC membranes.

MWCO is nominally defined for UF, NF, and dialysis membranes as the molecular weight of a neutral solute that is rejected by a 90% membrane.

2.3.2 Method of Permeation of Solutes

One of the methods to estimate this is the permeation of solutes dissolved in a solvent, rather than a pure solvent. Membrane manufacturers also accept this method, which evaluates the rejection of some molecules at different molecular weights, feeding the membrane with the solution, and is represented on a logarithmic scale as a function of molecular weights. Macromolecule solutions, such as poly(ethylene glycol) and dextran, are

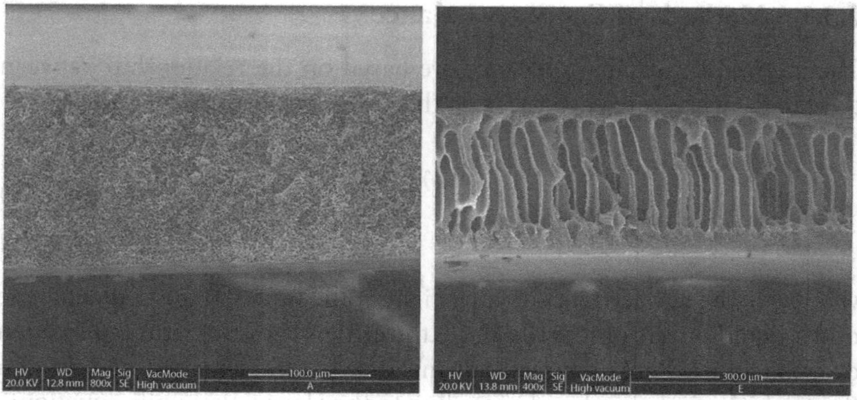

Figure 2.4 SEM images of different porous structure on cross section for PEEK-WC membranes, individual research.

generally used at relatively low concentrations, between 200 and 1000 mg/L, to minimize solute-solute interactions. Generally, it is necessary to foresee a possible blockage of the pores by molecules larger than their diameter, and to do this, the experiments are preceded by an estimate of the pore radius. Typically, a multi-component solution is employed and the feed and permeate compositions are determined by gel permeation chromatography (GPC).

2.3.3 Method of Bubble Pressure

The bubble pressure method was first employed by Bechhold in 1908 [9] and is based on the fact that the pressure (P) required to blow a gas through a liquid-filled capillary is inversely proportional to the capillary radius (r). The pore size can be calculated using Cantor's equation:

$$P = \frac{2\gamma}{r} cos\theta \tag{2.5}$$

where θ is the contact angle and γ the surface tension. With this test, it is possible to determine the radius of the largest pore. In fact, according to Cantor, the gas will pass through the largest pore first. This method has numerous advantages: it is not destructive, the procedure and apparatus are simple, and it provides information on integrity. On the other hand, it does not provide information on the pore size distribution.

2.3.4 Method of Liquid-Liquid Porosimetry

Liquid-liquid porosimetry is a method that provides information on the size distribution of small pores. The principle on which it is based is similar to that of the bubble point, since both use the correlation between the applied pressure and the radius of the pores open to the flow, as indicated by Washburn [10]:

$$\Delta P = \frac{2\gamma cos\theta}{r_P} \tag{2.6}$$

It is a very promising technique for evaluating active pores at the nano or subnanometric scale and the distribution of their dimensions. It can be used on membranes from UF and NF since the applied pressures are relatively low and do not risk causing compaction. Among the advantages include: the possibility of testing the membrane in the wet state, providing information that can be easily assimilated to the normal operating conditions of the membrane; it evaluates only the open pores, and does not use high pressures, thus avoiding rupture and compacting of the membrane; and it is a quick and easy method.

Using two immiscible liquids with low interfacial tension, the membrane is filled with one of the two and moved with the other. Monitoring of transmembrane pressure and flow allows for the determination of the radius of the open pores at a given applied pressure, according to the Cantor equation, assuming, however, that the contact angle between the liquid interface and the membrane can be considered equal to zero:

$$\Delta P = \frac{2\gamma}{r} \tag{2.7}$$

By gradually increasing the pressure, you will have corresponding radius and flow values, represented as membrane permeability. Assuming cylindrical pores, the Hagen-Poiseuille equation can be used to correlate the volume flow, J_V, and the number of pores, per unit area, N, with a given radius, r. For each pressure change, ΔP_i, the corresponding measured volumetric flow rate is related to the number of open pores:

$$J_{V_i} = \sum_{k=1}^{i} \frac{N_k \pi r_k^4 \Delta P_i}{8\gamma\delta} \tag{2.8}$$

where γ is the viscosity of the fluid, N_k is the number of pores in the kth class per unit area, and δ is the length of the pores. Obviously, the Hagen-Poiseuille equation assumes the cylindrical shape of the pores, but it is an applicable model to determine the distribution of dimensions in the number of pores, although it depends on it.

2.3.5 Method of Permeability

The permeability of a membrane for a certain liquid is a characteristic parameter, from which the hydraulic radius can be calculated. What is determined is the permeability, while the porosity (E), tortuosity (z), and thickness of the membrane (l) are known. The size of the pores is calculated from the Hagen-Poiseuille equation. This model does not discriminate between a system with many small pores and one with few large pores, although it can be improved by using a permeating gas rather than a liquid. In fact, the gas transport mechanism depends on the overall pressure of the system, so that, by measuring the permeability at different pressures, it is possible to discriminate the size of the pores between fine and coarse porous media. In literature, both gas permeation and solute techniques have been used in hollow polymeric fibers [11], obtaining similar information in the average pore size, but gas permeation has advantages. In fact, the polymer reduces the flow of the aqueous solution, while it has little influence on the permeability to gases and the surface porosity measured with respect to the length of the pores [12]. In the case of asymmetrical membranes, gas permeation has proven very useful in providing information on the active thin layer, while it says nothing about the underlying layer [13].

2.3.6 Method of Gas Adsorption/Desorption

The gas adsorption/desorption isotherms is another method used to determine the pore size distribution. Generally, nitrogen is used, subjected to adsorption and capillary condensation in the pores. BET analysis on nitrogen adsorption is generally used to determine the surface area and the distribution of the pore size of ceramic membranes, within the range of micro and mesopores. Traditional dense polymeric membranes, conventionally considered non-porous, are not studied with this technique, with the exception of glassy, semi-crystalline, or composite-amorphous polymer structures (e.g., TiO_2 - ZrO_2 - organic) [14].

2.3.7 Method of Mercury Intrusion Porosimetry

Porosimetry is an analytical technique that allows the measurement of some characteristics of a material, attributable to the presence of pores, including total pore volume, pore diameter, surface area, bulk density, and absolute density. This technique consists of introducing a high-pressure liquid into the material to be tested, which does not wet the material in question. The high pressure is necessary to counteract the surface tension of the liquid, which alone would not allow the liquid to penetrate inside the pores. Often, the liquid used is mercury, so we speak of mercury intrusion porosimetry.

The pore size can be determined starting from the pressure value of the liquid through the so-called Washburn equation, which represents the balance of forces for materials with cylindrical pores:

$$P_L - P_G = \frac{4\sigma \, cos\theta}{D_P} \tag{2.9}$$

where P_L is the pressure of the liquid, P_G is the pressure of the gas present inside the pores, σ is the surface tension of the liquid, θ is the contact angle of intrusion of the liquid (in general, for mercury it is included between 135° and 142°), and D_P is the diameter of the pores.

The pressure is increased, and the pressure required for mercury to overcome the capillary pressure and penetrate the pores is recorded with respect to the volumetric fraction of mercury that penetrates the pores and the distribution of pore sizes is thus determined, using the eq. (2.6). With subsequent extraction experiments, by reducing the pressure, other information can be obtained. In this case, pores ranging in size from 5 nm to 200 μm can be evaluated. Unlike other methods, in this case, there is a risk of membrane compaction, since it is performed dry, which may cause some pores to deform or collapse, without taking into account the fact that dead-end pores are also evaluated and, therefore, unsuccessful for the purposes of permeation [15].

2.3.8 Method of Thermometry

Thermometry is based on the decrease in the freezing or melting point of a liquid (usually water) with the pore size recorded using differential scanning calorimetry (DSC) [16]. A water-soaked membrane is placed in DSC, and the temperature is raised at a constant rate while the heat flux is measured. The distribution of the pore sizes is therefore deduced from the DSC thermogram.

2.3.9 Method of Perporometry

Permporometry is attributed to Eyraud, although the technique was modified by Katz *et al.* [17, 18]. The vapor pressure on the surface of a liquid depends on its curvature, which is the principle on which the technique is based. The vapor is capillary-condensed in the pores of the membrane, thus blocking the permeations of the incondensable gas. The filling of the pores follows the Kelvin equation:

$$d_K = \frac{-4v\sigma\ cos\theta}{RT\ln(\frac{P}{P_s})} \tag{2.10}$$

where θ is the contact angle, s the surface tension, v the molar volume, and T the temperature. The equation is valid in pores with dimensions smaller than the Kelvin diameter, dK. The latter is proportional to the vapor pressure (P) of the condensable gas and correlated to the true diameter of the pores, d_p, through the thickness (t) of the adsorbed layer: $d_P=d_K+t$. From the measurement of the permeabilities of the incondensable gas as a function of the relative pressure, P/P_S, an estimate of the distribution of the pore sizes is obtained. Non-active pores are excluded from the estimate, so permporometry is considered a precise technique in the study of pores. Consequently, nanopermporometry, based on the capillary condensation of steam (water or hexane) and the blocking of the permeation of a non-condensable gas (N_2), has found numerous applications to characterize hybrid organic/inorganic membranes with pores smaller than 1 nm, used in the nanofiltration of organic solutions [14, 19].

In addition to the techniques mentioned above, there are other newer methods, both direct and non, which are used to characterize the porosity and pore size of a membrane [20].

2.3.10 Method of Positron Annihilation Duration Spectroscopy

In positron annihilation duration spectroscopy (PALS), the sample is irradiated with positron generated using atoms that emit positrons such as ^{22}Na [21]. The emission of a ^{22}Na positron is detected by γ 1.28 MeV rays. To determine the positron duration, both the 1.28 MeV and 511 keV γ rays can be used. Moreover, by using many emitted positions, the duration distribution can be determined. To convert the duration distribution into pore size and porosity distribution on a molecular scale (fractional free volume), the cavities are assumed to be spherical. The usefulness of this method concerns the molecular pores in dense membranes, rather than the normal porous membranes from MF and UF. The most advanced PALS configurations allow the profiling of the fractional free volume depth within the upper layer of composite membranes, with a resolution of a few nanometers [22].

2.3.11 Methods of Scatter Radiation

Porous materials can scatter radiation of various types, such as visible light, X-rays, and neutrons, depending on the size of the pores. Therefore, X-ray diffraction (XRD), small angle X-ray diffusion (SAXS) techniques, Small angle neutron scattering (SANS), and light scattering can be used to examine porosity and pore size, at least in principle. Therefore scattering techniques are commonly applied to very regular materials, for example, zeolite and metal organic framework (MOF) [23, 24], which produce sharp XRD spectra, and to pseudoporous polymers separated by microphase, such as block copolymers and ionomers, particularly Nafion [25].

2.4 Microscopy Techniques

Microscopy techniques are a valid support to the morphological study of membranes, even though they too have relative limitations. The high energy of the electron beam used to observe the sample at maximum magnification can damage the surface of a polymer membrane. The membrane is cold fractured (under liquid N_2), to study its section, but the resolution of an SEM observation has a resolution around 10–20 nm, lower than that of the TEM, which reaches 0.2 nm. Therefore, TEM is more useful in investigating the skin of asymmetrical membranes. Despite this, the preparation of the samples for TEM is difficult and tedious, and the technique does not allow observation of the membrane surface directly. These are widely used analysis techniques but they are aimed at the study of the predominant morphology more than anything else. Atomic force microscopy (AFM) provides quantitative data in the study of membrane pores [26]. In this case, the preparation of the sample does not require special precautions, and in the case of using the tapping mode, there is no risk of damaging the sample. The basic principle of the technique involves a tip, with a known radius of curvature, fixed by a flexible arm and placed at a constant distance from the sample of interest at its resonance frequency. As the tip "runs" over the sample, the distance between the tip and the membrane remains constant, so when there are pores, the arm holding the tip flexes. The variations of the arm are processed by the microscope software, resulting in a 3D image of the surface. The subsequent processing and analysis of the images allow the measurement of the pores present. However, it should be pointed out that the radius of curvature of the tip represents a first limit; if it is too large, it invalidates the measurement of the pores. While, in any case, the doubt remains whether a pore is passing or not, precisely due to the physical limit of the tip and the technique. In Figure 2.5, it is possible to observe the porous surface structure of a PES membrane, both in AFM and SEM visualizations.

To quantify the size of the nanopores within the skin of the NF asymmetric membranes, Stawikowska and Livingston [27, 28] have devised an ingenious method. Large Os particles, a few nanometers in size, were injected into the interface opposite the skin by reflux and the region was visualized by TEM. The high-contrast nanoparticles trapped within the

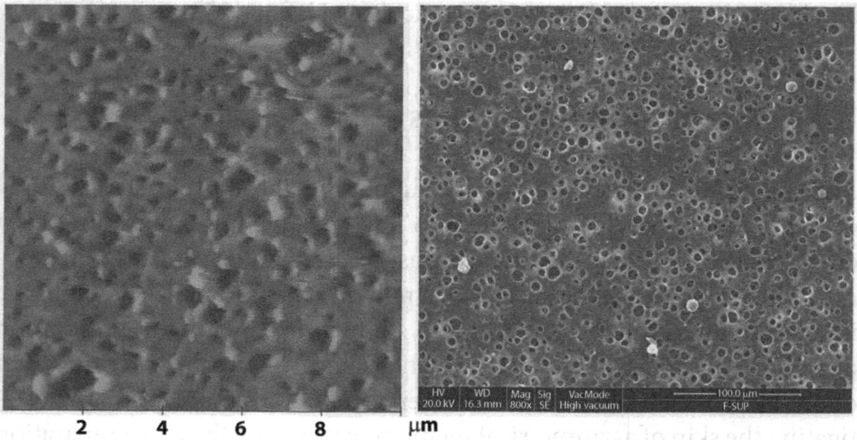

Figure 2.5 AFM and SEM porous PES membrane surface, individual research.

finest nanopores of the skin region could be easily observed and counted. Using particles of different sizes, it is possible to make consistent estimates of the pore size and compare different membranes. The porosity of the underbody region can also be determined using Rutherford backscattering spectrometry (RBS), an absolute atomic counting method, which produces the stoichiometric ratio of different atoms to the internal surface layers and can be profiled in depth. Zhang *et al.* estimated the porosity of the upper 500 nm of the polysulfone support (structurally identical to the UF membranes) in the RO composite membranes. After having equilibrated the membrane with a saline solution and measured the atomic composition of the substrate by RBS. Assuming that the concentration of salt, and therefore the Na and Cl atoms, in the pore was identical to that in the external solution, the porosity is 0.45±0.05 [29].

2.5 Conclusions

Membrane technology is growing day by day and becoming increasingly competitive, effective, and commercially attractive for separation and purification processes in the chemical and related industries, with a wide variety of both industrial and scientific applications [30].

Porosity measurement is a key factor in identifying the hydraulic performance of a porous material. The performance of the membranes depends on various factors, the morphology of the membrane, the characteristics of the material that composes it, the structure and size of the pores. Direct and indirect methods are used to evaluate the porosity of a membrane. The former measure the actual porosity by measuring the pore volume and the total volume of the sample. Using indirect methods, on the other hand, an estimate of the porosity is made by measuring the physical properties related to it. There is also an optical measurement of the pores of a membrane, which is examined under a microscope, and a variety of new and evolving techniques. In almost all cases, however, it is necessary to make some basic assumptions and, therefore, one necessarily runs into errors. For this reason, studies are constantly evolving, and again, for this reason, to have an estimate as accurate as possible, it is necessary to compare and cross the results of various types of tests. As the saying goes, "It is precisely when you think you know something that you have to look at it from another perspective" (cit. The Fleeting Moment).

References

1. Tylkowski, B., Trusheva, B., Bankova, V., Giamberini, M., Peev, G., Nikolova, A., Extraction of biologically active compounds from propolis and concentration of extract by nanofiltration. *J. Membr. Sci.*, 348, 124–130, 2010.
2. Fan, Q., Sirkar, K.K., Michniak, B., Iontophoretic transdermal drug delivery system using a conducting polymeric membrane. *J. Membr. Sci.*, 321, 240-249, 2008.
3. Szymanska, K., Zouboulis, A.I., Zamboulis, D., Hybrid ozonation–microfiltration system for the treatment of surface water using ceramic membrane. *J. Membr. Sci.*, 468, 163–171, 2014.
4. Tylkowski, B., Castelao, N., Giamberini, M., Garcia-Valls, R., Reina, J.A., Gumí, T., The importance of orientation in proton transport of a polymer film based on an oriented self-organized columnar liquid-crystalline polyether. *Mater. Sci. Eng.*, C 32, 105–111, 2012.
5. Park, J. and Kim, D., Effect of cerium/18-crown-6-ether coordination complex OH quencher on the properties of sulfonated poly(ether ether ketone) fuel cell electrolyte membranes. *J. Membr. Sci.*, 469, 238–244, 2014.
6. Mulder, M., *Basic Principles of Membrane Technology*, Kluwer Academic Publishers, Dordrecht, 1996.
7. Cuperus, F., Bargeman, D., Smolders, C., Permporometry. The determination of the size distribution of active pores in UF membranes. *J. Membr. Sci.*, 71, 57–67, 1992.
8. Torquato, S., *Random Heterogeneous Materials: Microstructure and Macroscopic Properties*, Springer, New York, 2001.
9. Cuperus, F.P. and Smolders, C.A., Characterization of UF membranes: Membrane characteristics and characterization techniques. *Adv. Colloid Interface Sci.*, 34, 135–173, 1991.
10. Calvo, J.I., Hernández, A., Prádanos, P., Martınez, L., Bowen, W.R., Pore Size Distributions in Microporous Membranes II. Bulk Characterization of Track-Etched Filters by Air Porometry and Mercury Porosimetry. *J. Colloid Interface Sci.*, 176, 467–478, 1995.
11. Khayet, M. and Matsuura, T., Determination of surface and bulk pore sizes of flat-sheet and hollow-fiber membranes by atomic force microscopy, gas permeation and solute transport methods. *Desalin.*, 158, 57–64, 2003.
12. Feng, C.Y., Khulbe, K.C., Matsuura, T., Ismail, A.F., Recent progresses in polymeric hollow fiber membrane preparation, characterization and applications. *Sep. Purif. Technol.*, 111, 43–71, 2013.
13. See-Toh, Y.H., Silva, M., Livingston, A., Controlling molecular weight cut-off curves for highly solvent stable organic solvent nanofiltration (OSN) membranes. *J. Membr. Sci.*, 324, 220–232, 2008.

14. Tsuru, T., Narita, M., Shinagawa, R., Yoshioka, T., Nanoporous titania membranes for permeation and filtration of organic solutions. *Desalin.*, 233, 1–9, 2008.
15. Yu, J., Hu, X., Huang, Y., A modification of the bubble-point method to determine the pore-mouth size distribution of porous materials. *Sep. Purif. Technol.*, 70, 314–319, 2010.
16. Brun, M., Lallemand, A., Quinson, J.F., Eyraud, C., A new method for simultaneous determination of the size and shape of the pores: the thermoporometry. *Thermochim. Acta*, 21, 59–88, 1977.
17. Mey-Marom, A. and Katz, M.G., Measurement of active pore size distribution of microporous membranes - a new approach. *J. Membr. Sci.*, 27, 119–130, 1986.
18. Katz, M.G. and Baruch, G., New insights into the structure of microporous membranes obtained using a new pore size evaluation method. *Desalin.*, 58, 199–211, 1986.
19. Tsuru, T., Nakasuji, T., Oka, M., Kanezashi, M., Yoshioka, T., Preparation of hydrophobic nanoporous methylated SiO_2 membranes and application to nanofiltration of hexane solutions. *J. Membr. Sci.*, 384, 149–156, 2011.
20. Tylkowski, B. and Tsibranska, I., Overview of main techniques used for membrane characterization. *J. Chem. Technol. Metall.*, 50, 1, 3–12, 2015.
21. Cahill, D.G., Freger, V., Kwak, S.Y., Encyclopedia of Membrane science and technology. *MRS Bull.*, 33, 27–32, 2008.
22. Chen, H., Hung, W., Lo, C., Huang, S., Cheng, M., Liu, G., Lee, K., Lai, J., Sun, Y., Hu, C., Suzuki, R., Ohdaira, T., Oshima, N., Jean, Y.C., Free-Volume Depth Profile of Polymeric Membranes Studied by Positron Annihilation Spectroscopy: Layer Structure from Interfacial. *Macromolecules*, 40, 7542–7557, 2007.
23. Kallus, S., Condre, J., Hahn, A., Golemme, G., Algieri, C., Timmins, P., Ramsay, J., Colloidal zeolites and zeolite membranes. *J. Mater. Chem.*, 12, 3343–3350, 2002.
24. Zornoza, B., Tellez, C., Coronas, J., Gascon, J., Kapteijn, F., Metal organic framework based mixed matrix membranes: An increasingly important field of research with a large application potential. *Microporous Mesoporous Mater.*, 166, 67–78, 2012.
25. Gebel, G. and Diat, O., Neutron and X-ray scattering: suitable tools for studying ionomer membranes. *Fuel Cells*, 5, 261–276, 2005.
26. Gordano, A., Arcella, V., Drioli, E., New Hyflon AD composite membranes and AFM characterization. *Desalin.*, 163, 127–136, 2004.
27. Stawikowska, J. and Livingston, A.G., Nanoprobe imaging molecular scale pores in polymeric membranes. *J. Membr. Sci.*, 413–414, 1–16, 2012.

28. Stawikowska, J. and Livingston, A.G., Assessment of Atomic Force Microscopy for Characterization of Nanofiltration Membranes. *J. Membr. Sci.*, 425–426, 58 –70, 2013.
29. Zhang, X., Cahill, D.G., Coronell, O., Mari˜nas, B.J., Partitioning of salt ions in FT30 reverse osmosis membranes. *Appl. Phys. Lett.*, 91, 181904–181904, 2007.
30. Bernestein, R., Kaufman, Y., Freger, V., Membrane Characterization, in: *Encyclopedia of Membrane Science and Technology*, E.MV. Hoek and V.V. Tarabara (Eds.), John Wiley & Sons, Inc, Hoboken, New Jersey, USA, 2013 Copyright.

PART II

SUSTAINABLE FABRICATION OF HIGH-DEFINED AND DYNAMIC MEMBRANE PORES

3

Smart Porous Membranes with Gating Pores for Responsive Separations

Zhuang Liu and Liang-Yin Chu*

School of Chemical Engineering, Sichuan University, Chengdu, Sichuan, China

Abstract
Smart porous membranes with responsive gates offer a novel approach to selective separations by dynamically adjusting their surface properties and pore sizes in response to environmental stimuli. In this chapter, the recent advancements in the design, fabrication, and separation application of smart membranes have been highlighted. The fundamental principle behind the smart membranes lies in the responsive nature of the gating materials integrated into the structure of membrane substrates. When the external environment, such as temperature, pH and/or ion/molecule changes, the gates undergo reversible transitions between hydrophilic and hydrophobic states, altering the pore size and surface properties. This feature enables the selective adsorption and desorption of guest molecules or ions which is crucial for applications such as protein separation, chiral resolution and toxic ion removal. Further addressing the challenges of the stability, scalability and long-term performances, responsive gating membranes have the potential to revolutionize various fields, offering efficient, environmentally friendly solutions for selective separations.

Keywords: Smart membranes, responsive membranes, advanced separation, temperature-response, ion-response, protein separation, chiral resolution

3.1 Introduction

Separation membranes are crucial for global sustainable development. Due to their versatility, adaptability, space-efficiency, and energy-saving

**Corresponding author*: chuly@scu.edu.cn

Annarosa Gugliuzza and Wanqin Jin (eds.) Porous Membranes: Breakthroughs in Manufacturing and Applications, (49–72) © 2025 Scrivener Publishing LLC

potential [1], they have found widespread applications in diverse fields, including water purification and desalination [2], hemodialysis, [3] bioprocessing, [4] pharmaceutical engineering [5, 6], and more. While significant advancements have been made in membrane technology, several persistent challenges remain. One of the primary issues is that traditional porous membranes often feature fixed pore sizes and surface properties, dictated by their immutable physical and chemical structures [7]. This results in predictable problems, such as a reduction in pore size caused by membrane fouling, which not only diminishes pore size but also obstructs the surface, interfering with the interaction between substances and the membrane.[8] Furthermore, although commercial adsorption-enantioselective membranes can effectively separate isomers through stereospecific interactions between enantiomers and chiral recognition sites, recycling the adsorbed isomers proves difficult due to the strong intermolecular forces involved [9]. Consequently, the development of innovative membranes capable of overcoming these constraints is crucial for advancing separation technologies.

Smart porous membranes with gating pores offer new possibilities for responsive separations [10, 11], enabling enhanced control over permeability and selectivity. These membranes reversibly alter the surface characteristics and pore sizes, which are triggered by changes in environmental stimuli, such as temperature, pH, specific ions or molecules, ionic strength, light, glucose concentration, and magnetic or electric fields [10]. To fabricate such membranes, responsive materials are integrated as functional gates onto stable and mechanically strong porous substrates. The conformation of these grafted functional polymers can be modified depending on environmental conditions, leading to the membrane's responsive behavior [12]. This responsive capability makes smart porous membranes valuable tools for precise size-based separations, affinity separations, and chiral separations. For example, the modulation of affinity via stimuli-responsive polymers can be used to control protein interactions with membrane surfaces, making these membranes suitable for protein separation applications [13–15].

For responsive separations, four key characteristics are essential for smart membranes: responsivity, stability, reversibility, and reproducibility [12]. Responsivity refers to both the extent and speed at which the membrane responds to environmental stimuli. Stability indicates the membrane's ability to maintain consistent performance, ensuring that its material properties and resistance remain unaffected during operation. The materials responsible for the membrane's responsiveness must be sufficiently durable and stable to withstand the operational conditions.

Reversibility is another important factor, ensuring that the smart membrane can be reused multiple times without loss of performance. The gating membrane should maintain their performances even after undergoing multiple cycles of environmental variation. Lastly, reproducibility is critical for the practical deployment of these membranes, as it ensures that both the materials and fabrication processes can be consistently replicated. Notably, reproducibility is influenced by the specific methods used in the fabrication of the smart porous membranes.

This chapter begins by examining various fabrication methods for smart membranes with gating pores, such as the "grafting from," "grafting to," and "blending" techniques. It then explores several important applications of these smart porous membranes, including responsive size-based separation, affinity separation, chiral resolution, and the removal of heavy metal ions. The chapter concludes by highlighting the challenges faced in the development of innovative smart gating membranes for industrial-scale responsive separations.

3.2 Fabrication Approaches of Smart Membranes with Gating Pores

Two main approaches are employed in the fabrication of smart porous membranes. The first approach involves the incorporation of gate materials onto the porous membrane substrates using "grafting from" or "grafting

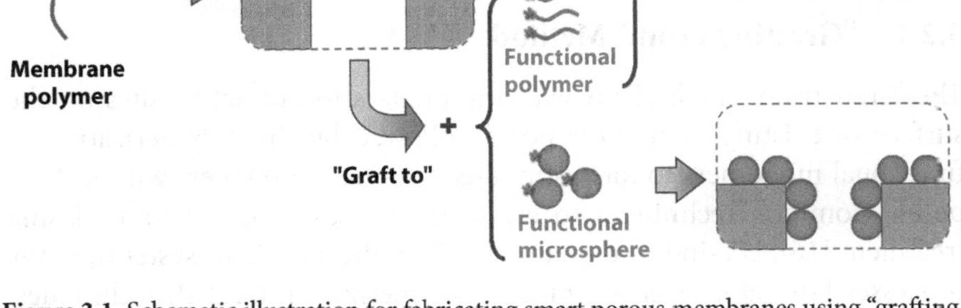

Figure 3.1 Schematic illustration for fabricating smart porous membranes using "grafting from" or "grafting to" technique.

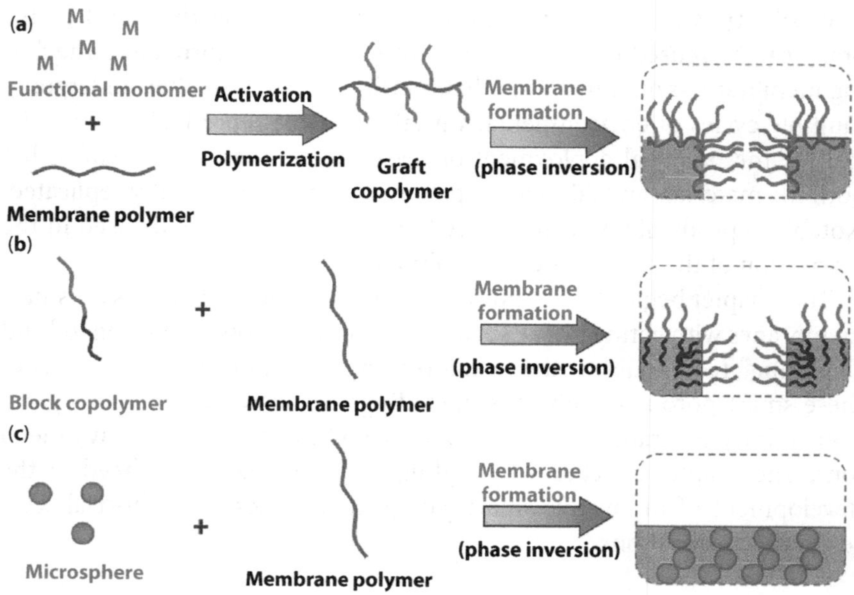

Figure 3.2 Schematic illustration for fabricating smart porous membranes with gating pores by blending stimuli-responsive polymer-additive mixtures such as grafted polymers (a), block copolymers (b), and microspheres (c) during membrane formation.

to" techniques (Figure 3.1). These methods allow for the precise attachment of functional polymers to the membrane surface, leading to highly stable gating structures that can exhibit robust and efficient gating behavior in response to environmental stimuli. The second approach blends stimuli-responsive polymers, block polymers, or polymer-additive mixtures like microspheres with membrane-forming polymers during the membrane fabrication process (Figure 3.2) [10, 11]. The blending technique allows for the incorporation of responsive functionalities within the membrane matrix, creating membranes with tunable characteristics.

3.2.1 "Grafting From" Method

The "graft-from" method involves the preparation of active sites on the surface of existing membrane pores, followed by the polymerization of functional monomers to form stimulus-responsive polymers within these pores. Common techniques to create these active sites include plasma treatment [16], UV-induced methods [17], radical initiator systems, redox processes [16], atom transfer radical polymerization (ATRP) techniques, and more [18].

The pore-filling polymerization approach with the assistance of plasma treatment has been successfully employed to graft membrane pores with responsive polymeric gates [19–21]. Plasma, a kind of partially ionized gas in excited state, would activate the surface of a porous material (e.g., membrane) and make it reactive. The treatment generates free radicals or other reactive species on the surface of the material, which can then serve as initiation sites for further reactions. This step also creates micro-roughness on the surface and introduces functional groups (such as -OH, -COOH, or -NH_2), which increase the surface reactivity. After plasma treatment, a monomer solution (or vapor) is introduced into the porous membrane. The plasma-modified surface provides reactive sites where the monomer molecules can adsorb and diffuse into the pores. The key challenge is ensuring that the monomer infiltrates deeply into the pores. The polymerization of the infiltrated monomers is typically initiated by the free radicals or reactive species created during the plasma treatment. This can occur through various polymerization techniques, including radical polymerization, cationic polymerization, or photo-polymerization, depending on the type of monomer and the plasma treatment conditions. The polymerization process fills the pores with polymer chains, which are grafted from the surface. The polymer chains grow and "fill" the pore space, altering the size, surface properties, and functionality of the material [21].

In the alternative, the ATRP method is a highly effective and versatile approach for grafting monomers onto membrane pores. Unlike other methods that rely on external power sources, such as γ-rays, plasma, or UV light, ATRP does not require such conditions [18]. Compared to these externally-powered grafting techniques, ATRP allows for precise control over grafting density and chain length, both of which significantly influence the performance and responsiveness of smart porous membranes. ATRP is a catalytic process, typically utilizing halogenated Cu^+/Cu^{2+} as a catalyst [18]. Considerable research has focused on modifying both inorganic and organic membranes using the ATRP technique. For example, the hydroxyl groups or the carboxylate groups on the surface of the membranes substrate are readily functionalized with ATRP initiators (–Br) [22, 23]. To initiate polymerization, the membrane anchored with -Br groups, is immersed in a degassed monomer solution with CuBr and $CuBr_2$ as catalysts. After polymerization, the grafted membranes are washed to remove any unreacted components, including residual catalysts [22]. The ATRP polymerization mothed is a versatile and effective grafting from technique for modifying porous materials, enhancing their performance in a wide range of applications.

3.2.2 "Grafting To" Method

The "grafting to" technique involves attaching pre-formed functional polymers, such as linear chains or microspheres, to the membrane surface through chemical or physical means. A key advantage of this method is the ability to precisely synthesize the polymers in advance, whether in the form of chains or microspheres. This enables precise control over their length or size. This leads to well-defined gating pore structures of the membrane by physical interactions and/or covalent bonds [10, 24]. The former, like van der Waals forces, are usually considered to be weak and less stable, while the latter, formed through chemical reactions, provide stronger and more durable attachments between the gates and the pore surfaces [25]. Functionalized reactive polymers can be synthesized using various polymerization techniques, such as controlled radical, anionic, or copolymerization methods, and then chemically linked to complementary reactive groups on the membrane surface [24]. However, the "grafting to" method is inherently limited, as the responsive polymeric chains should diffuse in the polymer membrane to access the reactive sites [24]. As the grafting reaction progresses, the surface of the porous membrane becomes coated with polymers, which hinders the diffusion of other polymers into the membrane pores. As a result, the grafting effect is typically limited to the surface. Common chemical reactions used in this process include those between epoxy and carboxyl groups [26, 27], epoxy and amino groups [27–29], carboxyl and amino groups [30, 31], as well as "click" chemistry reactions, such as the interaction between epoxy groups and maleic anhydride [32–35], among others.

Polydopamine is commonly employed as an interlayer for functionalizing membrane substrates due to its ability to interact with various surface materials [35]. For instance, amine-terminated polymers, such as amine-poly(*N*-isopropylacrylamide) (PNIPAM-NH_2), can be attached to a polydopamine layer through a Michael addition reaction (Figure 3.3). The resulting polydopamine coating is a thin, adhesive, and stable layer with a variety of functional groups (amine, hydroxyl, and quinone), which

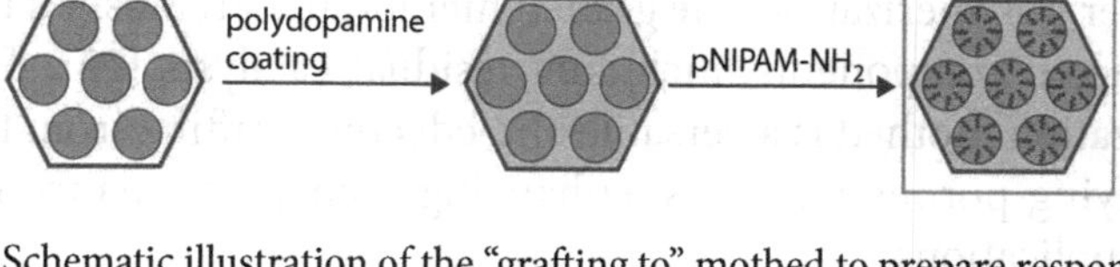

Figure 3.3 Schematic illustration of the "grafting to" mothed to prepare responsive membranes by polydopamine modification and Michael addition reaction [35].

can serve as reactive sites for further modifications. The Michael addition reaction is a type of conjugate addition where a nucleophile (often a thiol, amine, or enolate) adds to an electron-deficient α,β-unsaturated carbonyl compound. This reaction is commonly used in the synthesis of functionalized polymers, materials, and in bio-conjugation processes. The combination of polydopamine modification and Michael addition reactions provides a powerful strategy for modifying materials in a way that enables functionalization with polymer grafting, creating advanced surfaces and membranes for bioengineering separation.

3.2.3 "Blending" Method

The "blending" approach involves integrating responsive materials with membrane-forming polymers, enabling the simultaneous production of both the membrane structure and stimuli-responsive gates in a single step [10]. One of the key benefits of this method is its suitability for large-scale production using existing industrial equipment for membrane fabrication. This method can be easily scaled up as long as appropriate responsive functional polymers or domains are accessible. One variation of the blending technique involves grafting stimuli-responsive functional polymer side chains onto membrane-forming polymers [36, 37], and the resulting grafted copolymer, either on its own or combined with unmodified polymers, could form smart gating membranes via phase separation. Various methods, including UV or O_3 pre-treatment [17, 36], radical initiators, and self-driven ATRP [18, 37], can be employed to graft functional polymer side chains. Another alternative is to blend stimuli-responsive domains, such as block copolymers or functional microspheres, into the membrane matrix polymer, which would undergo phase separation induced by liquid or vapor to form smart membranes induced phase separation [38, 39]. Advances in controlled polymerization have made it increasingly feasible to synthesize amphiphilic block copolymers with well-defined structures, which, in turn, influence the morphology and stimuli-responsive properties of the resulting membranes.

The responsive nanogels as gates *in situ* assemble on the surfaces of three-dimensionally interconnected pores in the membrane during the vapor-induced phase separation (VIPS) process [40]. The VIPS process tends to proceed more slowly due to differences in kinetic parameters. The polymer membranes typically solidify over several minutes in the VIPS process, providing greater flexibility in modifying the membrane's morphology. Sometimes, the porous structure of the polyethersulfone (PES) membranes becomes symmetric and cellular-like during the VIPS process of a uniform membrane-forming solution (Figures 3.4a and 3.4b). However, these membranes often exhibit low transmembrane flux due to the closure of pores and a lack of interconnectivity [41].

The responsive nanogels, such as monodisperse poly(*N*-isopropylacrylamide) (PNIPAM), can be incorporated into a PES membrane-forming solution. During the VIPS process, the nanogels self-assemble at the interface between the growing pores and the PES matrix. When these nanogels are dispersed in a medium and encounter an interface (e.g., the boundary between a matrix and a pore phase), they can adsorb to the interface. The adsorption process is driven by a reduction in the interfacial energy (ΔG_1) [42]. This means that when the nanogels adhere to the interface, they lower the overall energy of the system by stabilizing the interface (Figure 3.4c). As the pore expands, there is a tendency for the nanogels to move away from the interface and into the expanding pore phase. However, the displacement of nanogels from the interface is hindered by an increase in interfacial energy (ΔG_2). This indicates that when the nanogels try to move from the interface into the expanding pore phase, the energy required to separate the nanogel from the interface increases, creating a barrier that resists this movement (Figure 3.4d). At 25°C, PNIPAM is hydrophilic, which encourages the nanogels at the interface to absorb additional water into the expanding pore spaces. This water absorption results in the enlargement of the membrane pores, which are subsequently modified by the *in situ*-assembled nanogels.

The membranes with responsive nanogels fabricated by the VIPS method feature enlarged, interconnected pores within the porous matrix (Figures 3.4e and 3.4f). The PNIPAM nanogels, which are assembled *in situ* at the interfaces of these interconnected pores, function as thermo-responsive gates (Figures 3.4g and 3.4h). Below the volume phase transition temperature (VPTT) of PNIPAM nanogels (T < VPTT), the nanogels remain swollen, causing the gate to close (Figure 3.4g). When the temperature exceeds the VPTT (T > VPTT), the nanogels shrink, leading to the opening of the gate (Figure 4h). Because of the interconnected nature of the pores, the thermo-responsive gates form a three-dimensional network that links the pores of the membrane (Figures 3.4i–3.4l). This three-dimensional structure, which integrates interconnected pores with responsive gates, significantly improves both the flux and the stimuli-responsive properties of the smart gating membranes. The *in situ* assembly of nanogels on pore surfaces via the VIPS method, combined with its simple fabrication process, effectively addresses the challenges of enhancing the flux, responsiveness, and mechanical strength of smart gating membranes.

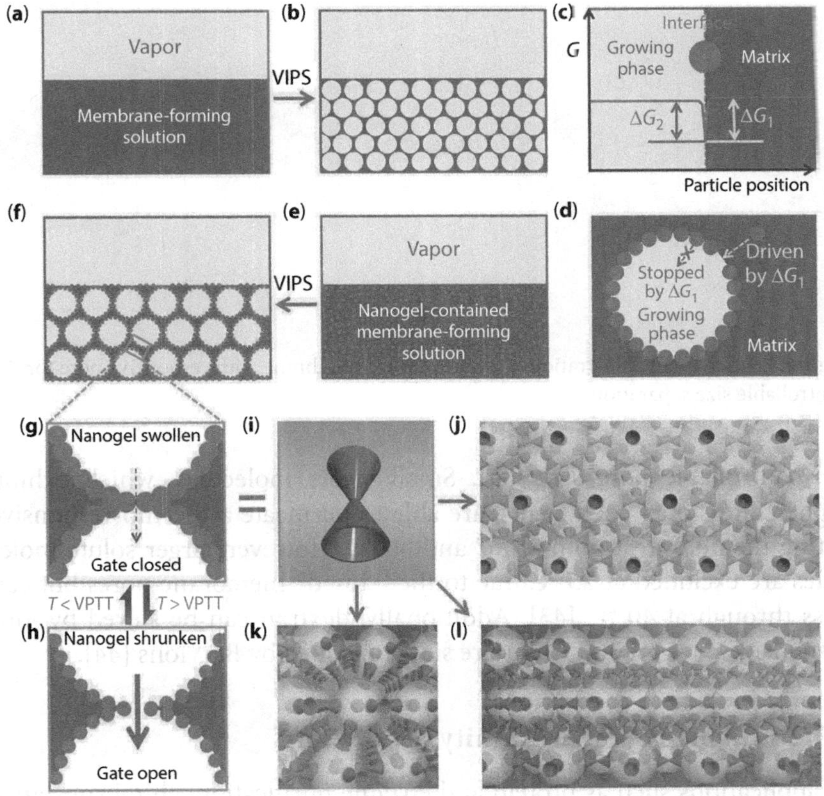

Figure 3.4 Schematic illustration of the VIPS process of membrane formation including the *in situ* assembly of responsive nanogels, three-dimensional interconnected pores, and the functional gating behavior [40].

3.3 Stimuli-Responsive Separations

3.3.1 Smart Pores for Size Separations

The smart porous membranes can facilitate graded sieving separations based on the size on the premise of optimal grafting or coating yield of responsive gates [12]. The swollen gates would diminish the membrane pores, thereby permitting only smaller molecules or particles to pass through, while larger ones are blocked (Figure 3.5). However, if exposed to environmental stimuli, the pore size increases, allowing large molecules or particles to permeate the membrane. For instance, a pH-responsive gating membrane can selectively reject fluorescein isothiocyanate (FITC)-dextran of different molecular weights under acidic or basic conditions due to the

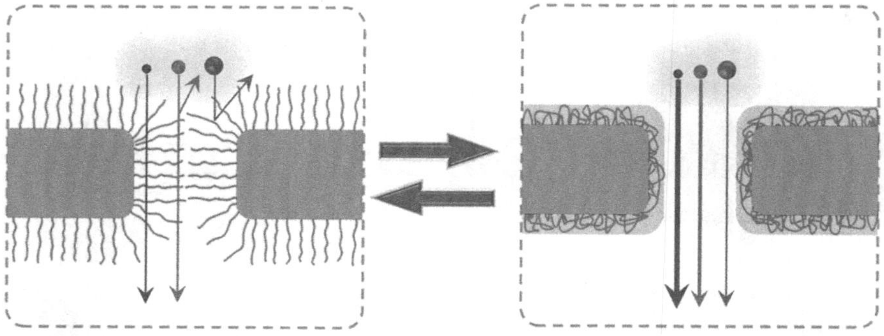

Figure 3.5 Schematic illustration of smart porous membrane with responsive pore for controllable size separation.

pH-controllable pore size [37]. Small solute molecules, which exhibit higher diffusional coefficients, are able to permeate a thermo-responsive gating membrane at both 25 °C and 40 °C. However, larger solute molecules are excluded at 25 °C due to the "closed" membrane pores but can pass through at 40 °C [43]. Additionally, dextran can be sieved by ion-responsive membranes, with pore size controlled by Ba^{2+} ions [44].

3.3.2 Smart Pores for Affinity Separation

In applications such as protein and enzyme purification, chiral substance separation, and the isolation of hydrophobic solutes, smart porous membranes with self-regulating surface properties have been widely studied. These membranes modulate the interactions between the pore surfaces and target substances, thereby facilitating efficient separation processes [12].

3.3.2.1 Hydrophobic Adsorption

Smart porous membranes featuring PNIPAM gates, which exhibit reversible transitions between hydrophilic and hydrophobic states in response to temperature changes, can be utilized for selective separation of hydrophobic substances like bovine serum albumin (BSA) through hydrophobic adsorption [45]. PNIPAM undergoes a clear and reversible phase transition at 32°C, which is defined as lower critical solution temperature (LCST). Below this temperature, PNIPAM forms extensive hydrogen bonds with water molecules at its amide groups, resulting in a swollen, hydrophilic state. When the temperature rises above the LCST, these hydrogen bonds break, causing PNIPAM to shrink and become hydrophobic.

This thermo-responsive behavior enables PNIPAM-functionalized membranes to effectively separate hydrophobic molecules. Additionally, the presence of micro- and/or nano-structured surfaces of PNIPAM-modified membranes can enhance the wettability shift between hydrophilic and hydrophobic states [46, 47].

A thermo-responsive affinity membrane for hydrophobic adsorption has been developed, which incorporates nano-structured pores and a PNIPAM surface layer [45]. As illustrated in Figure 3.6, the Shirasu porous glass (SPG) substrate alone is ineffective at adsorbing BSA molecules, regardless of temperature. The large size of BSA (68 kDa) means that the smooth pores of the SPG membrane are not suitable for trapping these macromolecules. In the case of the ungrafted nanostructured membrane formed by depositing SiO_2 nanoparticles onto the SPG membrane surface, BSA is increasingly adsorbed at first, about 0.5 mg/m^2, but no further increase is observed despite temperature changes. However, compared to the SPG substrate, BSA adsorption is enhanced on the nanostructured SPG membrane due to the concave-convex surface, which provides multiple steric sites for BSA molecules to interact, along with a larger surface area for adsorption. Despite this, the ungrafted nanostructured SPG membrane does not show thermo-responsiveness in the BSA adsorption process.

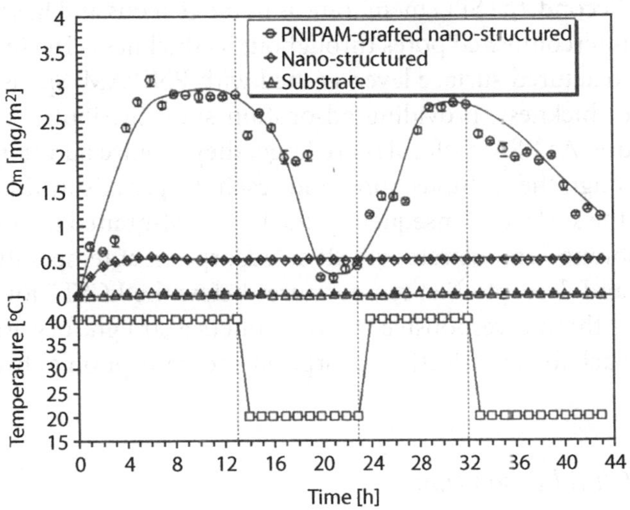

Figure 3.6 Temperature-driven dynamic adsorption and desorption of BSA on the SPG substrate membrane, nanostructured SPG membrane, and PNIPAM-grafted nanostructured SPG membrane [45].

When PNIPAM brushes are grafted onto these nano-structured surfaces, the PNIPAM-grafted nanostructured SPG membrane demonstrates significantly enhanced adsorption. At 40°C, the equilibrium adsorption (Q_m) reaches approximately 2.95 mg/m^2, much higher than the 0.5 mg/m^2 observed with the ungrafted membrane. This increase is attributed to the hydrophobic nature of the PNIPAM-grafted surface above the LCST, where BSA adsorption primarily occurs through hydrophobic interactions. This advanced membrane design offers a highly efficient solution for the thermo-responsive adsorption and desorption of BSA molecules, driven by PNIPAM grafting and nanostructuring. Below the LCST, the nanostructured surface enhances hydrophilicity, facilitated by capillary action within the nano-concave spaces formed between SiO_2 nanoparticles. This promotes strong hydration and adsorption of BSA molecules. The increased surface area from the nanostructuring further amplifies adsorption capacity by providing more active sites. Additionally, the concave-convex surface topology introduces steric interaction sites and enhances hydrophobic forces above the LCST, enabling selective and reversible adsorption. These synergistic features establish the membrane as a versatile tool for protein separation and other bioengineering applications.

From the result of the adsorption and desorption processes, it reaches equilibrium slowly. This can be attributed to two primary factors. First, the large molecular size of BSA molecules slows down their movement, making it take several hours for them to reach equilibrium on non-porous materials. Second, the SPG membrane is homogeneous and has nearly uniform and interconnected pores throughout its thickness. The introduction of a nanostructured surface layer, grafted with PNIPAM, spans the entire membrane thickness, providing adsorption sites distributed throughout the structure. As BSA molecules are large, they require additional time to diffuse through the tortuous pores and reach deeper adsorption sites, farther from the surface. Consequently, the PNIPAM-grafted nanostructured SPG membrane demonstrates effective "adsorption at temperatures above the LCST and desorption at temperatures below the LCST" for BSA molecules. This thermo-responsive behavior offers a straightforward and efficient approach for the selective adsorption and desorption of hydrophobic solutes.

3.3.2.2 Chiral Resolution

Enantiomers of chiral drugs often exhibit significant differences in their pharmacological properties, including metabolic effects, rates of metabolism, and toxicity levels, owing to the high stereo-selectivity of various biological and enzymatic processes. To minimize undesirable side effects or toxicity, smart porous membranes with chiral recognition capabilities are increasingly used to purify enantiomeric drug mixtures [9]. The working principle behind these responsive membranes for chiral resolution is depicted in Figure 3.7 [48]. These membranes incorporate thermo-responsive PNIPAM chains that are grafted with beta-cyclodextrin (β-CD) groups, which function as selective gates. The β-CD molecule, characterized by its hydrophobic cavities and hydrophilic outer surface, can selectively interact with a guest molecule. The species' intermolecular interactions, such as hydrophobic association electrostatic forces and hydrogen bonding, make β-CD an effective chiral selector for enantio-separation. Moreover, combined with the phase transition of PNIPAM, PNIPAM-modified β-CD systems would modulate the association constant between β-CD and the guest molecules via changing environmental temperature [20–24].

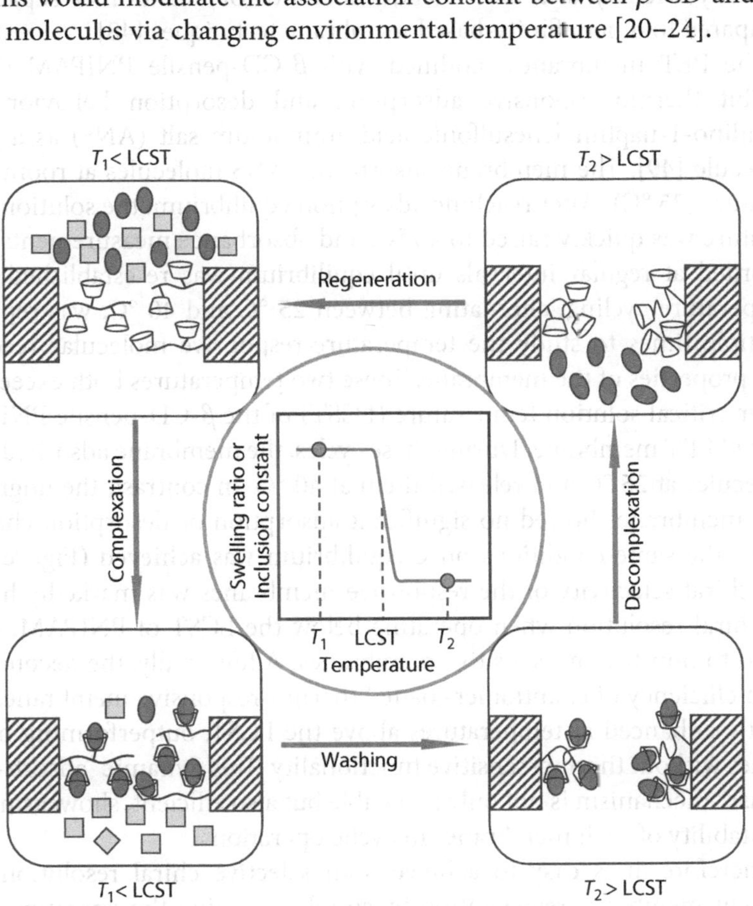

Figure 3.7 Schematic illustration of the chiral resolution process using a thermo-responsive membrane [48].

When the membrane operates chiral resolution at temperatures below the LCST, the grafted PNIPAM modified β-CD chains become swollen and hydrophilic. In this state, the β-CD enables selective capture of one enantiomer, while the other enantiomer passes through the membrane. Once the β-CD-enantiomer complex reaches equilibrium, a washing step is performed to remove any unbound molecules. Subsequently, raising the temperature above the LCST would cause the PNIPAM modified β-CD chains to shrink and transition to a hydrophobic state. This change significantly lowers the binding affinity between β-CD and the captured enantiomer, promoting the release of the bound enantiomer. As a result, the enantiomers are effectively separated, and the membrane is regenerated. This approach, especially the decomplexation step, offers a more environmentally friendly and user-friendly method for enantiomer separation compared to other affinity-based membrane techniques [48].

The PET membranes modified with β-CD-pensile PNIPAM chains exhibit thermo-responsive adsorption and desorption behavior with 8-anilino-1-naphthalenesulfonic acid ammonium salt (ANS) as a guest molecule [49]. The membrane adsorbs the ANS molecules at room temperature (25 °C). After reaching adsorption equilibrium, the solution temperature was quickly raised to 40 °C, and absorbance measurements were recorded at regular intervals until equilibrium was re-established. This temperature cycling, alternating between 25 °C and 40 °C, was repeated multiple times to study the temperature-responsive molecular recognition properties of the membrane. These two temperatures both exceed the lower critical solution temperature (LCST) of the β-CD-pensile PNIPAM grafted PET membrane. During these cycles, the membrane adsorbed ANS molecules at 25 °C and released them at 40 °C. In contrast, the ungrafted PET membrane showed no significant adsorption or desorption changes under the same conditions once equilibrium was achieved (Figure 3.8). The chiral selectivity of the responsive membranes was markedly higher for chiral resolution when operating below the LCST of PNIPAM, compared to non-thermo-sensitive membranes. Additionally, the decomplexation efficiency of enantiomer-loaded thermo-responsive membranes was greatly enhanced at temperatures above the LCST, outperforming membranes without thermo-sensitive functionality. This dynamic "adsorb-and-release" mechanism is not only reversible but also efficient, showcasing the adaptability of such membranes in cyclic operations.

Therefore, it is easy to achieve both selective chiral resolution and efficient membrane regeneration by simply adjusting the operating temperature. The enhanced chiral selectivity below the LCST suggests potential applications in enantiomeric separations where fine discrimination

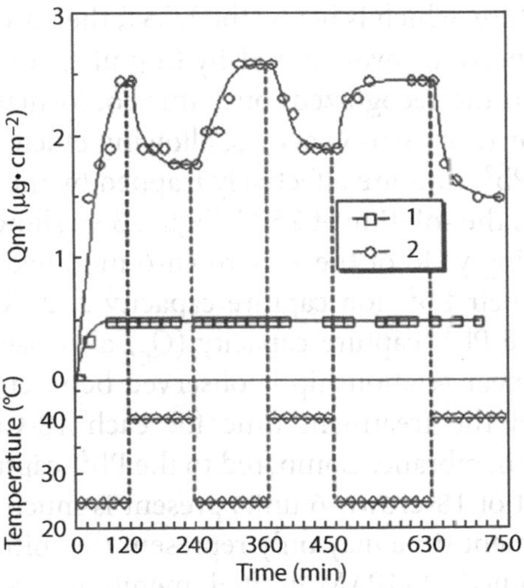

Figure 3.8 Temperature-responsive adsorption of ANS on ungrafted membrane (1) and β-CD-pensile PNIPAM grafted PET membrane (2) [49].

between chiral molecules is required. The superior decomplexation efficiency above the LCST further underscores the utility of these membranes in practical applications, as they enable easy regeneration and reuse without compromising performance. This dual capability of selective binding and efficient release highlights their promise for advanced separations in pharmaceutical, environmental, and chemical industries. This thermal-responsive membrane presents an innovative and effective solution to the complex decomplexation challenge faced by chiral solid membranes, making it especially promising for applications in chiral separation processes.

3.3.3.3 *Removal of Heavy Metal Ions*

18-Crown-6 is well known for its outstanding capacity to selectively interact with metal ions, forming highly stable 1:1 or 2:1 host-guest complexes. Among the various metal ions investigated, lead(II) (Pb^{2+}) exhibits one of the highest inclusion constants when binding with 18-crown-6, highlighting its exceptional affinity for this particular ion [50, 51]. Due to its high selectivity and strong complexation capabilities, polymers incorporating 18-crown-6 have been identified as promising materials for detecting and removing Pb^{2+}. Membranes functionalized with 18-Crown-6 modified PNIPAM and can be used in water treatment applications, where their Pb^{2+}-responsive gating pores efficiently eliminate trace amounts of lead ions.

At room temperature, which is below the LCST, the 18-Crown-6 modified PNIPAM chains become swollen and hydrophilic, resulting in minimal steric hindrance to the recognized ions. This conformation enhances the accessibility of the 18-crown-6 cavities, allowing efficient capture of Pb^{2+} ions. As a result, Pb^{2+} ions are effectively trapped by the 18-crown-6 units and removed from the solution at 25°C. Figure 3.9a shows the relationship between the grafting yield of the 18-Crown-6 modified PNIPAM grafted membranes and their Pb^{2+} ion capture capacity at 25°C. As the grafting yield increases, the Pb^{2+} capture capacity (Q_C) also rises over the course of treatment. A linear relationship is observed between the Pb^{2+} capture capacity (Q_C) and the treatment time for each 18-Crown-6 modified PNIPAM-grafted membrane. Compared to the Pb^{2+} captured by the membrane, the amount of 18-crown-6 units present is much larger. Therefore, the 34-hour treatment time may only represent the initial phase, with the 18-Crown-6 modified PNIPAM-grafted membrane requiring a longer period to reach saturation with Pb^{2+} ions.

The effect of grafting yield on the Pb^{2+} capture ratio (R_C) of the 18-Crown-6 modified PNIPAM-grafted membrane, with R_C defined as:

$$Rc = \frac{Qc}{VC_0} \times 100\%$$

where Q_C represents the amount of Pb^{2+} accumulated on the membrane surface (mmol/m^2), C_0 is the initial Pb^{2+} concentration in the solution (mmol/L), and V is the total volume of Pb^{2+} solution treated for 34 hours. As the grafting yield increases, the Pb^{2+} capture ratio (R_C) also increases, reaching approximately 75% when treating trace concentration of the Pb^{2+} (Figure 3.9b).

When the 18-Crown-6 modified PNIPAM-grafted membrane is used to treat the Pb^{2+}-contained solution mixed with trace amounts of K^+ and Ba^{2+}, Pb^{2+} is selectively removed with high efficiency (Figure 3.9c). The capture ratio (R_C) of Pb^{2+} exceeds 80%. Due to the relatively low inclusion constants of 18-crown-6 with K^+ and Ba^{2+} ions, the R_C values for K^+ and Ba^{2+} are both below 5%. Additionally, even though the concentration of K^+ ions is four orders of magnitude higher than the trace Pb^{2+} ions in the feed solution, the Pb^{2+} ions are still selectively adsorbed by the gating membrane (Figure 3.9d). The Pb^{2+} capture ratio (R_C) reaches about 78%, while the average capture ratio for K^+ is approximately 1.6% ± 0.8%. The experimental error in the K^+ capture ratio is likely due to the high concentrations of K^+ in both the feed and filtrate solutions, resulting in a very low

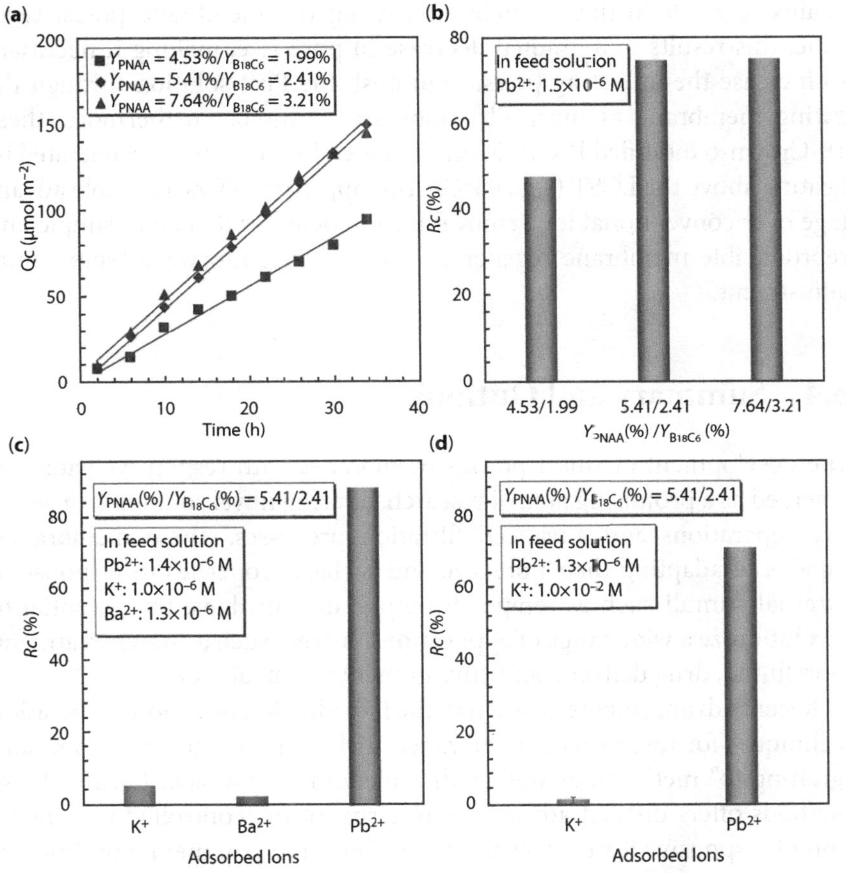

Figure 3.9 A, B) Effects of the grafting yield of 18-Crown-6 modified PNIPAM-grafted membranes on Pb^{2+} capture capacity (Q_c) (a) and Pb^{2+} capture ratio (R_C) (b) at 25°C. The Pb^{2+} solution has a concentration of 1.5×10^{-6} M. C, D) Capture ratios (R_C) for 18-Crown-6 modified PNIPAM-grafted membranes tested with mixed solutions containing trace concentrations of Pb^{2+}, K^+, and Ba^{2+} ions (c), and with a mixed solution containing high-concentration K^+ and trace Pb^{2+} (d) [52].

relative adsorption of K^+ ions on the membrane. These results demonstrate that gating membranes grafted with an adequate amount of 18-Crown-6 modified PNIPAM copolymers can effectively and selectively remove trace Pb^{2+} ions from environmental solutions, even in the presence of obstructive ions such as K^+ and Ba^{2+}.

When Pb^{2+} ions are adsorbed by the 18-crown-6 units on the gating polymer of the membranes at 25°C, the electrostatic repulsion between the charged complexes causes the 18-Crown-6 modified PNIPAM copolymer chains to swell further, thereby restricting the membrane pores. Over time, this results in a gradual decrease in pore size, making it necessary to increase the operating pressure to push the Pb^{2+} solution through the gating membrane as more Pb^{2+} ions are captured. Furthermore, these 18-Crown-6 modified PNIPAM-grafted membranes can be regenerated by heating above the LCST (e.g., 45°C). This approach offers a notable advantage over conventional ion removal technologies, as it enables simple and reproducible membrane regeneration via a straightforward temperature adjustment.

3.4 Summary and Outlook

The development of smart porous membranes with responsive gates has emerged as a promising area of research, offering novel solutions for selective separations and advanced filtration processes. These membranes, capable of adapting their pore size and surface properties in response to external stimuli such as temperature, pH, or ions, have the potential to revolutionize a wide range of applications in areas such as water treatment, biorefining, drug delivery, and environmental remediation.

Recent advancements have led to the introduction of various fabrication techniques for these smart membranes, including the "grafting from" and "grafting to" methods, as well as the "blending" approach. Each of these methods offers distinct advantages in terms of the controlled incorporation of responsive features, enabling the fine-tuning of membrane characteristics to meet specific needs. The incorporation of functional responsive elements, such as polymers, nanoparticles, or molecular gates, has significantly enhanced the ability of membranes to selectively separate compounds based on size exclusion or molecular recognition.

Despite these advancements, the field remains in the early stages, and several challenges must be addressed to unlock the full potential of smart membranes. These challenges include improving the scalability of fabrication methods, enhancing membrane stability under operational conditions, and optimizing the responsiveness of gating mechanisms. Moreover, while laboratory-scale success has been achieved, the translation of these membranes into large-scale industrial applications remains a significant hurdle. Issues related to cost, long-term durability, and process optimization require focused attention to facilitate the commercial viability of smart gating membranes.

The future of this field is promising, with many exciting directions for further research and development. Key areas for exploration include:

Process-Oriented Design: Developing smart membranes with tailored properties designed to meet the specific demands of industrial separation processes. This involves optimizing not only the functional properties but also the manufacturing processes to ensure reproducibility and scalability.

Micro/Nano-Structural Control: Achieving better control over the micro and nano-architecture of the membrane surface to enhance selectivity, permeability, and responsiveness. Advanced fabrication techniques such as molecular imprinting, electrospinning, and self-assembly can offer new avenues for structural optimization.

Membrane Stability: Addressing the long-term operational stability of responsive membranes, particularly under harsh conditions such as extreme pH, high pressure, or the presence of organic solvents. Enhanced material selection and surface modification strategies are needed to improve membrane durability.

Scalability and Industrial Application: This includes optimizing membrane performance in continuous operation, reducing manufacturing costs, and ensuring that large-scale production does not compromise the membrane's functional performance.

In conclusion, while the development of smart porous membranes with responsive gates has made substantial strides, much work remains to be done. Future research must focus on addressing the current limitations, expanding the scope of applications, and optimizing performance to meet the rigorous demands of industrial-scale operations. As these challenges are overcome, the potential of smart gating membranes to drive innovation across a variety of industries will continue to grow, offering transformative solutions to some of the most pressing separation and filtration challenges.

References

1. Wang, L., Boutilier, M.S.H., Kidambi, P.R., Jang, D., Hadjiconstantinou, N.G., Karnik, R., Fundamental transport mechanisms, fabrication and potential applications of nanoporous atomically thin membranes. *Nat. Nanotech.*, 12, 509, 2017.
2. Werber, J.R., Osuji, C.O., Elimelech, M., Materials for next-generation desalination and water purification membranes. *Nat. Rev. Mater.*, 1, 16018, 2016.
3. Wang, Z.G., Wan, L.S., Xu, Z.K., Surface engineerings of polyacrylonitrile-based asymmetric membranes towards biomedical applications: An overview. *J. Membr. Sci.*, 304, 8, 2007.
4. van Reis, R. and Zydney, A., Bioprocess membrane technology. *J. Membr. Sci.*, 297, 16, 2007.

5. Stamatialis, D.F., Papenburg, B.J., Gironés, M., Saiful, S., Bettahalli, S.N., Schmitmeier, S., Wessling, M., Medical applications of membranes: drug delivery, artificial organs and tissue engineering. *J. Membr. Sci.*, 308, 1, 2008.
6. Liu, Z., Ju, X.J., Wang, W., Xie, R., Jiang, L., Chen, Q., Zhang, Y.Q., Wu, J.F., Chu, L.Y., Stimuli-responsive capsule membranes for controlled release in pharmaceutical applications. *Curr. Pharm. Des.*, 23, 295–301, 2017.
7. Chu, L.Y., *Smart Membrane Materials and Systems*, Zhejiang University Press, Hangzhou and Springer-Verlag Berlin Herdelberg, 2011.
8. Rana, D. and Matsuura, T., Surface modifications for antifouling membranes. *Chem. Rev.*, 110, 2448, 2010.
9. Xie, R., Chu, L.Y., Deng, J.G., Membranes and membrane processes for chiral resolution. *Chem. Soc. Rev.*, 37, 1243, 2008.
10. Liu, Z., Wang, W., Xie, R., Ju, X.J., Chu, L.Y., Stimuli-responsive smart gating membranes. *Chem. Soc. Rev.*, 45, 460, 2016.
11. Darvishmanesh, S., Qian, X., Wickramasinghe, S.R., Responsive membranes for advanced separations. *Curr. Opin. Chem. Eng.*, 8, 98–104, 2015.
12. Chu, L.Y., Xie, R., Ju, X.J., Stimuli-responsive membranes: smart tools for controllable mass-transfer and separation processes. *Chin. J. Chem. Eng.*, 19, 891, 2011.
13. Zhao, Y.H., Wee, K.H., Bai, R., A novel electrolyte-responsive membrane with tunable permeation selectivity for protein purification. *ACS Appl. Mater. Inter.*, 2, 203, 2009.
14. Hsu, C.C., Wu, C.S., Liu, Y.L., Multiple stimuli-responsive poly (vinylidene fluoride) (PVDF) membrane exhibiting high efficiency of membrane clean in protein separation. *J. Membr. Sci.*, 450, 257, 2014.
15. Qiu, X., Yu, H., Karunakaran, M., Pradeep, N., Nunes, S.P., Peinemann, K.V., Selective separation of similarly sized proteins with tunable nanoporous block copolymer membranes. *ACS Nano*, 7, 768, 2012.
16. Khulbe, K.C., Feng, C., Matsuura, T., The art of surface modification of synthetic polymeric membranes. *J. Appl. Polym. Sci.*, 115, 855, 2010.
17. Deng, J., Wang, L., Liu, L., Yang, W., Developments and new applications of UV-induced surface graft polymerizations. *Prog. Polym. Sci.*, 34, 156, 2009.
18. Ran, J., Wu, L., Zhang, Z., Xu, T., Atom transfer radical polymerization (ATRP): a versatile and forceful tool for functional membranes. *Prog. Polym. Sci.*, 39, 124, 2014.
19. Yamaguchi, T., Nakao, S., Kimura, S., Plasma-graft filling polymerization: preparation of a new type of pervaporation membrane for organic liquid mixtures. *Macromolecules*, 24, 5522, 1991.
20. Xie, R., Chu, L.Y., Chen, W.M., Xiao, W., Wang, H.D., Qu, J.B., Characterization of microstructure of poly(*N*-isopropylacrylamide)-grafted polycarbonate track-etched membranes prepared by plasma-graft pore-filling polymerization. *J. Membr. Sci.*, 258, 157, 2005.

21. Liu, Z., Luo, F., Ju, X.J., Xie, R., Luo, T., Sun, Y.M., Chu, L.Y., Positively K+-responsive membranes with functional gates driven by host-guest molecular recognition. *Adv. Funct. Mater.*, 22, 4742, 2012.
22. Li, P.F., Xie, R., Jiang, J.C., Meng, T., Yang, M., Ju, X.J., Yang, L., Chu, L.Y., Thermo-responsive gating membranes with controllable length and density of poly (*N*-isopropylacrylamide) chains grafted by ATRP method. *J. Membr. Sci.*, 337, 310, 2009.
23. Alem, H., Duwez, A.S., Lussis, P., Lipnik, P., Jonas, A.M., Demoustier-Champagne, S., Microstructure and thermo-responsive behavior of poly (*N*-isopropylacrylamide) brushes grafted in nanopores of track-etched membranes. *J. Membr. Sci.*, 308, 75, 2008.
24. Zdyrko, B. and Luzinov, I., Polymer brushes by the "grafting to" method. *Macromol. Rapid Comm.*, 32, 859, 2011.
25. Menne, D., Pitsch, F., Wong, J.E., Pich, A., Wessling, M., Temperature-Modulated Water Filtration Using Microgel-Functionalized Hollow-Fiber Membranes. *Angew. Chem. Int. Ed.*, 53, 5706, 2014.
26. Luzinov, I. and Tsukruk, V.V., Ultrathin triblock copolymer films on tailored polymer brushes. *Macromolecules*, 35, 5963, 2002.
27. Motornov, M., Sheparovych, R., Katz, E., Minko, S., Chemical gating with nanostructured responsive polymer brushes: mixed brush versus homopolymer brush. *ACS Nano*, 2, 41, 2007.
28. Penn, L.S., Hunter, T.F., Quirk, R.P., Lee, Y., Deactivation of epoxide-derivatized surfaces. *Macromolecules*, 35, 2859, 2002.
29. Huang, H., Cammers, A., Penn, L.S., Grafting of free chains in the presence of preexisting polymer brushes. *Macromolecules*, 39, 7064, 2006.
30. Montagne, F., Polesel-Maris, J., Pugin, R., Heinzelmann, H., Poly (*N*-isopropylacrylamide) thin films densely grafted onto gold surface: preparation, characterization, and dynamic AFM study of temperature-induced chain conformational changes. *Langmuir*, 25, 983, 2008.
31. Crooks, R.M., Patterning of hyperbranched polymer films. *ChemPhysChem.*, 2, 644, 2001.
32. Ostaci, R.V., Damiron, D., Capponi, S., Vignaud, G., Léger, L., Grohens, Y., Drockenmuller, E., Polymer brushes grafted to "passivated" silicon substrates using click chemistry. *Langmuir*, 24, 2732, 2008.
33. Binder, W.H. and Sachsenhofer, R., 'Click' chemistry in polymer and materials science. *Macromol. Rapid Comm.*, 28, 15, 2007.
34. Wang, C., Wu, J., Xu, Z.K., High-density glycosylation of polymer membrane surfaces by click chemistry for carbohydrate-protein recognition. *Macromol. Rapid Comm.*, 31, 1078, 2010.
35. Clodt, J.I., Filiz, V., Rangou, S., Buhr, K., Abetz, C., Höche, D., Hahn, J., Jung, A., Abetz, V., Double stimuli-responsive isoporous membranes via post-modification of pH-sensitive self-assembled diblock copolymer membranes. *Adv. Funct. Mater.*, 23, 731, 2013.

36. Xue, J., Chen, L., Wang, H.L., Zhang, Z.B., Zhu, X.L., Kang, E.T., Neoh, K.G., Stimuli-responsive multifunctional membranes of controllable morphology from poly (vinylidene fluoride)-graft-poly [2-(*N, N*-dimethylamino) ethyl methacrylate] prepared via atom transfer radical polymerization. *Langmuir*, 24, 14151, 2008.
37. Luo, T., Lin, S., Xie, R., Ju, X.J., Liu, Z., Wang, W., Mou, C.L., Zhao, C.S., Chen, Q.M., Chu, L.Y., pH-responsive poly (ether sulfone) composite membranes blended with amphiphilic polystyrene-*block*-poly (acrylic acid) copolymers. *J. Membr. Sci.*, 450, 162, 2014.
38. Wang, G., Xie, R., Ju, X.J., Chu, L.Y., Thermo-responsive polyethersulfone composite membranes blended with poly(*N*-isopropyl acrylamide) nanogels. *Chem. Eng. Technol.*, 35, 2015, 2012.
39. Luo, F., Zhao, Q., Xie, R., Cao, Y., Ju, X.J., Wang, W., Liu, Z., Chu, L.Y., Effects of fabrication conditions on the microstructures and performances of smart gating membranes with *in situ* assembled nanogels as gates. *J. Membr. Sci.*, 519, 32, 2016.
40. Luo, F., Xie, R., Liu, Z., Ju, X.J., Wang, W., Lin, S., Chu, L.Y., Smart gating membranes with *in situ* self-assembled responsive nanogels as functional gates. *Sci. Rep.*, 5, 14708, 2015.
41. Venault, A., Chang, Y., Wang, D.M., Bouyer, D., A review on polymeric membranes and hydrogels prepared by vapor-induced phase separation process. *Polym. Rev.*, 53, 568, 2013.
42. Chen, L.Y., Xu, J.Q., Choi, H., Konishi, H., Jin, S., Li, X.C., Rapid control of phase growth by nanoparticles. *Nat. Commun.*, 5, 3879, 2014.
43. Chu, L.Y., Niitsuma, T., Yamaguchi, T., Nakao, S., II, Thermoresponsive transport through porous membranes with grafted PNIPAM gates. *AlChE J.*, 49, 896, 2003.
44. Ito, T., Hioki, T., Yamaguchi, T., Shinbo, T., Nakao, S.I., Kimura, S., Development of a molecular recognition ion gating membrane and estimation of its pore size control. *J. Am. Chem. Soc.*, 124, 7840, 2002.
45. Meng, T., Xie, R., Chen, Y.C., Cheng, C.J., Li, P.F., Ju, X.J., Chu, L.Y., A thermo-responsive affinity membrane with nano-structured pores and grafted poly (*N*-isopropylacrylamide) surface layer for hydrophobic adsorption. *J. Membr. Sci.*, 349, 258, 2010.
46. Sun, T., Wang, G., Feng, L., Liu, B., Ma, Y., Jiang, L., Zhu, D., Reversible switching between superhydrophilicity and superhydrophobicity. *Angew. Chem. Int. Ed.*, 116, 361, 2004.
47. Song, W., Xia, F., Bai, Y., Liu, F., Sun, T., Jiang, L., Controllable water permeation on a poly (*N*-isopropylacrylamide)-modified nanostructured copper mesh film. *Langmuir*, 23, 327, 2007.
48. Yang, M., Chu, L.Y., Wang, H.D., Xie, R., Song, H., Niu, C.H., A thermoresponsive membrane for chiral resolution. *Adv. Funct. Mater.*, 18, 652, 2008.

49. Xie, R., Zhang, S.B., Wang, H.D., Yang, M., Li, P.F., Zhu, X.L., Chu, L.Y., Temperature-dependent molecular-recognizable membranes based on poly (*N*-isopropylacrylamide) and β-cyclodextrin. *J. Membr. Sci.*, 326, 618, 2009.
50. Izatt, R.M., Bradshaw, J.S., Nielsen, S.A., Lamb, J.D., Christensen, J.J., Sen, D., Thermodynamic and kinetic data for cation-macrocycle interaction. *Chem. Rev.*, 85, 271, 1985.
51. Izatt, R.M., Pawlak, K., Bradshaw, J.S., Bruening, R L., Thermodynamic and kinetic data for macrocycle interactions with cations and anions. *Chem. Rev.*, 91, 1721, 1991.
52. Liu, Z., Luo, F., Ju, X.J., Xie, R., Sun, Y.M., Wang, W., Chu, L.Y., Gating membranes for water treatment: detection and removal of trace Pb^{2+} ions based on molecular recognition and polymer phase transition. *J. Mater. Chem. A*, 1, 9659, 2013.

4

Development of Anion Exchange Membranes *via* Click Chemistry

Binoy Maiti[1], Alex Abramov[1] and David Díaz Díaz[1,2,3]*

[1]*Institut für Organische Chemie, Universität Regensburg, Regensburg, Germany*
[2]*AFM-NANO, Instituto Universitario de Bio-Orgánica Antonio González (IUBO-AG), Universidad de La Laguna, Avda. Astrofísico Francisco Sánchez 2, La Laguna, Spain*
[3]*Departamento de Química Orgánica, Universidad de La Laguna, Avda. Astrofísico Francisco Sánchez 3, La Laguna, Spain*

Abstract
Anion exchange membrane fuel cells (AEMFCs) have gained increasing attention due to significantly reduced fuel cell device costs, particularly in the automotive industry and other energy-related applications. The alkaline anion exchange membrane (AAEM) is the key component of the anion exchange membrane fuel cell, which acts as an electrolyte in the fuel cell. In recent years, numerous studies have focused on the development of new anion exchange membranes for better fuel cell performance. In this contribution, we discuss the advantages and disadvantages of anion exchange membranes made from different polymer backbones using a click chemistry approach.

Keywords: Click chemistry, anion exchange membranes, fuel cells

Abbreviations

AAEM	Alkaline Anion Exchange Membrane
AEM	Anion Exchange Membrane
AEMFC	Anion Exchange Membrane Fuel Cell
IEC	Ion Exchange Capacity

**Corresponding author*: ddiazdiaz@ull.edu.es

Annarosa Gugliuzza and Wanqin Jin (eds.) Porous Membranes: Breakthroughs in Manufacturing and Applications, (73–96) © 2025 Scrivener Publishing LLC

PEMFC	Proton Exchange Membrane Fuel Cell
QA	Quaternary Ammonium
SR	Swelling Ratio
ED	Electrodialysis
SEM	Scanning Electron Microscopy
AFM	Atomic Force Microscopy
XPS	X-ray Photoelectron Spectroscopy
TGA	Thermogravimetric Analysis
DSC	Differential Scanning Calorimetry
PPO	Poly(2,6-dimethyl-1,4-phenylene oxide)
BPPO	Brominated poly(2,6-dimethyl-1,4-phenylene oxide)
PS	Polystyrene
PES	Polyether sulfone

4.1 Introduction

Over the last few decades, intensive research efforts have focused on alternative energy sources to reduce our dependency on fossil fuels. In this context, fuel cells have shown great potential for replacing conventional fossil energies due to their high efficiency and low emissions [1]. Recently, alkaline polymer electrolyte fuel cells (APEFCs) that utilize anion-exchange membranes (AEMs) as key components, separating oxidant and fuel chambers and acting as conductors of hydroxide ions, have drawn tremendous attention, mostly because of their advantages of enhanced oxygen reduction reaction kinetics and the utilization of nonprecious metal catalysts compared with acidic fuel cells. However, the most commonly reported AEMs displayed inferior hydroxide conductivity compared to proton exchange membranes because hydroxide ions (OH^-) inherently possess less mobility than protons. To further enhance the charge densities of AEMS, the ion exchange capacity (IEC) of the polyelectrolyte must increase. However, it is important to mention that increasing the IEC value will also significantly increase membrane swelling, which affects the dimensional durability or mechanical strength of the membrane, especially at elevated temperatures. Thus, the key to balance the trade-off between conductivity and mechanical strength is to keep the AEMs at a moderate IEC value. Besides, IEC value can be controlled by designing new AEMs based on block copolymers and side-chain-type polymers. In this sense, several methods have already been described in the literature to functionalize AEMs. Most commonly, nucleophilic aromatic substitution polycondensation reactions have been employed to introduce functional side groups at high temperatures [2]. Unfortunately, these traditional reactions have been unable to control the functionalization of AEMs and, hence, the IEC values. Thus, it is necessary to develop a new procedure to fabricate AEMs with new and tailorable functionalities. In this regard, a click chemistry-based approach is easy to carry out, broadly applicable for all functional groups, and provides convenient synthetic scalability at low cost. Over the last few years, the Cu(I)-catalyzed alkyne–azide cycloaddition (CuAAC), the prototype of click chemistry, has drawn tremendous research attention for constructing novel polymer architectures because of its high efficiency, functional group tolerance, and moderate reaction conditions [3, 4]. Additionally, triazole groups in the polymer backbone help to form a 3D network of hydrogen bonds with hydroxide ion and water, which further enhances the ionic conductivity value of the system. Furthermore, the triazole ring-containing polymer usually shows higher thermal stability compared to the polymer without the triazole moiety [5]. Moreover, alkylation of 1,2,3-triazole provides the 1,2,3-triazolium cation, which has high thermal stability, tunable solubility, and low flammability, and has also been used for the preparation of AEMs [6]. The thiol-ene reaction is another example of a click reaction, which has also been used by several groups to fabricate AEMs. In this report, we have focused on AEMs fabricated using specifically click chemistry. Therefore, the preparation of AEMs without involving click chemistry is out of the scope of this contribution and has been reported elsewhere [7, 8]. For the sake of clarity, only the click step in the synthetic schemes provided in this chapter is given with experimental details such as catalyst, reaction time, and temperature.

4.2 Poly(2,6-Dimethyl Phenylene Oxide) (PPO)-Based Anion Exchange Membranes

Recently, commercially available PPO has been explored as a potential candidate to produce AEMs [9], mainly due to its high glass transition temperature (T_g = 210°C) and high mechanical strength [10]. Although PPO itself is not conductive, its functionalization with ionic compounds (cationic or anionic) makes it suitable for the preparation of ion exchange membranes. Such functionalization is typically carried out through the nucleophilic substitution at the benzylic position of PPO.

Li *et al.* [11] prepared a series of AEMs from quaternized PPO materials containing 1,2,3-triazoles moiety using CuAAC (Figure 4.1). It was already established that 1,2,3-triazole moieties help to form a continuous 3D hydrogen network with water/hydroxide molecules, which facilitates the water uptake and hydroxide diffusion. In this particular example, the

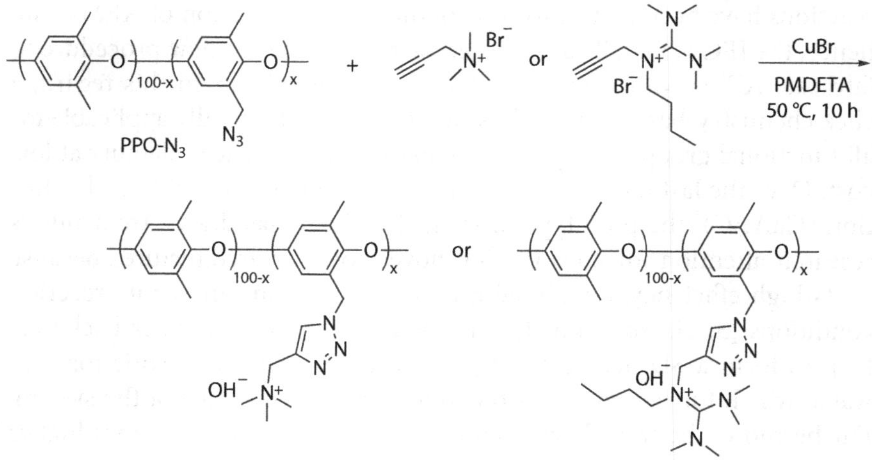

Figure 4.1 Synthesis of PPO copolymers bearing quaternary ammonium (QA) and guanidinium moieties through CuAAC. PMDETA = *N,N,N',N',N"*-pentamethylethylenetetramine [11].

observed hydroxide conductivity was 27.8–62 mS cm^{-1} at 20°C in water, which was much higher than that of a PPO-based AEM (TMA-20) made from trimethylamine (5 mS cm^{-1}) without the triazole moiety.

The authors also prepared a clicked PPO containing a guanidinium moiety, which is known as a super strong organic base, with an extra butyl group in the side chain to further confirm the beneficial effect of the triazole moiety on the conductivity of AEMs. The butyl chain was expected to reduce the interaction between water or hydroxides and the triazole ring. However, this membrane still absorbed more water and showed an almost seven-fold increase in hydroxide conductivity compared to the non-clicked TMA-20 membrane, regardless of the presence of the butyl chain.

Ge *et al.* [12] used CuAAC to prepare a series of AEM membranes based on PPO backbones bearing an alkyl side chain with different lengths (Figure 4.2). The resulting triazole groups and QA groups help to form a 3D hydrogen bond network of side-chain-type alkaline polymer electrolytes (APEs).

The long hydrophobic alkyl side chain with different chain lengths constitutes the key factor for the microphase separation and mechanical properties of AEMs. In this regard, different alkyl chains were attached to the cationic center in order to develop a better microphase-separated membrane that performs well in alkaline conditions. The results showed that the membranes having C14 chain length displayed better properties such as lower water uptake, higher OH^- conductivity, and more distinct ionic domains compared to those observed with C7 and C10 chain length membranes.

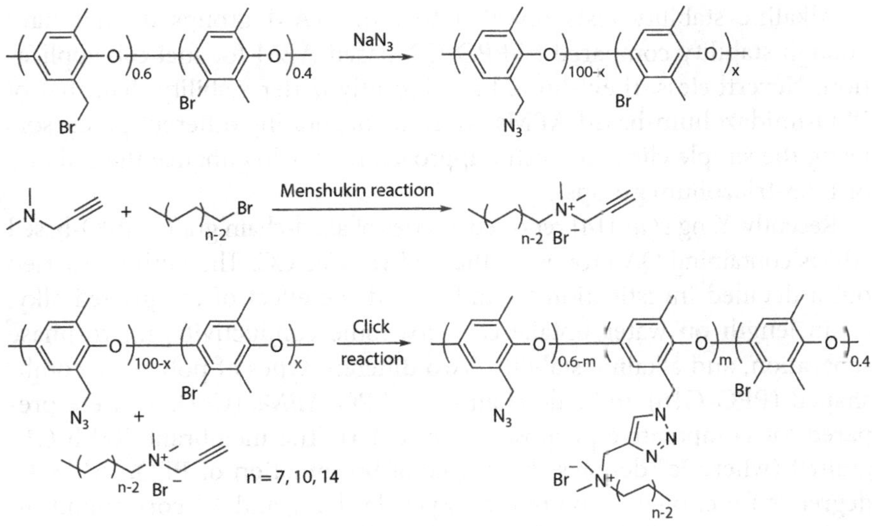

Figure 4.2 Synthesis of QA-containing PPO-based anion conductive copolymer [12].

Liu *et al.* [13] synthesized a series of anion exchange membranes from PPO *via* click chemistry followed by successive *N*-alkylation and ion exchange (Figure 4.3). Although the so-obtained 1,2,3-triazolium cation-functionalized membranes showed incomplete ion exchange in 1 M NaOH or $NaHCO_3$ for 24 h, even at elevated temperatures, they exhibited moderate hydroxide conductivity compared to other reported AEMs.

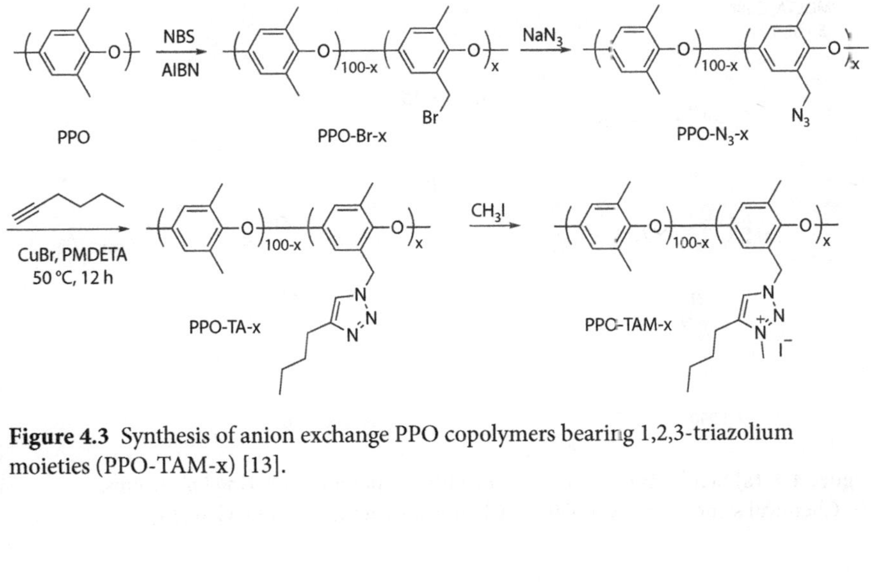

Figure 4.3 Synthesis of anion exchange PPO copolymers bearing 1,2,3-triazolium moieties (PPO-TAM-x) [13].

Alkaline stability tests revealed that the TAM groups do not have enough stability compared to PPO-QA-based AEM for fuel cell application. Nevertheless, they showed significantly better stability than that of PPO-imidazolium-based AEMs. Thus, incorporating different precursors using the simple click chemistry approach may help enhance the stability of 1,2,3-triazolium groups.

Recently, Yang *et al.* [14] reported a series of alkyl-chain grafted PPO-based AEMs containing QA groups synthetized *via* CuACC. The authors carried out a detailed investigation to understand the effect of the grafted alkyl chain length on water uptake capacity, ionic conductivity, micro-phase separation, and alkaline stability. Two different types of polymers, comb-shaped (PPO-C16) and side-chain-type PPO AEMs (C4D30), were prepared for comparative purposes (Figure 4.4). The membrane XaYb-C16 grafted (where "a" denotes the degree of bromination of PPO, "b" is the degree of functionalization of the alkyl side chain, and "n" corresponds to the length of the alkyl side chain) with the longest alkyl chain showed the

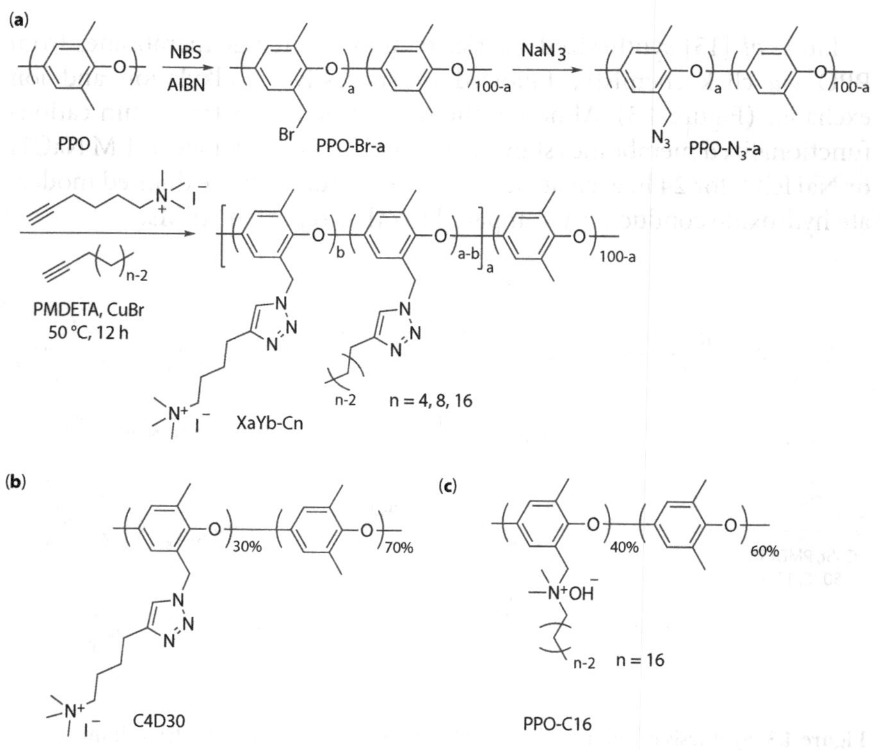

Figure 4.4 (a) Synthesis of anion exchange PPO copolymers bearing QA groups. (b) Chemical structures of C4D30. (c) Chemical structure of PPO–C16 [14].

highest bicarbonate/hydroxide conductivity compared to XaYb-C8 and XaYb-C4 membranes. In addition, the membrane X52Y10-C16 exhibited higher bicarbonate conductivity (14.1 mS cm^{-1}) than the C4D30 membrane (8.6 mS cm^{-1}) with the same IEC value. Furthermore, the membrane exhibited high mechanical strength and alkaline stability, as well as moderate water uptake. As mentioned earlier, the triazole groups in the polymer chain helped to form a 3D network of hydrogen bonds with hydroxide ions and water, which enhanced the ionic conductivity value. Such a hydrophobic alkyl chain is expected to induce well-defined microphase separation in the membrane and, hence, high ionic conductivity. In fact, the X40Y10-C16 membrane showed a higher power density of 67 mW cm^{-2} compared to X40Y10-C8 (52 mW cm^{-2}) and X40Y10-C4 membranes (36 mW cm^{-2}). The results suggested that a more efficient anion transport channel was formed in the grafted X40Y10-C16 membrane due to the presence of well-defined microphase separation in the membrane. Therefore, these grafted copolymers can be used for the production of AEMs, which will fulfill all the demands of an alkaline fuel cell.

Mondal *et al.* [15] synthesized a novel AEM using a combined approach based on thiol-ene click chemistry and the Menshutkin reaction. Specifically, the monomer 9-(2-((3-(triethoxysilyl)propyl)thio)ethyl)-9H-carbazole (TESPTEC) was synthesized by a thiol-ene click reaction, which was further incorporated into the brominated poly(2,6-dimethyl-1,4-phenylene oxide) (BPPO) backbone through the Menshutkin reaction (Figure 4.5).

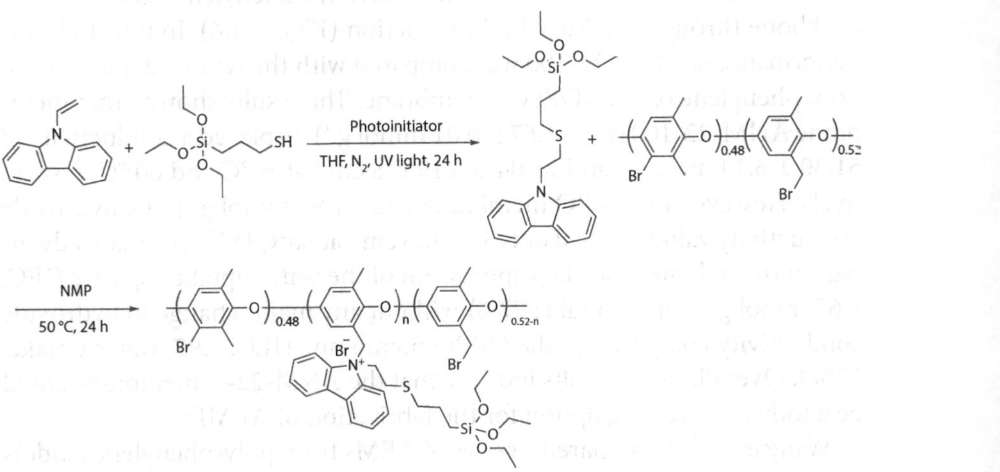

Figure 4.5 Synthesis of thiol-ene click-mediated monomer and their corresponding PPO-based copolymer. Photoinitiator = 2,2-dimethoxy-2-phenylacetophenone [15].

The so-prepared polymeric membrane was fully characterized by NMR, XRD, X-ray photoelectron spectroscopy (XPS), FTIR, AFM, FE-SEM, TGA, etc. It is important to mention that water uptake and conductivity are two major properties that must be considered while designing a high-performance anion exchange membrane. In this context, AEMs need low water uptake capacity and reasonable conductivity values (at least 10 mS cm^{-1}) to be effective in fuel cell applications, especially because higher water uptake capacity decreases mechanical strength. Therein, a (3-mercaptopropyl)triethoxysilane (MPTES) silica-based network was incorporated inside the membrane to maintain a lower water uptake capacity. The highest water uptake capacity of the synthesized membrane was 18.36 wt.% having a higher conductivity (19.84 + 1.81 mS cm^{-1}). In contrast, the reference QPPO membrane, which showed higher ionic conductivity than the synthesized membrane, displayed extremely high water uptake capacity (141.6 ± 6.8%). A series of experiments showed that the alkaline stability was quite moderate compared to the reference membrane, and after 12 days (288 h), the synthesized membrane retained almost 60% of its initial conductivity. However, the QPPO membrane retained only 38% of its initial conductivity after 260 h of alkaline stability test. The reasonable conductivity with moderate swelling ratios made these membranes good candidates for AEM applications.

In 2017, the same group [16] reported another AEM using a similar strategy. In this case, the monomer 1-(2-(hexylthio)ethyl)-1H-imidazole (HLTEI) was synthesized *via* thiol-ene click reaction and further incorporated into the brominated poly(2,6-dimethyl-1,4-phenylene oxide) (BPPO) backbone through the Menshutkin reaction (Figure 4.6). In this study, the performance of ANM-22-100 was compared with the reported quaternized poly(phenylene oxide) (QPPO) membrane. The results showed that membrane ANM-22-100 (IEC: 1.67 ± 0.01 mmol g^{-1}) displayed a conductivity of 51.30 ± 6.14 mS cm^{-1} and 78.04 ± 6.19 mS cm^{-1} at 30°C and 60 °C, respectively. However, for QPPO membrane (IEC: 1.97 mmol g^{-1}), the hydroxide conductivity value reached only 18.2 mS cm^{-1} at 60°C [17]. The main advantage of the polymer was the suppression of the water uptake capacity (IEC: 1.67 mmol g^{-1}, water uptake: 77%) without any major change in hydroxide conductivity compared to the QPPO membrane (IEC: 1.97, water uptake: 77%). Overall, these results indicate that the ANM-22-X membrane could be another interesting option for the fabrication of AEMFCs.

Wang *et al.* [18] prepared a series of AEMs from poly(phenylene oxide)s (PPOs), containing a flexible pendant imidazolium (Im) cation linked *via* flexible spacer units using CuAAC (Figure 4.7). In this example, the water uptake capacity of the PPO-PIm membrane increased with increasing the

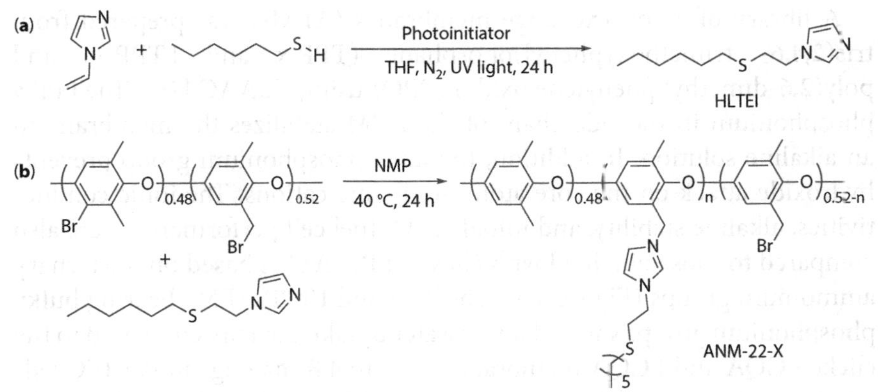

Figure 4.6 Schematic representation of the (a) photoinitiated thiol–ene click reaction for the preparation of HLTEI; (b) Menshutkin reaction between BPPO and HLTEI to produce the ANM-22-X polymer. Photoinitiator = 2,2-dimethoxy-2-phenylacetophenone [16].

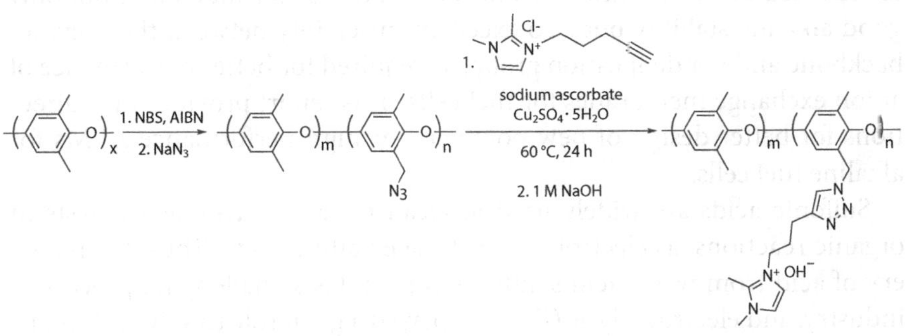

Figure 4.7 Synthetic route for the preparation of PPO-imidazolium AEMs [18].

triazole group in the side chain, which helps to absorb more water molecules *via* hydrogen bonding with the nitrogen atoms [11]. These AEMs showed higher hydroxide conductivity (95 mS cm^{-1}, 80°C, 100% RH) than the traditional benzylic dimethylimidazolium AEM (55 mS cm^{-1}, 80°C, 100% RH), with a similar IEC value due to the formation of a significant microphase separated structure. The alkaline stability of the membranes was further tested by soaking the membrane in 1 M NaOH solution at 80°C and monitoring the variation in conductivity over time. A slight loss of conductivity (5%) was observed after seven days of aging test. Additionally, the incorporation of a flexible spacer unit and the 1,2,3-triazole moiety into the polymer side chain did not deteriorate its thermal and alkaline stability.

A library of anion exchange membranes (AEMs) was prepared from tris(2,4,6- trimethoxyphenyl)phosphane (TPP-x and LTPP-x) and poly(2,6-dimethyl phenylene oxide) (PPO) using CuAAC [19]. The bulky phosphonium in the side chain of the AEM stabilizes the membrane in an alkaline solution. In addition, the large phosphonium group prevents hydroxide attack on the core atom in organic cations. The ionic conductivities, alkaline stability, and initial H_2/O_2 fuel cell performance were also compared to those obtained with clicked PPO AEMs based on quaternary ammonium groups (Figure 4.8). The TPP and LTPP AEMs bearing bulky phosphonium groups showed lower water uptake capacity compared to the clicked CQA and LCQA membranes (Figure 4.8) having similar IEC values. Lower ion conductivities were observed for all of the AEMs due to the lower water uptake. In addition, due to the presence of highly hydrophobic bulky phosphonium groups, the miscibility between the catalyst layer and membrane decreased and, therefore, failed to serve as a membrane for fuel cell devices. From this study, the authors concluded that not only good alkaline stability but also excellent miscibility between the polymer backbone and pendent cation groups is required for better performance of anion exchange membranes for fuel cells. This report provides new directions for better design of new polymers for high-performance AEMs for alkaline fuel cells.

Sulfonic acids are widely used to clean metal surfaces, as catalysts in organic reactions, as electrolytes in storage batteries, etc. Thus, the recovery of acid from waste acid solutions constitutes a challenging process in industry, and electrodialysis (ED) has played a great role to solve this issue. ED is an electro-membrane driven process where a specific ion can be transported from one solution to another through an ion exchange membrane under the influence of an applied electric field. Wang *et al.* [20] synthesized an AEM by incorporating tertiary ammonium groups into PPO using click chemistry (Figure 4.9). The authors demonstrated the use of this polymer in the fabrication of proton blockage membranes to maintain sulfonic acid concentration in ED processes. Considering that proton leakage or transport depends mainly on water uptake and swelling ratio, the authors compared values such as water uptake, swelling ratio, and IECw between two different AEMs bearing tertiary ammonium groups and quaternary ammonium groups. The proton blockage membranes or AEMs made of tertiary amine groups showed lower water uptake and swelling ratio than that of the AEMs containing quaternary ammonium groups, probably due to the weaker hydration effect of tertiary amine compared to quaternary ammonium groups. The highest concentration limitation of AEMs with tertiary ammonium groups for H^+ was 3.02 mmol/L with an

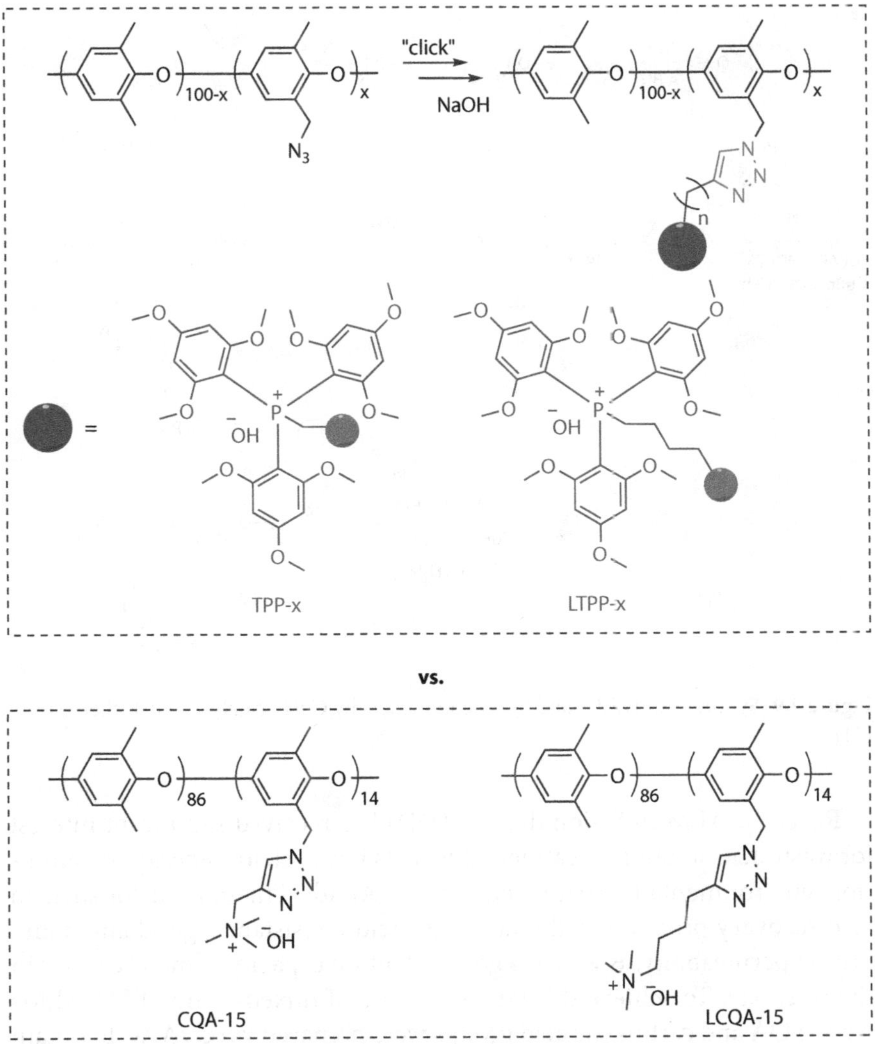

Figure 4.8 Synthetic route of triphenyl phosphonium functionalized PPO polymers (*top*) and chemical structures of CQA-15 and LCQA-15 (*bottom*) [19].

IECw value of 1.67 mmol/g, which is slightly higher than the 3.00 mol/L reported for the commercial AEM (Neosepta ACM). The concentration limitation of membranes with tertiary amine groups was higher compared to those of membranes with QA groups, which indicated that the weak base groups introduced into AEMs could effectively prevent proton leakage, supporting the potential of these membranes as candidates for acid recovery by the ED process.

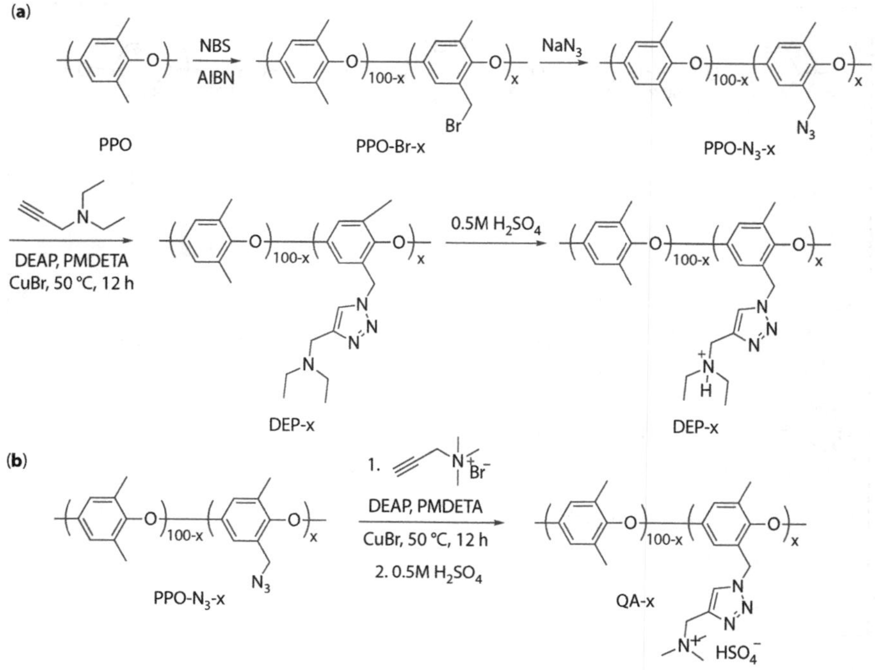

Figure 4.9 Synthesis of DEP-x and QA-x copolymers by Cu(I)-catalyzed click chemistry [21].

Recently, AEM diffusion dialysis (DD) has received significant interest for waste acid solutions treatment due to its low cost, low energy consumption, environmentally friendly nature, etc. An ideal membrane for such an acid recovery process should have high acidic resistance, good anion and proton permeability, as well as a good retention capacity of metal cations in the retentate. To achieve this target, a series of mixed-charge PPO-AEMs having QA and carboxylic acid groups were prepared *via* CuAAC to obtain high H^+ dialysis coefficients and highly selective AEMs (Figure 4.10) [21].

This polymer was used in DD for acid recovery applications. The water uptake, volumetric IECv, gravimetric IECw, swelling ratio, and H^+ dialysis coefficients (UH+) of AEMs were thoroughly examined, and the obtained values were compared with previously reported AEMs lacking carboxylic acid groups. The carboxylic acid groups were incorporated to improve H^+ transport in AEMs over metal ions. The results showed that both H^+ dialysis coefficients (UH+) and H^+/Fe^{2+} selectivity increased with increasing carboxylic acid groups into the AEM, unlike other AEMs. The highest H^+ dialysis coefficient, 20.37 m/h, and H^+/Fe^{2+} selectivity 34.52, were achieved with PPO-mixed-charge AEMs. This report suggests that introducing

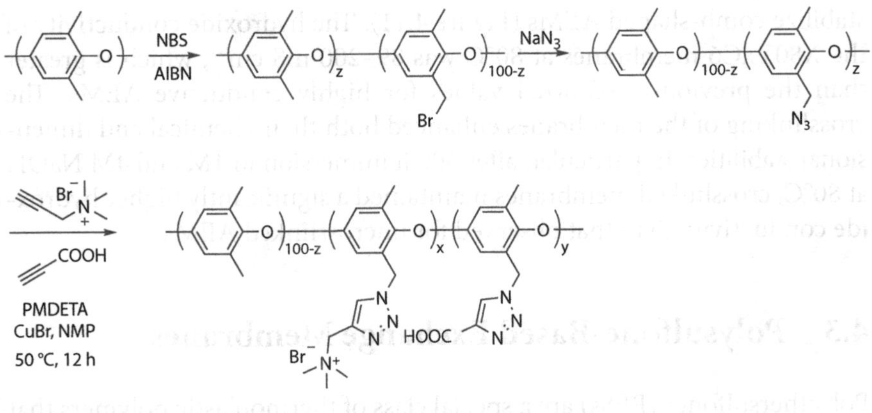

Figure 4.10 Synthesis of mixed-charge containing PPO-AEMs [21].

weak (carboxylic) acid groups into the polymer backbone increases acidic resistance and provides good anion and H^+ transport in AEMs over multi-charge metal ions. Therefore, they can be considered a promising candidate for acid recovery applications.

It has been well-established that comb-shaped polymers based on PPO backbones with different pendant alkyl side chain lengths show good ion conductivity under alkaline conditions [14]. However, to further improve the ion conductivity with the enhancement of the dimensional stability of the membrane, and decrease fuel crossover in the preparation of new AEMs, Zhu *et al.* [22] introduced a thiol-ene crosslinking strategy to

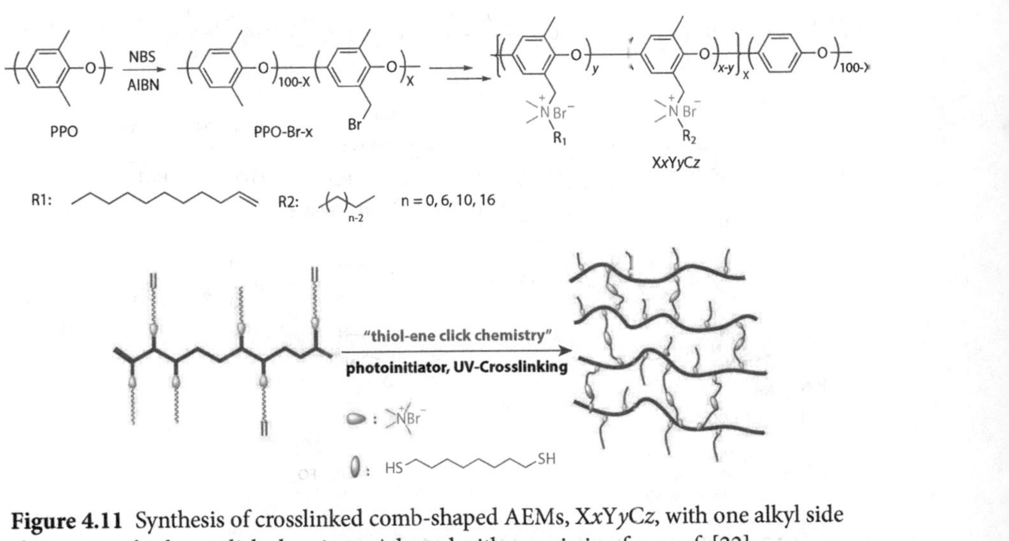

Figure 4.11 Synthesis of crosslinked comb-shaped AEMs, X*x*Y*y*C*z*, with one alkyl side chain using thiol-ene click chemistry. Adapted with permission from ref. [22].

stabilize comb-shaped AEMs (Figure 4.11). The hydroxide conductivity of the X80YxC6 membranes at 80°C was 89–200 mS cm^{-1}, which is greater than the previously reported values for highly conductive AEMs. The crosslinking of the membranes enhanced both their chemical and dimensional stabilities. In particular, after 500 h immersion in 1M and 4M NaOH at 80°C, crosslinked membranes maintained a significantly higher hydroxide conductivity than that observed for uncrosslinked AEMs.

4.3 Polysulfone-Based Exchange Membranes

Polyethersulfones (PESs) are a special class of thermoplastic polymers that display excellent mechanical, thermal, and chemical stabilities, making them mostly used in membrane separation processes. However, PESs possess low IEC and high ionic resistance, which are critical factors for AEMs development. Thus, a significant amount of research has been focused on the modification of PES to improve those properties. In this context, Chen *et al.* [23] used CuAAC for the fabrication of a series of functional PES-AEMs bearing guanidinium moieties in the side chains with different spacers (Figure 4.12). Briefly, azidomethylated PES was reacted with terminal alkyne molecules containing di-Boc guanidine precursor *via* click reaction. Then, the di-Boc guanidine groups were deprotected, affording the corresponding functional guanidinium moiety. For comparative purposes, hydrophobic alkyl

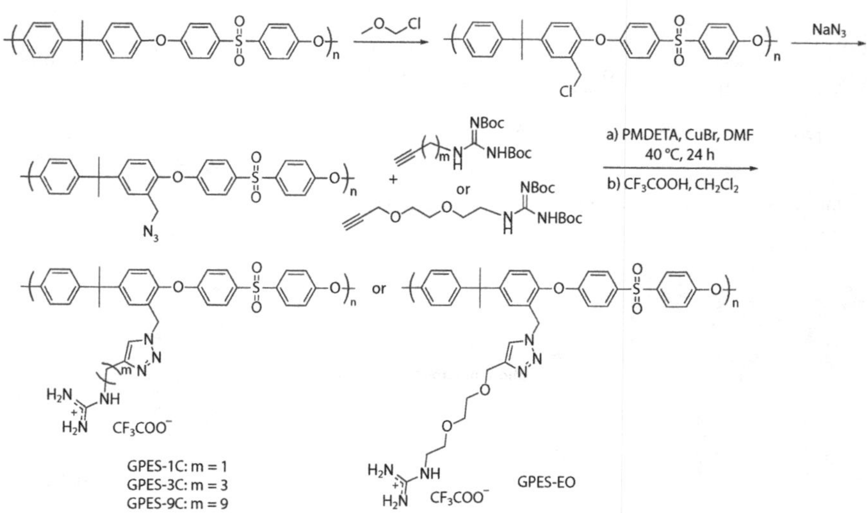

Figure 4.12 Synthesis of PES-based AEMs with different spacers [23].

spacers with different lengths and hydrophilic ethylene oxide (EO) spacer were introduced into the polymer structure. Interestingly, the hydroxide conductivity decreased as the length of flexible methylene increased, which may be attributed to the low IEC value and high ratio of hydrophobic alkyl spacer. In comparison, the GPES-EO AEM having EO spacers displayed higher conductivity than GPES-nC membranes (81 mS cm^{-1} at 80°C). The high ionic conductivity of GPES-EO may be interpreted by the connected ionic transport channels and higher water absorption. Thus, the incorporation of a flexible spacer with a proper length is essential for improving ionic conductivity.

Hu *et al.* [24] used CuAAC to synthesize a series of new fluorenyl-containing polysulfone-based crosslinked AEMs. In this study, the azide group was first introduced into the side chains of fluorenyl-containing polysulfone *via* a quaternization reaction followed by a crosslinking click reaction with different bifunctional alkynes at 120°C (Figure 4.13). The obtained crosslinked membranes were fully characterized by IR, SEM, and AFM. Furthermore, the fuel cell permeability, alkaline resistance, oxidative stability, as well as the mechanical strength of the membranes were improved by increasing the degree of crosslinking. Additionally, the water uptake capacity and swelling ratio were also tuned by variation of crosslinking agent's polarity and chain length. The ionic conductivity of the AEMs was mainly dependent on the water uptake capacity of membranes,

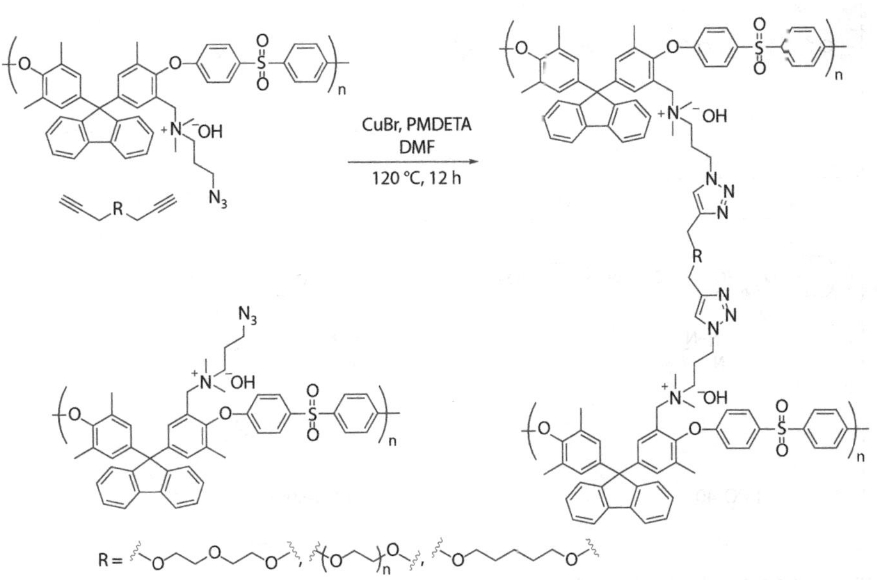

Figure 4.13 Schematic representation of the click-crosslinking process of PES [24].

which depends on the crosslinking density (i.e., the higher the crosslinking density, the lower the water uptake is). The hydroxide conductivity of the membranes was reduced after crosslinking. However, the conductivity of the high hydrophilic crosslinked membranes exhibited a relatively high value as compared to the comparatively hydrophobic membranes because of the compensation of water uptake. This leads to the conclusion that for crosslinked AEMs, a structure with uniform nano/microphase separation would avoid the decrease of ionic conductivity even after crosslinking.

4.4 Polystyrene-Based Anion Exchange Membranes

In 2012, Liu *et al.* [25] reported the synthesis of a series of crosslinked, comb-shaped polystyrene (PS) AEMs having a C-16 alkyl side chain *via* CuAAC (Figure 4.14). Small-angle X-ray scattering (SAXS) studies indicated a nanophase separation between the hydrophilic and hydrophobic domains in these membranes, similar to PPO-AEMs. The obtained crosslinked PS-AEMs were found to be very tough and transparent, probably due to the good compatibility between the polystyrene backbone and the

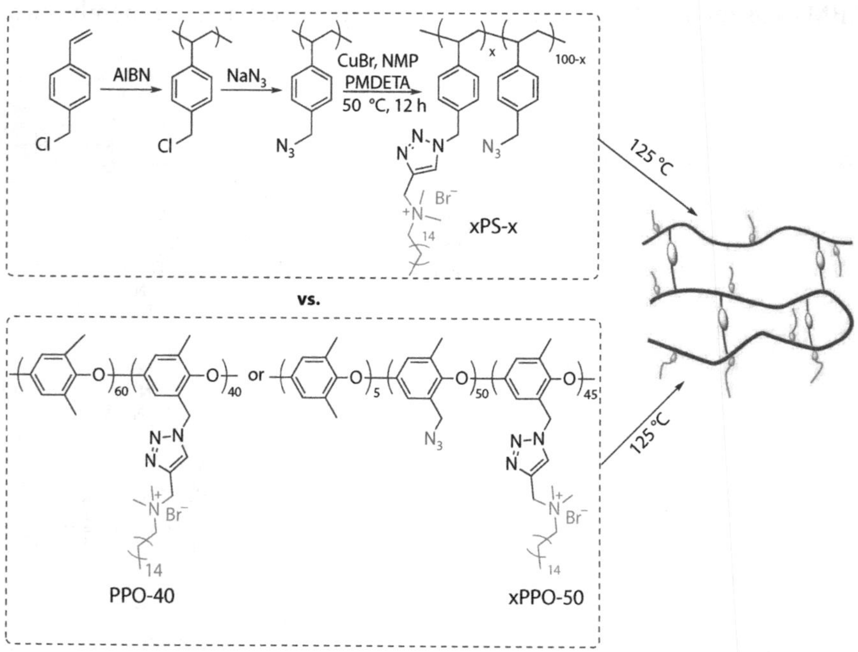

Figure 4.14 Synthesis of crosslinked comb-shaped xPS-x, xPPO-50, PPO-40 by click chemistry and azide-assisted self-crosslinking. Adapted with permission from ref. [25].

long alkyl side chains. In addition, these crosslinked membranes showed higher hydroxide conductivities than uncrosslinked PS-x membranes due to their lower water uptake capacity. The hydroxide conductivity values of crosslinked PS membranes were comparable to or even higher than those observed for crosslinked PPO-AEMs with similar comb-shaped architecture reported elsewhere [26]. The authors pointed out that the good miscibility between the alkyl and PS chains favors the formation of an ionic domain suitable for anion transport. In general, degradation of tetraalkylammonium ions under alkaline conditions took place through Hoffmann elimination of β-hydrogen, direct nucleophilic substitution, or nitrogen ylide formation. Consequently, the alkaline stability of the crosslinked AEMs was studied in 2 M, 5 M, and 10 M NaOH at 80°C for 400 h. The degradation behavior showed that the crosslinked membranes have much better alkaline stability than the uncrosslinked membranes. The initial fuel cell performance with the peak power density of 88 mW/cm^2 was observed for the H_2/O_2 alkaline fuel cell with xPS-65 membrane. In general, these results indicate that these tough and transparent membranes are promising candidates for fuel cell and other energy-related applications.

In 2018, a new series of crosslinked PS-based AEMs were prepared via CuAAC with different crosslinking densities and trimethylammonium cationic groups to investigate the toughness, water uptake, swelling, and conductivity of the membranes (Figure 4.15) [27]. It is important to note that high ion content present in the AEM can increase the hydroxide conductivity and water uptake but decrease the mechanical properties of the membranes. Additionally, when membranes having different amounts of crosslinker but the same IEC value were studied, the less crosslinked membrane showed

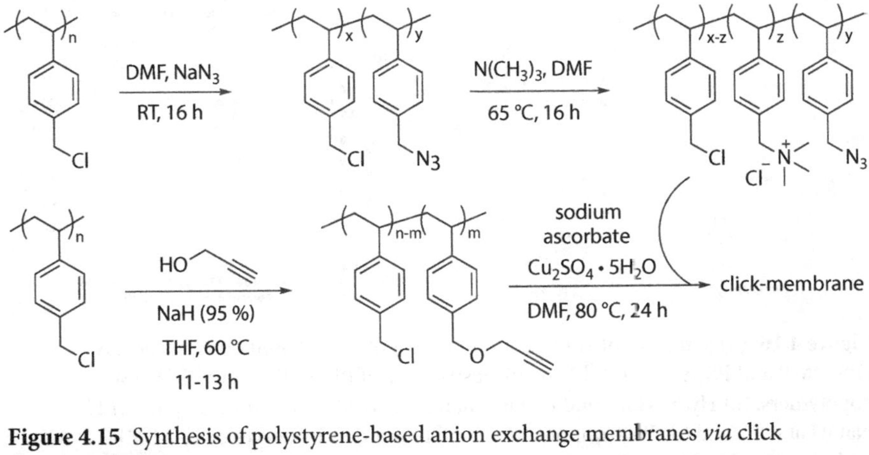

Figure 4.15 Synthesis of polystyrene-based anion exchange membranes *via* click chemistry [27].

higher conductivity than the highly crosslinked membranes. Specifically, the chemical stability of AEMs was tested by soaking the membranes in 8 mol L^{-1} KOH solution at 60°C and their conductivity was measured periodically. The results showed that almost 33–62% of the initial conductivity was retained after 49 days. In addition, the fuel cell performance with the peak power density of 157 mW/cm^2 was observed for the H_2/O_2 alkaline fuel cell, which was comparable to that of other AEM-based fuel cells reported.

Very recently, Wang *et al.* [28] prepared a series of PS-based AEMs bearing six-membered (DMP) and bis-six-membered *N*-cyclic QA (ASU) cations in the side chain using CuAAC, and compared their properties (water uptake, alkaline stability, etc.) with PPO-based AEMs (Figure 4.16). This study showed that the PS-ASU copolymer with bis-six-membered *N*-cyclic QA exhibited higher alkaline stability than the PS-DMP copolymer, presumably due to the higher ring strain associated to the ASU groups. The membrane formation ability of PS-QA was reduced due to the presence of bulky *N*-cyclic QA cations in the PS-AEM, resulting in the formation of a brittle material. In this study, styrene-ethylene butylene-styrene triblock copolymer (SEBS), a commercial thermoplastic elastomer, was blended with PS-ASU or PS-DMP copolymers by solution casting method and named as

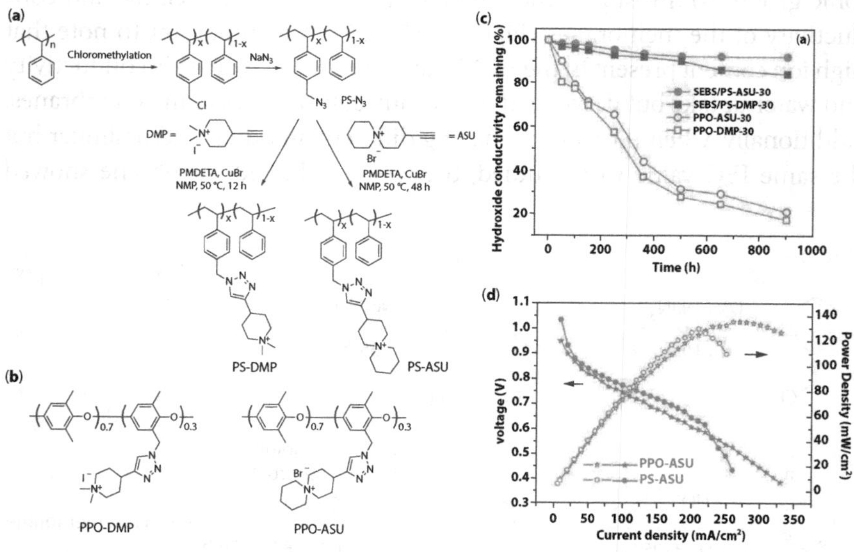

Figure 4.16 (a) Synthesis of anion-conductive copolymers based on *N*-cyclic QAs (PS-DMP and PS-ASU) and (b) chemical structure of PPO-DMP and PPO-ASU copolymers. (c) Hydroxide conductivity change of AEMs after immersing in (a) 1M NaOH at 80°C and (d) Polarization curves and power density curves of an AEMFC at 60°C with the PS-ASU-30 and PPO-ASU-30 ionomer. Adapted with permission from ref. [28].

SEBS/PS-DMP and SEBS/PS-ASU, respectively. Alkaline stability was determined by conductivity measurements at regular time intervals. After 900 h, the SEBS/PS-DMP and SEBS/PS-ASU AEMs displayed a slow decrease in conductivity (< 20%), whereas PPO-based AEMs showed 79–83% loss of their initial conductivity under the same testing period and conditions. These results indicated that the polymer backbone has a great influence on the alkaline stability of the membranes. Furthermore, fuel cell performance with a peak power density of 130 mW/cm^2 was observed at a current density of 210 mA/ for the H_2/O_2 alkaline fuel cell, which was comparable to that of the PPO-ASU membrane. Therefore, these *N*-cyclic QA cations-functionalized PS copolymers are also new options for the fabrication of stable AEMs for alkaline fuel cells and other energy-related applications.

Yang *et al.* [29] designed and synthesized a series of polystyrene (PS)-grafted PPO-based AEMs having pendent quaternary ammonium groups using the "grafting onto" method *via* a combination of atom transfer radical polymerization (ATRP) and CuACC (Figure 4.17). The ATRP method was used to control the length of the PS grafting chain, which can eventually regulate water uptake and thus improve the dimensional stability of AEMs in water. Due to excellent miscibility between PS grafting chains and PPO, a nanophase

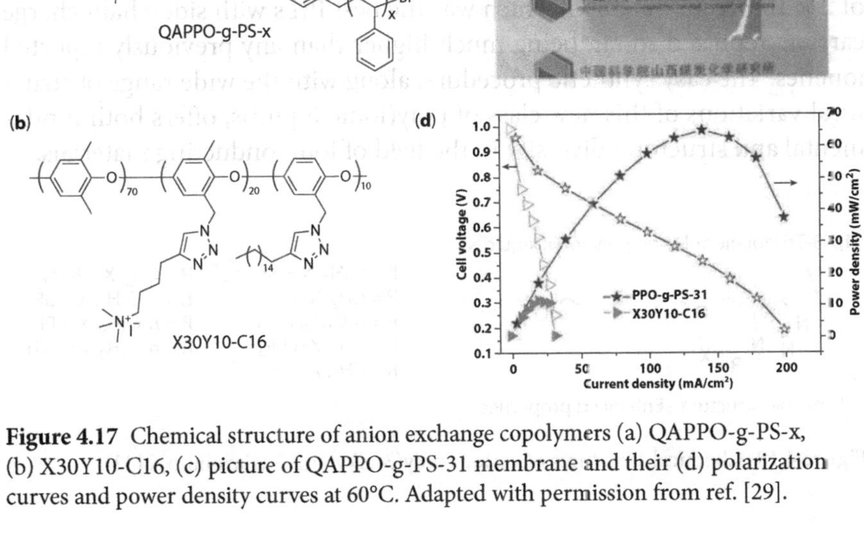

Figure 4.17 Chemical structure of anion exchange copolymers (a) QAPPO-g-PS-x, (b) X30Y10-C16, (c) picture of QAPPO-g-PS-31 membrane and their (d) polarization curves and power density curves at 60°C. Adapted with permission from ref. [29].

segregated structure was formed, confirmed by small-angle X-ray scattering (SAXS), which further helps control water uptake. The observed hydroxide conductivity was 15.9 mS/cm at 20°C, for QAPPO-g-PS-21 membranes with an IEC value of 1.21 meq./g in water. Moreover, the PS-grafted PPO-AEMs showed excellent alkaline stability in 1 M NaOH at 80°C, and the QAPPO-g-PS-45 membrane showed only a 10% drop in their initial ion conductivity after 500 h alkaline stability testing. Finally, fuel cell performance with a peak power density of 64.4 mW/cm^2 was observed at a current density of 140 mA/cm^2 for the H_2/O_2 alkaline fuel cell, which was comparable to that of the PPO-ASU membrane, but higher than that of typical side-chain-type AEMs without PS side chains (X30Y10-C16). This was probably due to the formation of a well-defined hydrophobic-hydrophilic microphase separated structure.

4.5 Poly(ionic Liquid)s-Based Anion Exchange Membrane

Poly(ionic liquid)s (PILs), a special class of polyelectrolytes that bears an electrolyte group either cation or anion in the repeating unit [30], has both the properties of ionic liquids (such as good ionic conductivity, thermal stability, tunable solution properties, chemical stability, etc.) and versatile polymer (mechanical strength, processability, etc.) [31]. Drockenmuller and co-workers reported the synthesis of 1,2,3-triazolium-based poly(ionic liquid)s containing a triethylene glycol spacer through CuAAC (Figure 4.18) [32]. These poly(ionic liquid)s bearing a methyl substituent and bis(trifluoromethylsulfonyl)imide anion displayed an ionic conductivity of 2 × 10^{-5} S cm^{-1} at 30°C, which was the best PILs with side-chain charge carriers reported so far, being much higher than any previously reported ionenes. The easy synthetic procedure, along with the wide range of structural variations of this new class of poly(ionic liquid)s, offers both fundamental and structural diversity in the field of ion-conducting materials.

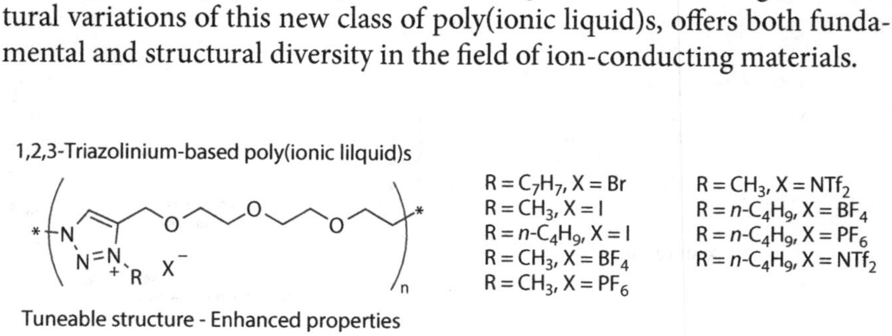

Figure 4.18 Chemical structures of various poly(3-alkyl-1,2,3-triazolium)s [32].

4.6 Conclusion

The main goal of AEMFC research is to enhance the long-term chemical and mechanical stability of AEMs in high-pH and high-temperature conditions. To achieve this target, research was initially focused on tuning the polymer backbone and side-chain structures for stability improvement. During the last few years, researchers have paid more attention to improving AEM head groups by incorporating quaternary ammonium groups. But still, degradation remains an unavoidable phenomenon under hot, alkaline conditions. Non-ammonium cation groups such as imidazolium and guanidinium have also been incorporated in AEMs by several research groups to enhance the stability, albeit the results were almost comparable. Inorganic head group such as phosphane group have been also incorporated into the polymer backbone with the same objective. The current contribution describes how the presence of triazole groups formed via CuAAC in the polymer mainchain/or sidechain increases thermal and chemical stability with IEC values compared to the polymer without triazole moieties. Additionally, the crosslinking strategy by thiol-ene click reaction improved ion conductivity with the enhancement of the dimensional stability. Nevertheless, AAEM is still a challenging topic and more research, especially focused on structure-property relationships, is needed to solve the stability issues for practical applications in fuel cells.

Acknowledgment

D.D.D. acknowledge Cátedra Fundación CEPSA of the ULL, Cátedra de Medioambiente y Sostenibilidad (Cabildo de Tenerife-ULL), Fundación Ramón Areces and the Ministry of Science and Innovation (project PID2022-142118OB-I00/AE/10.13039/501100011033/UE) for financial support, and Nanotec, INTech, Cabildo de Tenerife and ULL for laboratory facilities.

References

1. Merle, G., Wessling, M., Nijmeijer, K., Anion exchange membranes for alkaline fuel cells: A review. *J. Membr. Sci.*, 377, 1–35, 2011.
2. Hwang, G.J. and Ohya, H., Preparation of anion exchange membrane based on block copolymers. Part II. The effect of the formation of macroreticular structure on the membrane properties. *J. Membr. Sci.*, 149, 163–169, 1998.

4. Meldal, M. and Tornøe, C.W., Cu-catalyzed azide-alkyne cycloaddition. *Chem. Rev.*, 108, 2952–3015, 2008.
5. Wu, P., Feldman, A.K., Nugent, A.K., Hawker, C.J., Scheel, A., Voit, B., Fokin, V.V., Efficiency and fidelity in a click-chemistry route to triazole dendrimers by the copper (I)-catalyzed ligation of azides and alkynes. *Angew. Chem. Int. Ed.*, 43, 3928–3932, 2004.
6. Kantheti, S., Narayan, R., Raju, K.V.S.N., The impact of 1, 2, 3-triazoles in the design of functional coatings. *RSC Adv.*, 5, 3687–3708, 2015.
7. Obadia, M.M. and Drockenmuller, E., Poly(1, 2, 3-triazolium)s: a new class of functional polymer electrolytes. *Chem. Commun.*, 52, 2433–2450, 2016.
8. Maurya, S., Shin, S.H., Kim, Y., Moon, S.H., A review on recent developments of anion exchange membranes for fuel cells and redox flow batteries. *RSC Adv.*, 5, 37206–37230, 2015.
9. Noh, S., Jeon, J.Y., Adhikari, S., Kim, Y.S., Bae, C., Molecular Engineering of Hydroxide Conducting Polymers for Anion Exchange Membranes in Electrochemical Energy Conversion Technology. *Acc. Chem. Res., 52*, 2745–2755, 2019.
10. Ran, J., Wu, L., Ru, Y., Hu, M., Din, L., Xu, T., Anion exchange membranes (AEMs) based on poly (2, 6-dimethyl-1, 4-phenylene oxide)(PPO) and its derivatives. *Polym. Chem.*, 6, 5809–5826, 2015.
11. Hay, A.S., Polymerization by oxidative coupling: Discovery and commercialization of PPO® and Noryl® resins. *J. Polym. Sci., Part A: Polym. Chem.*, 36, 505–517, 1998.
12. Li, N., Guiver, M.D., Binder, W.H., Towards High Conductivity in Anion-Exchange Membranes for Alkaline Fuel Cells. *ChemSusChem.*, 6, 1376–1383, 2013.
13. Ge, Q., Ran, J., Miao, J., Yang, Z., Xu, T., Click chemistry finds its way in constructing an ionic highway in anion-exchange membrane. *ACS Appl. Mater. Interface*, 7, 28545–28553, 2015.
14. Liu, L., He, S., Zhang, S., Zhang, M., Guiver, M.D., Li, N., 1, 2, 3-Triazolium-based poly (2, 6-dimethyl phenylene oxide) copolymers as anion exchange membranes. *ACS Appl. Mater. Interfaces*, 8, 4651–4660, 2016.
15. Yang, C., Liu, L., Han, X., Huang, Z., Dong, J., Li, N., Highly anion conductive, alkyl-chain-grafted copolymers as anion exchange membranes for operable alkaline H_2/O_2 fuel cells. *J. Mat. Chem. A*, 5, 10301–10310, 2017.
16. Mondal, A.N., He, Y., Ge, L., Wu, L., Emmanuel, K., Hossain, M.M., Xu, T., Preparation and characterization of click-driven N-vinylcarbazole-based anion exchange membranes with improved water uptake for fuel cells. *RSC Adv.*, 7, 29794–29805, 2017.
17. Mondal, A.N., He, Y., Wu, L., Khan, M.I., Emmanuel, K., Hossain, M.M., Xu, T., Click mediated high-performance anion exchange membranes with improved water uptake. *J. Mater. Chem. A*, 5, 1022–1027, 2017.

18. Li, Q., Liu, L., Miao, Q., Jin, B., Bai, R., A novel poly (2, 6-dimethyl-1, 4-phenylene oxide) with trifunctional ammonium moieties for alkaline anion exchange membranes. *Chem. Commun.*, 50, 2791–2793, 2014.
19. Wang, J., Wei, H., Yang, S., Fang, H., Xu, P., Ding, Y., Constructing pendent imidazolium-based poly (phenylene oxide) s for anion exchange membranes using a click reaction. *RSC Adv.*, 5, 93415–93422, 2015.
20. Tang, H., Li, D., Li, N., Zhang, Z., Zhang, Z., Anion conductive poly (2, 6-dimethyl phenylene oxide) s with clicked bulky quaternary phosphonium groups. *J. Membr. Sci.*, 558, 9–16, 2018.
21. Wang, L., Li, Z., Xu, Z., Zhang, F., Efome, J.E., Li, N., Proton blockage membrane with tertiary amine groups for concentration of sulfonic acid in electrodialysis. *J. Membr. Sci.*, 555, 78–87, 2018.
22. Wang, L., Zhang, F., Li, Z., Liao, J., Huang, Y., Lei, Y., Li, N., Mixed-charge poly(2,6-dimethyl-phenylene oxide)anion exchange membrane for diffusion dialysis in acid recovery. *J. Membr. Sci.*, 549, 543–549, 2018.
22. Zhu, L., Zimudzi, T.J., Li, N., Pan, J., Lin, B., Hickner, M.A., Crosslinking of comb-shaped polymer anion exchange membranes via thiol–ene click chemistry. *Polym. Chem.*, 7, 2464–2475, 2016.
23. Chen, Y., Tao, Y., Wang, J., Yang, S., Cheng, S., Wei, H., Ding, Y., Comb-shaped guanidinium functionalized poly (ether sulfone) s for anion exchange membranes: Effects of the spacer types and lengths. *J. Polym. Sci., Part A: Polym. Chem.*, 55, 1313–1321, 2017.
24. Hu, B., Miao, L., Bai, Y., Lü, C., Facile construction of crosslinked anion exchange membranes based on fluorenyl-containing polysulfone via click chemistry. *Polym. Chem.*, 8, 4403–4413, 2017.
25. Liu, W., Liu, L., Liao, J., Wang, L., Li, N., Self-crosslinking of comb-shaped polystyrene anion exchange membranes for alkaline fuel cell application *J. Membr. Sci.*, 53, 6, 133–140, 2017.
26. Li, N., Yan, T., Li, Z., Thurn-Albrecht, T., Binder, W.H., Comb-shaped polymers to enhance hydroxide transport in anion exchange membranes. *Energy Environ. Sci.*, 5, 7888–7892, 2012.
27. Tuli, S.K., Roy, A.L., Elgammal, R.A., Zawodzinski, T.A., Fujiwara, T., Polystyrene-based anion exchange membranes *via* click chemistry: improved properties and AEM performance. *Polym. Int.*, 67, 1302–1312, 2018.
28. Wang, X., Sheng, W., Shen, Y., Liu, L., Dai, S., Li, N., N-cyclic quaternary ammonium-functionalized anion exchange membrane with improved alkaline stability enabled by aryl-ether free polymer backbones for alkaline fuel cells. *J. Membr. Sci.*, 587, 117135, 2019.

29. Yang, C., Liu, L., Huang, Y., Dong, J., Li, N., Anion-conductive poly (2, 6-dimethyl-1, 4-phenylene oxide) grafted with tailored polystyrene chains for alkaline fuel cells. *J. Membr. Sci.*, 573, 247–256, 2019.
30. Qian, W., Texter, J., Yan, F., Frontiers in poly (ionic liquid)s: syntheses and applications. *Chem. Soc. Rev.*, 46, 1124–1159, 2017.
31. Yuan, J., Mecerreyes, D., Antonietti, M., Poly (ionic liquid) s: An update. *Prog. Polym. Sci.*, 38, 1009–1036, 2013.
32. Mudraboyina, B.P., Obadia, M.M., Allaoua, I., Sood, R., Serghei, A., Drockenmuller, E., 1, 2, 3-Triazolium-based poly (ionic liquid)s with enhanced ion conducting properties obtained through a click chemistry polyaddition strategy. *Chem. Mater.*, 26, 1720–1726, 2014.

5

Supercritical Fluid-Assisted Porous Membrane Formation: Mechanisms and Applications

Lucia Baldino and Stefano Cardea*

Department of Industrial Engineering, University of Salerno, Via Giovanni Paolo II, Fisciano (SA), Italy

Abstract

Porous polymeric membranes are largely used in numerous fields: from environmental separation processes to reactors, pharmaceuticals, and electronics. The morphological characteristics of membranes influence their final application; therefore, selecting the proper polymeric system, polymer concentration, solvent, and production technique is relevant to ensure successful device performance. Supercritical CO_2 phase separation is a green and innovative alternative production process that can produce solvent-free membranes in a faster way than traditional ones. Moreover, the possibility of obtaining regular and interconnected porous morphology at micro- and nanoscale opens the range of applications for these polymeric membranes to other fields.

Keywords: Membranes, microporous, supercritical CO_2, supercritical phase separation, loaded membranes

5.1 Introduction

Porous polymeric membranes are largely used in numerous fields [1]: from environmental separation processes to reactors, pharmaceuticals, and electronics. Notably, 69% of the scientific papers on membrane production and applications have been published in the last 10 years (see Scopus: porous

**Corresponding author*: scardea@unisa.it

Annarosa Gugliuzza and Wanqin Jin (eds.) Porous Membranes: Breakthroughs in Manufacturing and Applications, (97–110) © 2025 Scrivener Publishing LLC

membranes, phase inversion), demonstrating this is a dynamic research field around which increasing interest is involved.

To obtain a membrane suitable for different applications, specific porosity and pore size are generally required. For instance, membranes used for ultrafiltration should have a mean pore size in the range of 2–100 nm [2], whereas membranes used as scaffolds in tissue engineering should have a mean pore size around 100 μm [3] for cell inoculation, proliferation, and migration. These morphological characteristics influence membranes' mechanical properties and resistance [1] and depend on several factors. Of particular relevance are the kind and concentration of the selected polymer and the membrane formation technique, specifically whether traditional or supercritical phase separation is adopted.

The chemico-physical properties of membranes can be varied by loading a filler [1, 4] and/or an active compound [1, 5] into the starting polymeric system. By operating in this way, numerous smart porous devices can be designed and produced, improving performance and enlarging the applications of the final membranes.

Therefore, in this chapter, the main operative parameters influencing membrane characteristics are critically reviewed and discussed, with particular attention to membrane formation processes. Traditional phase separation and supercritical phase separation will be compared to highlight the advantages of the supercritical process as an innovative and flexible tool to generate regularly porous, solvent-free polymeric membranes.

5.2 Membranes Morphological Characteristics

The morphological characteristics of membranes influence their final application. For this reason, the selection of the proper polymeric system, polymer concentration, solvent, and production technique is relevant to ensure a successful device performance.

Pore structures can be identified as finger, sponge, and dense layer, according to the composition of the polymer solution and the velocity of the precipitation front moving through the polymer solution during the solvent/non-solvent exchange. The shape of the pores is influenced by directional diffusive mass transport at the precipitation front. For example, if diffusive mass transport perpendicular to the precipitation front dominates over diffusive mass transport in other directions, dense lamellar structures are formed. Conversely, if diffusive mass transport parallel to the precipitation front dominates over other directions, finger pores are formed. When diffusive mass transport is of similar magnitude in every direction, sponge pores are formed [6].

In the case of porous membranes, a further classification is related to the kind of pores: open or closed. According to the definition proposed by Rouquerol *et al.* [7], when pores are totally isolated from their neighbors, they are described as closed pores (see Figures 5.1A, 5.1B, 5.1C). On the other hand, pores with a continuous channel of communication until the external surface of the body are described as open pores (see Figures 5.1D, 5.1E). In general, this classification refers to the pores' availability to an external fluid.

The formation of open or closed pores is influenced by the membrane production method. For instance, interconnected porous structure is considered to originate from the spinodal decomposition mechanism during phase separation [8–12], whereas closed pores are formed by the nucleation and growth mechanism [8–11]. However, the coalescence of droplets of the polymer-lean phase before solidification of the polymer-rich phase can lead to the formation of an open porous structure [11, 13].

The addition of polymeric additives to the casting solution can affect membrane structure and properties, in terms of porosity, pore size, roughness, and hydrophilic-lipophilic behavior [14–16]. Moreover, the presence of additives can favor a spongy membrane structure, preventing finger-like macrovoid formation [1]. Inorganic particle loading in membranes, forming nanocomposites, can also be used for various applications to improve membrane stiffness, toughness, chemical stability, electrical conductivity, and resistance [17].

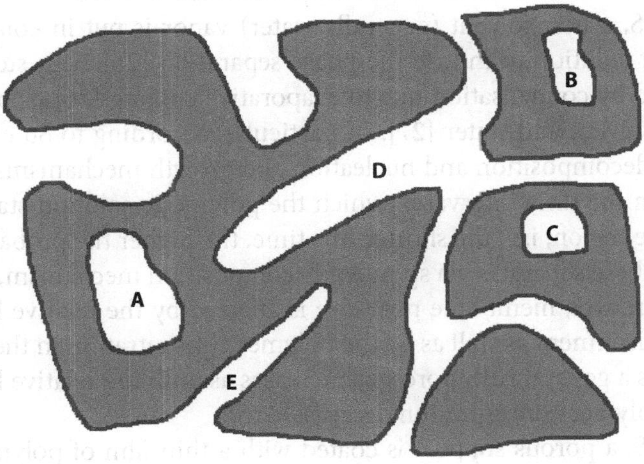

Figure 5.1 Schematic representation of a membrane porous section with open and closed pores.

5.3 Brief Overview on Traditional Membranes Formation Mechanisms and Applications

Numerous techniques are available to prepare porous polymeric membranes, such as sintering, stretching, track etching, sol–gel process, vapor deposition, solution coating, and phase separation [18]. However, phase separation is the most frequently used method and it can be induced by vapor (VIPS) or using a non-solvent (NIPS) [19]. As previously mentioned, pore structure and dimension influence mechanical resistance and membrane performance; therefore, their control is crucial and strictly depends on the mechanisms involved during membrane production.

In general, phase separation mechanisms can be of two types: nucleation and growth [20], and spinodal decomposition [21], depending on where the phase separation occurs in the thermodynamic phase diagram. In a ternary phase diagram, the binodal and spinodal curves divide the diagram into three regions: homogeneous, meta-stable, and unstable. The binodal separates the homogeneous and the metastable regions, and the spinodal separates the metastable and the unstable regions [22]. In the metastable region, solution phase separation occurs via nucleation and growth mechanism, producing cellular pores in the polymeric matrix. In the unstable region, solution phase separation occurs via the spinodal decomposition mechanism, resulting in a bi-continuous structure [21, 23]. Mass transfer related to the mutual diffusion coefficients between polymer, solvent, and non-solvent also plays a relevant role on the position of the phase separation composition inside the ternary phase diagram [24]; therefore, it has to be taken into account when these components are selected.

In VIPS, a non-solvent (generally water) vapor is put in contact with a polymer solution to induce the phase separation [25]. This step can be performed by condensation due to evaporative cooling [26] or by affinity between solvent and water [27]. In particular, according to Su *et al.* [24], spinodal decomposition and nucleation and growth mechanisms in VIPS depend on the time interval in which the polymeric solution stays in the metastable region, i.e., the shorter this time, the higher the probability the solution phase separates via spinodal decomposition mechanism, and vice versa. Moreover, membrane pore size is affected by the relative humidity in the environment as well as by the polymer concentration in the starting system. As a general rule, pore size increases as both the relative humidity and the polymer concentration decrease [28].

In NIPS, a porous support is coated with a thin film of polymer solution. This system is immersed into a coagulation bath, which consists of a poor solvent, i.e., the non-solvent. The polymer film solidifies through the exchange of solvent and non-solvent; therefore, the main requisite to perform NIPS is that the solvent/non-solvent system must be miscible [29]. The factors influencing this kind of phase separation are the choice of solvent/non-solvent, the composition of the polymer solution, the composition of the coagulation bath, and the film casting conditions [30]. Since NIPS is the common way to perform the traditional phase separation, some application examples are reported below.

Wu *et al.* [2] produced hydrophobic polyvinylidene fluoride (PVDF) membranes with hydrophilicity and antifouling properties using an aqueous solution of graphene oxide (GO) as the coagulation bath. Operating in this way, an increment in pore size and surface roughness were obtained. Zhou *et al.* [31] used cellulose nanocrystals (CNCs) as an additive to improve the hydrophilicity, permeability, antifouling, and mechanical properties of cellulose acetate membranes. These authors observed that these membranes became more porous and interconnected after the addition of CNCs, as well as the tensile strength of the composite membranes increased by 47%. Basile *et al.* [3] proposed microporous membranes based on poly(ε-caprolactone) (PCL) and PCL functionalized with amine (PCL-DMAEA) or anhydride groups (PCL-MAGMA). Nanowhiskers of hydroxyapatite (HA) were also incorporated into the polymer matrix to realize nanocomposite membranes. Compared to PCL alone, hydrophilicity increased for functionalized PCL, whereas the addition of HA significantly influenced the hydrophilic characteristics. PCL-DMAEA and the relative nanocomposite membranes were the most promising in terms of cell-biomaterial interactions. Hassan Kiadeh *et al.* [32] prepared asymmetrical polycaprolactone membranes for wound dressing. In particular, a composite membrane was composed of two layers: an asymmetrical poly(ε-caprolactone) layer coated with a drug-loaded chitosan-silica matrix. The results showed that the polycaprolactone-modified membranes were suitable for fibroblast cell adhesion and proliferation. Sabri *et al.* [33] blended polysulfone (PS) membranes with different loadings of arabic gum (AG). AG addition to the dope polymer solutions changed the porous structure, hydrophilicity, and zeta potential of the cast PS/AG membranes. The inhibition of bacterial attachment and growth on PS/AG membranes was observed for both Gram-positive and Gram-negative bacterial models.

These are some examples of polymeric membranes, also loaded with active agents, produced by NIPS and potentially useful in different fields: from filtration to tissue engineering to pharmaceuticals. However, some common limitations characterize these membranes: an irregular, asymmetrical and prevalently closed morphology; toxic organic solvents residue are rarely checked, and purification post-treatments are in all cases required.

Therefore, to overcome these drawbacks, an alternative and advanced process has been proposed: the supercritical CO2-assisted phase separation. This process is attracting growing interest not only in the scientific literature but also in the industry, thanks to the possibility of improving membrane morphology and properties, using a one-shot process that is faster with respect to the traditional ones.

5.4 Supercritical Phase Separation

Porous polymeric membranes can be obtained using an innovative and eco-friendly process called supercritical CO_2 phase separation. In this case, a supercritical fluid (SCF) is used as the non-solvent for the polymer [18, 34, 35]. Therefore, it can be defined as an evolution of the traditional NIPS.

The main advantages of using a SCF over an organic solvent are related to its physico-chemical properties, i.e., liquid-like density, gas-like diffusivity, and negligible surface tension. Thanks to these properties, processes assisted by SCF are generally superior to the corresponding traditional ones, in terms of selectivity [36], kinetics [37], mass transfer (much lower viscosities than organic solvents) [38], capacity to remove all the organic solvent residues [35], and flexibility [39]. Among SCFs, supercritical CO_2 (SC-CO_2) is the most frequently used since it is characterized by mild critical conditions ($P_c \approx$ 73 bar, $T_c \approx$ 31°C), is inert, cheap, non-toxic, and generally recognized as safe (GRAS) [35–39].

Since supercritical phase separation is similar to NIPS (the main difference is the kind of non-solvent), a ternary phase diagram can also be used in this case to explain the thermodynamics of the process. In Figure 5.2, the binodal and spinodal curves determine the phase separation regions and CP is the critical point in which the two curves overlap. In particular, if nucleation and growth of droplets of the polymer lean-phase with further solidification of the polymer rich-phase occur (liquid-liquid binodal demixing), a membrane cellular structure is formed; if spinodal decomposition, with subsequent solidification of the polymer rich-phase occurs, a bi-continuous membrane structure is obtained; if nucleation and growth of droplets of the polymer rich-phase, followed by solidification of the polymer rich-phase occurs, a bead-like (particulate) membrane structure is realized [40].

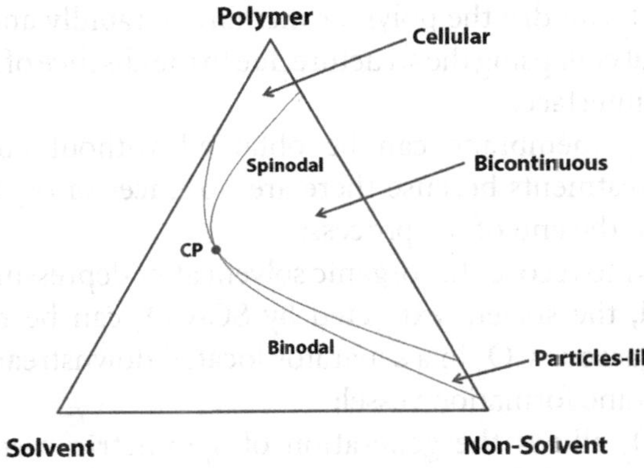

Figure 5.2 Qualitative ternary phase diagram used to describe the supercritical phase separation, in which the non-solvent is SC-CO_2.

Moreover, due to the flexibility of this process, changing the operative pressure and temperature that influence CO_2 density and solvent power, membrane mean pore size and, in some cases, membrane morphology, can be modified using the same polymer system. This is possible since the hole area and shape change in the ternary diagram when the SC-CO_2 characteristics are modified through the selection of different operative parameters (i.e., pressure and temperature). For example, from a thermodynamic viewpoint, decreasing the operative pressure, membrane mean pore size is generally larger, as the miscibility hole enlarges. From a kinetic viewpoint, when the operative pressure increases, the supercritical phase separation becomes faster, as the density and, thus, the solvent power of SC-CO_2 increase. The resulting membrane morphology and pore size will depend on the region of the miscibility hole where the composition path will fall, according to the classical theory on phase separation. Moreover, according to Shi *et al.* [41], mass transfer kinetics is also a relevant factor for membrane morphology. In particular, they demonstrated that the high instantaneous velocity of SC-CO_2 during phase separation led to a cellular pore structure in the membrane, whereas a decrease in the difference between the average solvent out-diffusion velocity and non-solvent in-diffusion velocity favored an increase in pore size and porosity of the membrane.

Thanks to these features, supercritical phase separation has several advantages as compared to NIPS:

- the use of toxic organic solvents, volatile and dangerous for human health and environment, is avoided;

- $SC\text{-}CO_2$ can dry the polymer membrane rapidly and totally without collapsing the structure due to the absence of a liquid-vapor interface;
- the dry membrane can be obtained without additional post-treatments because there are no traces of organic solvents at the end of the process;
- it is easy to recover the organic solvent after depressurization; indeed, the solvent extracted by $SC\text{-}CO_2$ can be removed from gaseous CO_2 in a separator located downstream of the membrane formation vessel;
- $SC\text{-}CO_2$ allows the generation of symmetric membranes and the modulation of membrane morphology and pore size by simply changing process pressure and temperature, which produce different CO_2 solvent power and diffusivities.

A possible supercritical phase separation plant layout is shown in Figure 5.3. The main components are the high-pressure pump (P-1), the high-pressure vessel (V-3), and the separator (S-1). This process is batch with respect to the membrane formation and works using a continuous flux of CO_2 [40], favoring a complete and fast removal of the organic solvent required for the preparation of the starting polymeric solution.

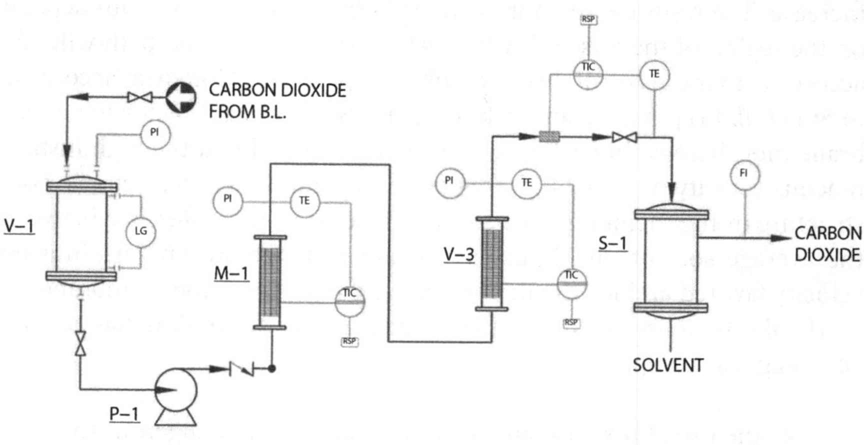

Figure 5.3 Schematic plant layout of the supercritical phase separation process.

5.5 Main Application Fields of Membranes Produced by Supercritical Phase Separation

The advantages of supercritical phase separation allow for the production of polymeric membranes with more regular and reproducible morphologies than the ones produced by traditional methods. Moreover, this process is faster than the traditional ones and it does not require post-treatments to purify the obtained membranes, since they are already safe (no final organic solvent residues) for various applications. However, in the literature, there are relatively few papers related to the production of membranes using this innovative process, even if they have been proposed for different applications: filtration, biomedical, pharmaceutical, electronics, and so on.

Barroso *et al.* [42] realized poly(ethylene glycol) methyl ether acrylate (PEGA) membranes grafted to poly(acrylonitrile) (PAN) by supercritical phase separation. These membranes showed enhanced hydrophilicity, larger protein and starch solution permeabilities, and good resistance to irreversible fouling, indicating that the technology is an efficient process to prepare fouling resistant membranes for biomacromolecule separations. Morgado *et al.* [43] proposed poly(vinylalcohol)/chitosan (PVA/CS) membranes for wound dressings in skin regeneration. SC-CO_2-assisted phase separation was performed at 200 bar and 45 °C. SEM images showed a regular porous morphology that, coupled with the biocompatibility and biodegradation properties, made these membranes suitable for skin wound healing. Baldino *et al.* [40] produced cellulose acetate (CA)/graphene oxide (GO) nanocomposite membranes by SC-CO_2-assisted phase separation, operating at 200 bar and 40°C. GO loading did not interfere with the cellular morphology of CA membranes; however, the presence of GO influenced pore sizes that became larger. The same research group [44] produced poly(vinyl alcohol) (PVA) membranes loaded with silver-hydroxyapatite nanoparticles by SC-CO_2 phase separation. Different operative conditions were studied to determine the influence of SC-CO_2 density on membrane morphology: temperature was varied from 35 to 55°C and pressure from 150 to 250 bar. PVA membranes with a uniform cellular morphology and nanoporous walls were obtained; silver ions were released from PVA membranes mainly by diffusion, showing a significant *Escherichia coli* inhibition. In another work, Baldino *et al.* [45] proposed cellulose acetate membranes loaded with curcumin for active food packaging. A uniform distribution at the nanometric level of curcumin was obtained, producing practically solvent-free membranes (5 ppm acetone residue). A very long continued release of curcumin was assured, maintaining more than

90% of the active compound's native antioxidant activity. Cardea *et al.* [46] produced PVDF-HFP membranes loaded with 3% w/w curcumin (Cm) using the same technique. Various operative conditions (from 250 bar/35°C to 150 bar/55°C) and polymer concentrations (from 10 to 30% w/w) were tested. A cellular membrane morphology was observed by SEM, and porosity values up to 89% were measured. Cm release was prolonged from 20 h to 26 h changing the PVDF-HFP membrane concentration from 20 to 30% w/w. Tabernero *et al.* [47] suggested the inclusion of a natural origin polymer (levan) in membranes of cellulose acetate, produced by SC-CO_2 phase separation. These membranes were characterized by interconnected pores; due to levan addition, composite membranes increased their mechanical resistance and cell adhesion from 8% to 30%. In a further work, these authors [48] applied supercritical phase separation to prepare cellulose acetate plus Laponite® membranes for tissue engineering applications. These composite membranes were characterized by a porosity of around 85%, with an average pore size ranging from about 9 to 13 μm, depending on the initial weight ratio cellulose acetate/Laponite®. Moreover, Laponite® increased material cell adhesion: thanks to its homogeneous inclusion into the membranes, a cell adhesion six times higher than the original material was obtained. Sizov *et al.* [49] produced porous poly(2,5-benzimidazole) membranes for vanadium redox battery applications, by SC-CO_2-assisted phase separation. Significant improvement in proton conductivity was observed (values up to 27 mS cm^{-1} were achieved), while VO^{2+} permeability was preserved at a low level (4.1×10^{-8} cm^2 min^{-1}). An improved coulombic efficiency without significant deterioration of the voltage efficiency compared to Nafion membranes was also demonstrated.

These examples confirm the flexibility of the supercritical phase separation in obtaining membrane systems tailored for a specific application, changing the operative parameters governing the process. Operating in this way, smart devices can be produced, also loaded with active compounds, ensuring a homogeneous distribution in the membrane matrix. This is favored by the kinetics of this kind of phase separation, which avoids cluster formation and active compound deposition on the bottom of the membrane. These advantages allow a general improvement of the membrane performance in the various application fields, in a reproducible, safe, and eco-friendly way.

5.6 Conclusions

Supercritical phase separation is a green and innovative process for the production of porous polymeric membranes. Solvent-free products can be obtained using a faster technique that, together with the regular and interconnected porous morphology at micro- and nanoscale, opens the range of applications for these membranes to other fields.

From a scientific research viewpoint, other experimentations can be performed to enlarge the knowledge and the complexity of the polymeric systems generated, starting from the basic theory of NIPS, and pointing towards an advanced process that is more flexible and environmental friendly.

References

1. Guillen, G.R., Pan, Y., Li, M., Hoek, E.M.V., Preparation and characterization of membranes formed by nonsolvent induced phase separation: A review. *Ind. Eng. Chem. Res.*, 50, 3798, 2011.
2. Wu, T., Zhou, B., Zhu, T., Shi, J., Xu, Z., Hu, C., Wang, J., Facile and low-cost approach towards a PVDF ultrafiltration membrane with enhanced hydrophilicity and antifouling performance via graphene oxide/water-bath coagulation. *RSC Adv.*, 5, 7880, 2015.
3. Basile, M.A., Gomez d'Ayala, G., Malinconico, M., Laurienzo, P., Coudane, J., Nottelet, B., Della Ragione, F., Oliva, A., Functionalized PCL/HA nanocomposites as microporous membranes for bone regeneration. *Mater. Sci. Eng.*, C, 48, 457, 2015.
4. Grasso, G., Galiano, F., Yoo, M.J., Mancuso, R., Park, H.B., Gabriele, B., Figoli, A., Drioli, E., Development of graphene-PVDF composite membranes for membrane distillation. *J. Membr. Sci.*, 604, 118017, 2020.
5. Sawada, I., Fachrul, R., Ito, T., Ohmukai, Y., Maruyama, T., Matsuyama, H., Development of a hydrophilic polymer membrane containing silver nanoparticles with both organic antifouling and antibacterial properties. *J. Membr. Sci.*, 1, 387–388, 2012.
6. HoppHirschler, M. and Nieken, U., Modeling of pore formation in phase inversion processes: analysis of pore formation mechanism. *J. Therm. Anal. Calorim.*, 139, 2831, 2020.
7. Rouquerol, J., Avnir, D., Fairbridge, C.W., Everett, D.H., Haynes, J.H., Pernicone, N., Ramsay, J.D.F., Sing, K.S.W., Unger, K.K., Recommendations for the characterization of porous solids. *Pure Appl. Chem.*, 66, 1739, 1994.

8. van de Witte, P., Dijkstra, P.J., van den Berg, J.W.A., Feijn, B.J., Phase separation processes in polymer solutions in relation to membrane formation. *J. Membr. Sci.*, 117, 1, 1996.
9. Nishi, T., Wang, T.T., Kwei, T.K., Thermally induced phase separation behavior of compatible polymer mixtures. *Macromolecules*, 8, 227, 1972.
10. Nunes, S.P. and Inoue, T., Evidence for spinodal decomposition and nucleation and growth mechanisms during membrane formation. *J. Membr. Sci.*, 111, 93, 1996.
11. Tsai, F.J. and Torkelson, J.M., Roles of phase separation mechanism and coarsening in the formation of poly(methylmethacrylate) asymmetric membranes. *Macromolecules*, 23, 775, 1990.
12. Song, S.W. and Torkelson, J.M., Coarsening effects on the formation of microporous membranes produced via thermally induced phase separation of polystyrene±cyclohexanol solutions. *J. Membr. Sci.*, 98, 209, 1995.
13. Mulder, M., *Basic principles of membrane technology*, Kluwer Academic Publishers, Dordrecht, 1991.
14. Basri, H., Ismail, A.F., Aziz, M., Polyethersulfone (PES)–silver composite UF membrane: effect of silver loading and PVP molecular weight on membrane morphology and antibacterial activity. *Desalin.*, 273, 72, 2011.
15. Urkiaga, A., Iturbe, D., Etxebarria, J., Effect of different additives on the fabrication of hydrophilic polysulfone ultrafiltration membranes. *Desalin. Water Treat.*, 56, 3415, 2015.
16. Bildyukevich, A.V., Plisko, T.V., Isaichykova, Y.A., Ovcharova, A.A., Preparation of high-flux ultrafiltration polyphenylsulfone membranes. *Petrol. Chem.*, 58, 747, 2018.
17. Alexandre, M. and Dubois, P., Polymer-layered silicate nanocomposites: Preparation, properties and uses of a new class of materials. *Mater. Sci. Eng., R*, 28, 1, 2000.
18. Temtem, M., Casimiro, T., Aguiar-Ricardo, A., Solvent power and depressurization rate effects in the formation of polysulfone membranes with CO_2-assisted phase inversion method. *J. Membr. Sci.*, 283, 244, 2006.
19. Plisko, T.V., Penkova, A.V., Burts, K.S., Bildyukevich, A.V., Dmitrenko, M.E., Melnikova, G.B., Atta, R.R., Mazur, A.S., Zolotarev, A.A., Missyul, A.B., Effect of Pluronic F127 on porous and dense membrane structure formation via non-solvent induced and evaporation induced phase separation. *J. Membr. Sci.*, 580, 336, 2019.
20. Lifshitz, I.M. and Slyozov, V.V., The kinetics of precipitation from supersaturated solid solutions. *J. Phys. Chem. Solids*, 19, 35, 1961.
21. Cahn, J.W., Phase separation by spinodal decomposition in isotropic systems. *J. Chem. Phys.*, 42, 93, 1965.
22. Mulder, M.H.V., *Preparation of synthetic membranes basic principles of membrane technology*, 2nd, pp. 54–95, Kluwer Academic Publishers, Dordrecht, 1996.

23. Chung, T.S. and Teoh, S.K., Breaking the limitation of composition change during isothermal mass-transfer processes at the spinodal. *J. Membr. Sci.*, 130, 141, 1997.
24. Su, Y.S., Kuo, C.Y., Wang, D.M., Lai, J.Y., Deratani, A., Pochat, C., Bouyer, D., Interplay of mass transfer, phase separation, and membrane morphology in vapor-induced phase separation. *J. Membr. Sci.*, 338, 17, 2009.
25. Wang, I.F., Ditter, J.F., Zepf, R., Highly porous polyvinylidene difluoride membranes, 2000, US Patent 6,146,747.
26. Caquineau, H., Menut, P., Deratani, A., Dupuy, C., Influence of the relative humidity on film formation by vapor induced phase separation. *Polym. Eng. Sci.*, 43, 798, 2003.
27. Lee, H.J., Jung, B., Kang, Y.S., Lee, H., Phase separation of polymer casting solution by nonsolvent vapor. *J. Membr. Sci.*, 245, 103, 2004.
28. Park, H.C., Kimb, Y.P., Kimb, H.Y., Kang, Y.S., Membrane formation by water vapor induced phase inversion. *J. Membr. Sci.*, 156, 169, 1999.
29. Koenhen, D.M., Mulder, M.H.V., Smolders, C.A., Phase separation phenomena during formation of asymmetric membranes. *J. Appl. Polym. Sci.*, 21, 199, 1977.
30. Ghosh, A.K., Jeong, B.H., Huang, X.F., Hoek, E.M.V., Impacts of reaction and curing conditions on polyamide composite reverse osmosis membrane properties. *J. Membr. Sci.*, 311, 34, 2008.
31. Zhou, J., Chen, J., He, M., Yao, J., Cellulose acetate ultrafiltration membranes reinforced by cellulose nanocrystals: Preparation and characterization. *J. Appl. Polym. Sci.*, 43946, 1, 2016.
32. Hassan Kiadeh, S.Z., Ghaee, A., Mashak, A., Mohammadnejad, J., Preparation of chitosan–silica/PCL composite membrane as wound dressing with enhanced cell attachment. *Polym. Adv. Technol.*, 28, 1396, 2017.
33. Sabri, S., Najjar, A., Manawi, Y., Omer Eltai, N., Al-Thani, A., Ali Atieh, M., Kochkodan, V., Antibacterial properties of polysulfone membranes blended with arabic gum. *Membranes*, 29, 9, 2019.
34. Huang, S., Wu, G., Chen, S., Preparation of microporous poly (vinylidene fluoride) membranes via phase inversion in supercritical CO_2. *J. Membr. Sci.*, 293, 100, 2007.
35. Tsivintzelis, I., Pavlidou, E., Panayiotou, C., Porous scaffolds prepared by phase inversion using supercritical CO_2 as antisolvent. I. Poly(l-lactic acid). *J. Supercrit. Fluids*, 40, 317, 2007.
36. Baldino, L., Della Porta, G., Sesti Osseo, L., Reverchon, E., Adami, R., Concentrated oleuropein powder from olive leaves using alcoholic extraction and supercritical CO_2 assisted extraction. *J. Supercrit. Fluids*, 133, 65, 2018.
37. Baldino, L., Cardea, S., Reverchon, E., Natural aerogels production by supercritical gel drying. *Chem. Eng. Trans.*, 43, 739, 2015.
38. Ruivo, R., Cebola, M.J., Simões, P.C., Nunes da Ponte, M., Fractionation of edible oil model mixtures by supercritical carbon dioxide in a packed column. Part II. A mass transfer study. *Ind. Eng. Chem. Res.*, 41, 2305, 2002.

39. Baldino, L., Reverchon, E., Della Porta, G., An optimized process for SC-CO_2 extraction of antimalarial compounds from *Artemisia annua* L. *J. Supercrit. Fluids*, 128, 89, 2017.
40. Baldino, L., Sarno, M., Cardea, S., Irusta, S., Ciambelli, P., Santamaria, J., Reverchon, E., Formation of cellulose acetate–graphene oxide nanocomposites by supercritical CO_2 assisted phase inversion. *Ind. Eng. Chem. Res.*, 54, 8147, 2015.
41. Shi, W., He, G., Zhao, W., Zhang, L., Du, L., Wang, Y., Qin, X., Effects of solvent and nonsolvent diffusion velocities on themorphology of cellular polyetherimide membranes prepared using supercritical CO_2 phase inversion. *J. Supercrit. Fluids*, 83, 6, 2013.
42. Barroso, T., Temtem, M., Casimiro, T., Aguiar-Ricardo, A., Antifouling performance of poly(acrylonitrile)-based membranes: From green synthesis to application. *J. Supercrit. Fluids*, 56, 312, 2011.
43. Morgado, P., Lisboa, P.F., Ribeiro, M.P., Miguel, S.P., Simões, P.C., Correia, I.J., Aguiar-Ricardo, A., Poly(vinylalcohol)/chitosan asymmetrical membranes: Highly controlled morphology toward the ideal wound dressing. *J. Membr. Sci.*, 469, 262, 2014.
44. Baldino, L., Aragón, J., Mendoza, G., Irusta, S., Cardea, S., Reverchon, E., Production, characterization and testing of antibacterial PVA membranes loaded with HA-Ag_3PO_4 nanoparticles, produced by SC-CO_2 phase inversion. *J. Chem. Technol. Biotechnol.*, 94, 98, 2019.
45. Baldino, L., Cardea, S., Reverchon, E., Biodegradable membranes loaded with curcumin to be use d as engineered independent devices in active packaging. *J. Taiwan Inst. Chem. Eng.*, 71, 518, 2017.
46. Cardea, S., Baldino, L., Reverchon, E., Comparative study of PVDF-HFP-curcumin porous structures produced by supercritical assisted processes. *J. Supercrit. Fluids*, 133, 270, 2018.
47. Tabernero, A., Baldino, L., González-Garcinuño, A., Cardea, S., Martín del Valle, E.M., Reverchon, E., Supercritical CO_2 assisted formation of composite membranes containing an amphiphilic fructose-based polymer. *J. CO2 Utiliz.*, 34, 274, 2019.
48. Tabernero, A., Cardea, S., Masa, R., Baldino, L., Martín del Valle, E.M., Reverchon, E., Preparation and characterization of cellulose acetate-Laponite® composite membranes produced by supercritical phase inversion. *J. Supercrit. Fluids*, 155, 104651, 2020.
49. Sizov, V.E., Kondratenko, M.S., Gallyamov, M.O., Stevenson, K.J., Advanced porous polybenzimidazole membranes for vanadium redox batteries synthesized via a supercritical phase-inversion method. *J. Supercrit. Fluids*, 137, 111, 2018.

6

Advanced Fabrication of Porous Membranes for Membrane Contactors Processes

M. Frappa, F. Macedonio, E. Drioli and A. Gugliuzza*

Research Institute on Membrane Technology-National Research Council (CNR-ITM), Rende, Italy

Abstract

This chapter provides new insights into the fabrication of porous membranes, which moves from traditional phase inversion to a set of advanced templating procedures. To get a high-defined structure in membranes is explained to provide interesting solutions for the advancement of membrane contactor processes. Benefits in productivity and efficiency can be from reproducible and amplified events, which take place in every region of the membrane. A mathematical description of mass and energy transfer through porous membranes focuses on the relationship between transport mechanisms and structural/chemical features of membranes. So, the basics are described for some membrane contactor processes, including membrane distillation (MD), osmotic distillation (OMD), membrane crystallization (MCr), membrane emulsification (ME), gas-liquid membrane contactor (GLMC), and membrane condenser (MeC). A discussion is provided on materials and membrane specificities, with a special emphasis on green materials and fabrication procedures. Manufacturing techniques that enable one to get lithographic precision in polymeric architectures are reviewed, while compelling MD water desalination case studies are examined in relation to productivity-efficiency trade-offs.

Keywords: Porous membranes, phase separation, molding, breath figure, self-assembly, electrospinning, track etching, 3D printing

**Corresponding author*: a.gugliuzza@itm.cnr.it

Annarosa Gugliuzza and Wanqin Jin (eds.) Porous Membranes: Breakthroughs in Manufacturing and Applications, (111–172) © 2025 Scrivener Publishing LLC

Nomenclature

a	exponent of r
C	concentration
D	diameter
D	diffusion coefficient
F	gas/liquid contact area
H	Henry's law coefficient
J	Trans-membrane flux
K	molecular free path for water vapor
k	overall mass transfer coefficient based on inner surface of the hollow fiber
M	molecular weight
P	pressure
P_{alm}	the log mean of the partial pressure of the air within the pores
Q	flow rate
r	Pore size
R	gas constant, resistance
t	time
T	temperature

Greek symbols

ε	porosity
δ	thickness
θ	contact angle
μ	dynamic viscosity
χ	nominal constant given by operating condition
σ	surface free energy
τ	tortuosity
ρ	packing density

Subscript

a	air
avg	average value
f	feed
g	gas
i	inner, inside, at the interface, i component

Kn	Knudsen
k_B	Boltzmann's constant
l	liquid
lm	liquid-membrane interface
m	membrane
o	outer, outside
P	pore
p	permeate
w	water

6.1 Introduction

Membrane contactor (MC) technology is regarded as a valuable alternative to conventional energy-intensive techniques for the purification and recovery of added-values products [1]. This technology uses porous membranes as physical interfaces between two adjacent phases, i.e., liquid-liquid, liquid-gas, liquid/gas/liquid [2]. The membrane keeps two phases in contact preventing their mixing while mass and/or energy transfer is induced under a specific driving force, i.e., the gradient of concentration, temperature, pressure, or partial pressure (Scheme 6.1). To play a key role, the membrane needs to fulfill specific morphological and chemical requirements. The pores have to be numerous and sufficiently small so that a wide interfacial area is provided for mass transfer while liquid intrusion and wetting need to be disallowed [3, 4]. Surface free energy along with pore size are critical factors to contrast wetting. Therefore, the structure and chemical properties of the membranes are

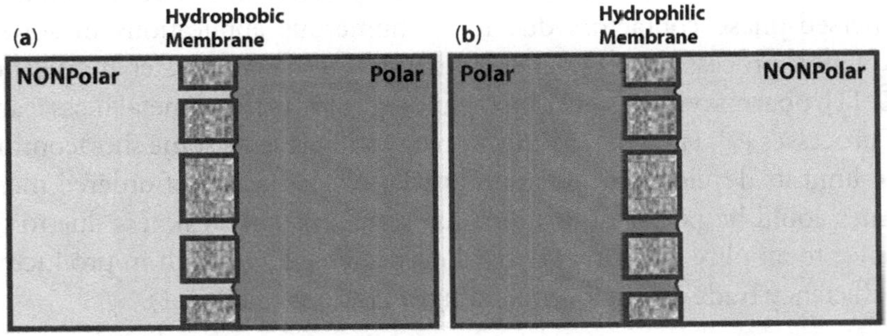

Scheme 6.1 Basic concept of a membrane contactor: (a) hydrophobic membrane and (b) hydrophilic membrane with polar and non-polar solution on the two membrane sides.

to be considered when a membrane contactor operation is implemented. Scheme 6.1 provides two concepts of membrane contactors based on hydrophobic and hydrophilic membranes wherein immiscible and miscible phases are faced.

In a situation of immiscibility, two effects come into play: a) a hydrophobic membrane is wet with the organic phase while a compound can be transported from the aqueous to the organic phase or *vice versa*; b) a hydrophilic membrane is filled with the aqueous phase, usually containing the solute so that the transport is promoted from the aqueous to the organic or gaseous phase. If the two phases are miscible, a hydrophobic membrane is then chosen to keep them at the external sides while a vapor-liquid (V-L) interface is formed at the pore entrance. Surface free energy is decisive for controlling wetting processes, but pore number, size, and distribution along with membrane thickness are pivotal to achieve a stable interface between the two phases. Based on these concepts, it becomes clear that there is a necessity to design membranes with suitable features not yet available in commercial membranes. High-defined structures can be crucial to extend desired events over different lengths of scale. Advanced fabrication procedures could fit this target making MC operations more proficient and competitive than conventional separation processes.

6.2 Membrane Contactors Technology: An Overview

A membrane contactor (MC) has several advantages over traditional contactors, i.e., packed and spray columns, including no flooding, channeling, and foaming problems, and low capital and operating costs are required [5, 6]. The use of a membrane as a physical barrier between two phases represents a powerful approach to improve the performance over conventional dispersed-phase contactors due to its numerous applications in solvent extraction processes [7, 8] semiconductors [9], air and water purification [10, 11], pharmaceuticals [12], biotechnology [13, 14], and metal mass transfer processes [9]. However, MC technology still presents some shortcomings that limit its development on an industrial scale. The use of ordered membranes could be powerful to overcome some current weakness due to the chance to amplify and direct desired events, enabling a shift in productivity-efficiency trade-offs at low operating conditions (Table 6.1).

Table 6.1 Main advantages and disadvantages of membrane contactors process.

Advantages	Disadvantages
High surface area per volume unit	Need for pre-treatment
Reduced size of modules	Limited operating pressure (breakthrough value)
Wide range of operating flowrates	Membrane stability (hydrophobic/hydrophilic character)
Low operating temperatures and pressure	Additional resistance offered by the membrane
Low equipment costs and increased process safety	Slow mass-transfer (the membrane may retard the mass transfer between the aqueous feed and the organic solvent)
High rejection (100% (theoretical) of non-volatile constituents)	Presence of an additional mass transport resistance and to the quite limited range of the operating pressures below the breakthrough threshold
Possibility to carry out simultaneously reaction and separation	Polymeric membrane replacement costs
Flexibility, control and automatization	Limited operating pressure below the breakthrough Value
Modular design, no moving parts	Relatively low membrane lifetime
Robust membranes (such as, PTFE, PP and PVDF; new amorphous perfluoro polymers (such as Hyflon, Teflon, etc.).	
Less membrane fouling	
Low sensitivity to concentration polarization phenomenon	

(*Continued*)

Table 6.1 Main advantages and disadvantages of membrane contactors process. (*Continued*)

Advantages	Disadvantages
High modularity and compatibility	
Easy scale up	
Independence of the fluid phases in contact	
Possibility to use of plastic equipment (absence of corrosion)	

6.2.1 Membrane Distillation

Membrane distillation (MD) is a type of membrane contactor-based process applied in desalination and wastewater treatment [15]. Porous, no-wetting membranes are required to recover freshwater with a theoretical rejection of 100% for salts and non-volatile components.

This process works at low operating temperatures compared to conventional distillation columns, reduced concentration polarization compared to pressure-driven membrane processes, and low operating pressure compared to reverse osmosis (RO). MD mainly involves two-phase changes: water evaporation and condensation (Figure 6.1). The first step occurs

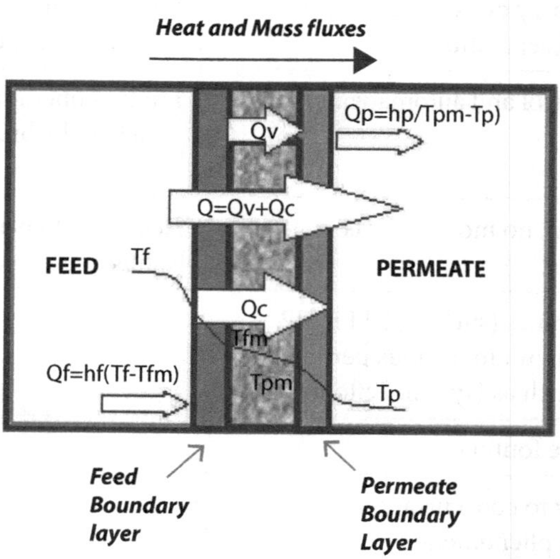

Figure 6.1 Representative scheme of a direct contact membrane distillation showing mass and heat transfer through a porous hydrophobic membrane.

at the L-V interface (i.e., feed side); successively, the water vapor diffuses through the no-wetted pores of hydrophobic membranes and then condenses on the opposite side (i.e., permeate side).

Depending on the approach used to induce water evaporation and condensation, different MD configurations can be assembled:

- Direct contact membrane distillation (DCMD) is obtained when a cold pure water solution is circulated tangentially to the permeate side of the membrane. A temperature difference is applied across the membrane to induce a vapor pressure difference necessary to promote mass transfer [16].
- Air gap membrane distillation (AGMD) is a configuration wherein an air gap is interposed between the membrane and a cold condensing surface placed inside the membrane module. The air gap increases the conductive heat transfer resistance while decreasing the amount of heat lost by conduction through the membrane [17, 18].
- Vacuum membrane distillation (VMD) is a configuration wherein a vacuum or low pressure is applied at the permeate side, while an external condenser is used to collect water as a liquid on the permeate side [19–21].
- Sweep gas membrane distillation (SGMD) is a configuration where air or nitrogen is used to sweep volatile compounds outside the membrane (at the permeate side) for condensation. The gas is used in place of the vacuum [20, 22, 23].

More recently other MD configurations have been developed and proposed:

- Liquid gap membrane distillation (LGMD) is a merging of DCMD and AGMD configurations. LGMD shows the same aspect as AGMD except for the gap, which is kept filled by a stagnant cold liquid solution (distilled water) [24].
- Permeate gap membrane distillation (PGMD) involves the addition of a third channel with a further non-permeable foil and a separation of the distillate from the coolant. In general, the coolant can be any other liquid, such as cold feed water [25]. One significant advantage of PGMD is the efficient heat recovery system. The presence of the distillate channel reduces heat losses due to an additional heat transfer

resistance and reduces the temperature difference across the membrane, which slightly lowers the permeation rate.

- Thermostatic sweeping gas membrane distillation (TSGMD) is a combination of SGMD and AGMD where the temperature of the sweeping gas is minimized. [26] The temperature of sweeping gas may be increased close to the membrane module due to the heat transfer from the feed side through the membrane to the permeate side. The presence of the cold wall on the permeate side reduces the increase in the sweeping gas temperature, resulting in an enhanced driving force and a subsequent increase in the water production rate.
- Material gap membrane distillation (MGMD) involves filling the gap between the membrane and the condensation plate with materials such as polyurethane (sponge), polypropylene mesh, sand, and deionized water [27]. Regarding heat transfer, MGMD shows performance comparable to DCMD mainly due to the high heat loss through conduction. The heat developed in the materials at higher temperatures decreases the mass transfer with a reduction of the flux.

In every case, the success of the separation depends on the capacity of the membrane to keep a stable and durable interface between two phases, so that water diffusion is allowed while heat loss is reduced. The trans-membrane flux can be measured with *eq. 6.1*

$$J = C * (P_F - P_p) = C\Delta P \tag{6.1}$$

where J is the water flux, P_f and P_p the vapor pressure at the feed and permeate side, respectively; C the membrane distillation coefficient, which depends on morphological parameters of the membrane described in *eq. 6.2*:

$$C \propto (r^a \varepsilon) / (\delta\tau) \tag{6.2}$$

where r is the nominal pore size of membrane, a the exponent of r in the range of 1 to 2 (a=1 for Knudsen diffusion, a =2 for viscous flux), τ the pore tortuosity and δ the membrane thickness.

Membrane pores and thickness play a relevant role in heat transfer as well. For a feed with a certain temperature T_f, the heat is first transferred from the heated feed solution across the thermal boundary layer to the membrane surface at a rate expressed as:

$$Q = h_f \cdot \Delta T_f \tag{6.3}$$

where h_f is the heat latent of feed and ΔT_f the difference of temperature between feed and membrane. The surface of the membrane liquid is vaporized, and heat is transferred across the membrane at a rate:

$$Q_v = h_v \cdot \Delta T_m = N \cdot \Delta H_v \tag{6.4}$$

where N is the rate of mass transfer and ΔH_V the heat of vaporization. Additionally, heat is conducted through the membrane material and the vapor that fills the pores at a rate of:

$$Q_m = h_m \cdot \Delta T_m \tag{6.5}$$

where h_m is the heat transfer coefficient of the vapor within the membrane. The total heat transfer across the membrane is measured as:

$$Q = Q_v + Q_m \tag{6.6}$$

After the passage of vapor across the membrane, the condensation takes place at the V-L interface, where heat is removed from the cold-side membrane surface through the thermal boundary layer at a rate of:

$$Q = h_p \cdot \Delta T_p \tag{6.7}$$

A limitation for MD operation is the heat transfer across the boundary layer, as a large quantity of heat must be supplied to the membrane surface to vaporize the liquid, resulting in thermal energy consumption. Concerning the membrane, it is hence important to choose materials with low thermal conductivity to suitably control pore size distribution, porosity, and thickness to reduce heat loss without affecting mass transfer.

Another important issue is the polarization phenomenon, which is mainly related to fluid dynamics. One way to reduce it relies on the generation of mixing and turbulence in the flow, thus reducing the thickness of the boundary layer. It is necessary to distinguish between the temperature polarization coefficient (TPC) and the concentration polarization coefficient (CPC). These two parameters are used to quantify heat and mass transport resistance, respectively, in the boundary layer with respect to the total transfer resistance of the system.

$$TPC = \frac{T_{fm} - T_{pm}}{T_f - T_p} \tag{6.8}$$

$$CPC = \frac{C_{Bm}}{C_{Bbf}} \tag{6.9}$$

where C_{Bm} is the concentration of the non-volatile solutes at the membrane surface, whereas C_{Bbf} the concentration of the non-volatile solutes at the bulk feed. To limit these events, it is conceivable to increase the flow rates - the increase has not to generate an operating pressure above the liquid entry pressure (LEP) - or to include the spacers for promoting turbulence.

6.2.2 Osmotic Distillation

Osmotic membrane distillation (OMD) is an evaporative membrane contactor process that involves the contact of two liquid streams with a hydrophobic microporous membrane, as illustrated in Figure 6.2 [28, 29]. It is regarded as a powerful operation for processing liquid foods such as fruit or vegetable juices [30].

The driving force for mass transport is the difference in concentration across the membrane. No high temperature is required, thus preserving the quality of the fresh juices. OMD is suitable for concentrating aqueous

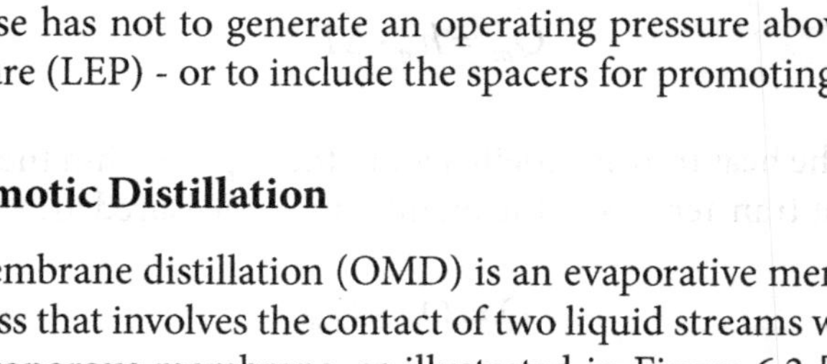

Figure 6.2 Schematic profile of temperature in the osmotic distillation process.

solutions containing non-volatile solutes and can concentrate a lot of amounts of solutes at low temperature and pressure [31]. Unlike RO, no osmotic pressure limitation takes place. Due to the hydrophobicity of the membrane, the juice and the brine are not able to enter the membrane pores, so a thin gas layer—basically air—is entrapped within the pores, forming the so-called gas membrane. Water vapor diffuses through the gas membrane, and the vapor pressure difference between the membrane sides drives the process. The difference with MD is due to the fact that the water transport is sustained by a difference in concentration rather than a difference in temperature. Owing to this strict analogy, the process has also been referred to as 'isothermal membrane distillation'. The mass transfer mechanism in OMD is similar to that in MD: at the pore entrance V-L equilibria are established giving rise to a partial pressure difference of water vapor, which diffuses through the gas immobilized within the pores. The water vapor pressures at the pore inlet are related to the temperature and activities prevailing in the liquids facing the membrane by:

$$Pwf = P_{wf} \cdot a_{wf} \tag{6.10}$$

$$Pwp = P_{wp} \cdot a_{wp} \tag{6.11}$$

where P_w represents the vapor pressure of pure water and a_w the water activity in the solutions of feed (*f*) and permeate (*p*) side, respectively.

In OMD, the difference of vapor pressure ($P_{wf}-P_{wp}$) for water transport is sustained by an activity difference, i.e., $a_{wp}<a_{wf}$. In membrane distillation, the activity difference represents a counter-driving force to be overcome by a temperature difference across the membrane, *i.e.*, in that case, $a_{wf}<a_{wp}$, $P_{wf}>P_{wp}$. There exists the possibility that both the temperature and activity differences act synergistically. In any case, the driving force allows the diffusion of the water vapor through the stagnant gas phase.

For quite large pore size with respect to the mean molecular free path of water vapor, molecular diffusion becomes the controlling mechanism and the water mass flux can be written as [31]:

$$v = KDln\frac{PAp}{PAf} = KD\frac{Pwf - Pw2}{Palm} \tag{6.12}$$

where P_{alm} is the log mean of the partial pressure of the air within the pores and KD expresses a sort of membrane permeability depending on the membrane properties, temperature, and pressure:

$$KD = \frac{\varepsilon M}{X\delta}\frac{PD}{RT} \tag{6.13}$$

where R is the gas constant, δ membrane thickness, T the temperature inside the membrane, ε the membrane porosity, M the molecular weight of water, P the total pressure inside the pore, and D the diffusion coefficient of vapor.

For pore sizes lower than the mean molecular free path, vapor transport is limited by the collision against the pore wall rather than against the molecules of the gas present in the pores. Therefore, the water mass flux can be conveniently written by the Knudsen model:

$$v = K_K(P_{w1} - P_{w2}) \tag{6.14}$$

in which K_k is expressed as:

$$K_k = \frac{1}{3}\frac{\varepsilon d_p}{X\delta}\sqrt{\frac{8M}{\pi RT}} \tag{6.15}$$

where K is the molecular free path for water vapor at 25°C and atmospheric pressure.

In this case, the diffusing molecules of water vapor are slowed down by molecules of air as well as by pore walls, resulting in a lower flux with the same driving force. Considering the ideal pore size used in OMD and MD, both the molecular diffusion and Knudsen models are expected to be valid.

6.2.3 Membrane Crystallization

Membrane crystallization (MCr) is an emerging excellent technique for the recovery of raw materials from high-concentrated brine [32]. The concentration polarization does not significantly affect the driving force of the process and, therefore, it is possible to reach high recovery factors and high concentration values. In addition with MD or/and reverse osmosis (RO) processes, it is possible to increase the recovery factor with potential zero brine production transforming the traditional problem of disposal of the brines into a potential new profitable market [33, 34]. The technique is based on MD principles, where the evaporation of water from a solution in contact with a hydrophobic membrane causes a concentration of the solution to reach super-saturated conditions. The operational principle is based on the vapor pressure gradient created by a different temperature across a microporous hydrophobic membrane [35–38]. Figure 6.3 shows optical microscope pictures of cubic sodium chloride crystals as a final output of MCr operations [36].

In MCr, controlled crystallization is expected to minimize the deposition and growth of crystals on the membrane and maximize crystal production through a broad set of available kinetic trajectories in the thermodynamic phase diagram. A suitable choice of the membrane and specific process parameters—temperature, concentration, flow rate, etc.—address the crystallization process to produce well-sized polymorph crystals in relatively short times. Different polymorphic forms of the same substance, which are considered different materials with their own physical, chemical, and biological properties, could be separated [39–41]. Inorganic salts, as well as protein and organic (bio)molecules, could be obtained in crystalline form to be reused in agriculture, cosmetics, hygiene and personal care, pharmaceuticals, fine chemicals, pigments, and foods [37, 41]. In the organic semiconductor field, the crystalline phase is utilized for the preparation of single-crystal-based devices and photonic crystals due to their wide diffusion in heterogeneous catalysis. MCr might be also applied for the preparation of new structure-based drugs in medical devices, where crystal forms are required to obtain the highest surface-volume ratio and greater catalytic efficiency [42, 43]. Functional membranes have been proposed to cause localized supersaturation; heterogenous nucleation and short formation of crystals have been successfully induced differently from what was observed with crystallization in a homogeneous phase [36, 41].

Heat and mass transport through the membrane can be described under non-thermodynamic equilibrium conditions [34]. Mass transport takes

Figure 6.3 Optical images @20X of NaCl crystals recovered from feed solution by using membranes filled with graphene.

place in the feed boundary layer, through the membrane pores, and in the permeate boundary layer [35]. To analyze the mass transfer in boundary layer, the Dusty-Gas Model (DGM) may be applied. This model also describes mass transfer in porous media by four possible mechanisms: viscous flow, Knudsen diffusion, molecular diffusion, and surface diffusion. Considering the simplest configuration (DC), the molar trans-membrane flux (*J'*) can be determined as follows [44]:

$$J' = \frac{\varepsilon PD_{ij}}{\tau\delta RT} ln\left(\frac{p_a^{2\frac{2r}{3}\left(\frac{8RT}{\pi M_i}\right)^{1/2}} + PD_{ij}}{p_a^{1\frac{2r}{3}\left(\frac{8RT}{\pi M_i}\right)^{1/2}} + PD_{ij}}\right) \tag{6.16}$$

where P is the total pressure, D_{ij} the diffusivity, R the ideal gas constant, T the temperature, M_i the molecular weight, and p_a^1 and p_a^2 the partial pressure of air at feed and membrane surface, respectively. It is clear that the trans-membrane flux is proportional to the membrane porosity ε and inversely proportional to the membrane thickness δ, confirming again how membrane morphologies strongly influence the process performance. Another critical parameter is the critical size of crystals n^*, which can be estimated from *eq. 6.17*:

$$n^* = \frac{32\pi v_0 \sigma^3}{3(k_B T)^2 ln^3 S} \tag{6.17}$$

where v_0 is the molecular volume, κ_B the Boltzmann's constant, and S the supersaturation. This equation yields an indication of the relationship between a high operating level of supersaturation and crystal size.

6.2.4 Membrane Emulsification

Membrane emulsification (ME) is a simple method to prepare stable and complex emulsions with potential applications in many fields such as dairy and food industry [45], formulations of particulate products for treating wastewater containing dyes [46, 47], enzyme immobilization [48], and biomedical applications as drug delivery systems [49, 50]. An emulsion is a mixture of two or more liquids that are usually immiscible (unmixable or unblendable). The process of membrane emulsification exploits porous membranes filled with the phase to disperse, while the continuous phase flows along the membrane surface (Figure 6.4). The droplets grow at pore outlets until they are detached once a certain size is reached. This event depends on the balance between the drag force on the droplet and the flowing continuous phase, as well as the buoyancy of the droplet, surface tension forces, and driving pressure [51, 52]. An advantage of this technique is its energy consumption is lower than traditional methods such as homogenization and rotor-stator system. At the same time, the difficulty of industrial scale up is one of the main limitations of this process.

The droplet at the pore outlet tends to form a spherical shape to obtain the least free energy under the action of the surface tension, but some distortion may occur depending on the flow rate of the continuous phase and the contact angle between the drop and the surface of the membrane [53]. Therefore, membrane pore size and distribution as well as porosity, roughness, and chemistry can affect the final droplet size; crossflow velocity, transmembrane pressure and emulsifier-type can influence the dimensions and distributions of emulsions. By choosing appropriate process parameters, complex emulsions can be prepared with droplets sized within the range of 2 to 10 times the nominal pore size of the membrane.

The transmembrane pressure ΔP can be determined as the difference between the pressure of the dispersed phase outside the membrane P_d, and the mean pressure of the continuous phase:

$$\Delta P = \frac{Pd-(Pc,in+Pc,out)}{2} \tag{6.18}$$

where $P_{c,in}$ and $P_{c,out}$ are the pressures at the inlet and outlet of the membrane module, respectively. The minimum emulsification pressure (or

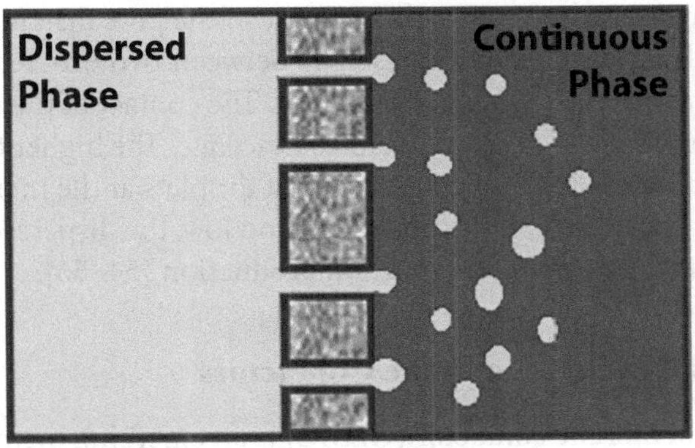

Figure 6.4 Schematic diagram of membrane emulsification process.

capillary pressure), required to make the dispersed phase flow outside the membrane, can be calculated from the Laplace equation (*eq. 6.19*):

$$\Delta P = \frac{4\sigma cos\theta}{r} \tag{6.19}$$

where σ is the surface tension of water, θ the contact angle, and r the pore size of the membrane. From capillary pressure, it is possible to define the dispersed phase flux J_d related to the transmembrane pressure according to Darcy's law:

$$J_d = \frac{K\Delta P}{\mu\delta} \tag{6.20}$$

where K is the membrane permeability, δ the membrane thickness, and μ the viscosity of the dispersed phase. To obtain the formation of an emulsion with the desired droplet dimension and narrow size distribution, it is necessary to find equilibrium between all the parameters of the process, including membrane morphology. The relationship between the droplet size of an emulsion and the pore size of the membrane is given by:

$$d_d = xr' \tag{6.21}$$

where d_d is the droplet size of an emulsion, r the droplet size of the membrane pore size, and x a nominal constant typically ranging from 2 to 10. The higher the narrow pore size, the higher the quality of monodisperse emulsion. The surface porosity of a membrane is another critical issue because it inversely determines the increase of the pore distance from each other. This distance decides the distance between two droplets, providing the space necessary to prevent coalescence. The contact of the droplets can result in adhesion, which may lead to coalescence. The higher the porosity, the higher the probability of coalescence of droplets at the membrane surface. On the other hand, if the porosity is too low, the dispersed phase flux may be insufficient for viable emulsion production [54–56].

6.2.5 Gas–Liquid Membrane Contactors

Gas–liquid membrane contactor (GLMC) is an emerging membrane technology, which is of particular interest in acid [57] or organic gas absorption, such as CO_2 [58] absorption into aqueous solutions [59], oxygen removal in semiconductors [60], ozonolysis [61], water treatment [62], dehumidification process, and concentration processes at low temperatures [63, 64]. In principle, a gas–liquid membrane reactor is referred to as a membrane contactor in which gas and liquid flow independently on the two sides of porous or dense membranes. Thanks to the use of membranes, it is possible to have a wide and fixed gas–liquid interfacial area, which leads to an advantageous mass transfer, process design, and scale up. In addition, this process overcomes problems frequently faced in dispersive reactors such as floods, emulsions, and foaming. For porous membranes, the mass transfer is strongly dependent on the position of the gas–liquid interface in the membrane pores, which is determined by the membrane wettability and pressure difference between the two phases (Figure 6.5).

GLMC can be divided into three categories:

- A dispersive gas–liquid membrane reactor utilizes porous membranes to disperse gas into micro-bubbles with a variety of unique properties. This kind of configuration finds widespread use in chemical and biochemical reactions, such as the decomposition of organic compounds.
- A non-dispersive gas–liquid membrane reactor allows gas and liquid to flow independently on the two sides of the membranes. In this case, both porous and dense membranes can be used to provide a very large and fixed specific interfacial area. This configuration has been applied in non-catalytic reactions, e.g., chemical absorption, preparation of nanoparticles, ozonolysis, hydration, and catalytic reactions, e.g., nitrate reduction, hydrogenation, direct methanol fuel cell.

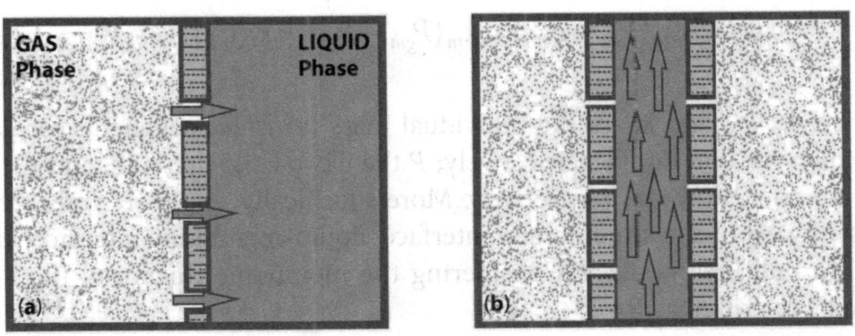

Figure 6.5 Schematic diagram of membrane gas–liquid reactors: (a) disperse membrane reactor; (b) non-disperse membrane reactor.

- A pore flow through reactor is a configuration where the reaction solution is pumped quickly through the catalytic membrane pores to reduce the contact time and eliminate pore diffusion.

Membrane wetting resistance is normally characterized using the Laplace–Young equation (*eq. 6.19*). Partial wetting in porous membranes could happen due to an excessive pressure drop and/or the surface free tension of the liquid. In dense membranes, a much higher pressure difference can be used. The mass transfer is determined by consecutive steps across three phases: diffusion of gaseous component *i* from the bulk gas to the membrane wall, diffusion through the totally gas-filled pores of the membrane to the membrane–liquid interface, dissolution into the liquid absorbent, followed by liquid phase diffusion and/or chemical reaction. Hence, the total resistance can be defined as the reciprocal of the overall mass transfer coefficient:

$$R_{\text{total}} = \frac{1}{K_{\text{ol}}} = \frac{H}{k_g} + \frac{H}{k_m} + \frac{1}{k_l} = R_g + R_m + R_l \tag{6.22}$$

where *Kol* is the overall mass transfer coefficient based on the liquid phase, *H* Henry's law coefficient, $k_{l\,g}$ the gas phase mass transfer coefficient, k_m the membrane mass transfer coefficient, k_l the liquid phase mass transfer coefficient, and *R* the resistance.

In the case of a non-wetted membrane, R_m expresses the resistance to mass transfer in the membrane pores filled by gas. The trans-membrane flux of gas (*J*) from gaseous phase to liquid phase passing through the porous membrane can be expressed by:

$$J = K_g(P_{gb} - P_{gm}) = k_m(P_{gm} - P_{lm}) = k_l(C_{lm} - C_{lb}) \tag{6.23}$$

where k_g, k_m, and k_l are the individual mass transfer coefficients for gas, membrane, and liquid, respectively; *P* the gas partial pressure, and *C* the liquid concentration, respectively. More specifically, g_b, g_m, l_m and l_b represent gas bulk, gas-membrane interface, liquid-membrane interface, and liquid bulk, respectively. Considering the membrane thickness, the flux can be rewritten as:

Table 6.2 Properties of membranes used in gas-liquid membrane.

Membrane	ID (μm)	OD (μm)	Pore Size (μm)	Porosity %	Process	Ref.
Polysulfone (PS)	200	400	0.05	n.a.	SO_2 absorption with NaOH, Na_2CO_3, Na_2SO_3 and $NaHCO_3$ solutions	[65, 66]
Polypropylene (PP)	600	1000	0.265	0.79	CO_2 absorption with CORAL 20 solution	[67]
PP	344	442	0.02-0.2	>0.45	CO_2 absorption with PG, MEA and MDEA solutions	[68]
Poly(vinylidene fluoride) (PVDF)	300	514	n.a.	0.698	CO_2 absorption with pure water	[69
PVDF	607	907	0.04	n.a.	H_2S and CO_2 absorption with Na_2CO_3 solution	[70]
PP	244	300	n.a.	0.35	CO_2 absorption with MEA and AMP	[71]
PTFE	1000	2000	n.a.	0.50	CO_2 absorption with MEA and AMP	[71]
EPTFE	1000	2000			CO_2 in CH_4 adsorption with Distilled water/aqueous sodium hydroxide/amine solutions	[72]

(Continued)

Table 6.2 Properties of membranes used in gas-liquid membrane. (*Continued*)

Membrane	ID (μm)	OD (μm)	Pore Size (μm)	Porosity %	Process	Ref.
PVDF	300	500	0.094	71.65	CO_2 adsorption with water	[73]
PSF			0.03-0.1	71.05-75.3	CO_2 adsorption with water	[74]
PAI	1100	1400		65.8-79	CO_2 adsorption with water/sodium taurinate	[75]
PEI	450	800	0.039-0.093		CO_2 adsorption with water	[4]
Aluminum oxide (α-Al2O3)	3000	4000	0.8		0.15-4.8 vol.% SO_2, 10 vol.% Coe and air with N,N-Dimethylaniline/1-ethyl-3-methylimidazolium ethylsulfate	[76]

$$J = k_g(P_{gb} - P_{gm})(\pi d_o \delta) = k_m(P_{gm} - P_{lm})(\pi d ln \delta) = k_l(C_{lm} - C_{lb})(\pi diL)$$
$$J = K_g(P_{gb} - P*)(\pi d_o L) = K_l(C* - C_{lb})(\pi diL) \quad (6.24)$$

where d_o, d_i, and d_{ln} are the outer, inner, and logarithmic mean diameters of the membrane, respectively. K_g the overall mass transfer coefficient in the gas phase, and K_l the liquid phase. These can be respectively expressed as:

$$\frac{1}{K_g} = \frac{1}{k_g} + \frac{do}{k_m d_{ln}} + \frac{Hdo}{k_l d_i} \quad (6.25)$$

$$\frac{1}{K_l} = \frac{di}{Hdok_g} = \frac{di}{Hdokg} + \frac{di}{Hd_{ln}k_m} + \frac{1}{Ek_l} \quad (6.26)$$

where E is the enhancement factor (E) due to chemical reaction.

In the case of a very thin catalytic membrane layer, mass transport limitation by pore diffusion can occur if the reaction is very fast. This is possible by setting the flow rate and other parameters. The pore structure, catalyst amount, and convective flow rate are the main parameters affecting the conversion rate and selectivity of consecutive reactions as reported in Table 6.2.

6.2.6 Membrane Condenser

The membrane condenser (MeC) process consists of condensing and blocking water vapor at the hydrophobic membrane surface from streams containing a saturated gaseous stream (Figure 6.6). Once dehydrated, the latter passes through the pores of the membrane to be collected at the permeate side [77, 78].

Different sources of gaseous streams (e.g., flue gas) can be considered such as power plants, cooling towers, stacks, etc. The gaseous stream, at a certain temperature and water-saturated, is inlet to the membrane condenser, which is kept at a lower temperature to cool down the gas to a supersaturation state. This difference in temperature is the real driving force of this process. The saturated streams coming into contact with the hydrophobic surface of the membrane condense and cannot pass through the membrane. Differently, the dehydrated stream can pass through. Therefore, the

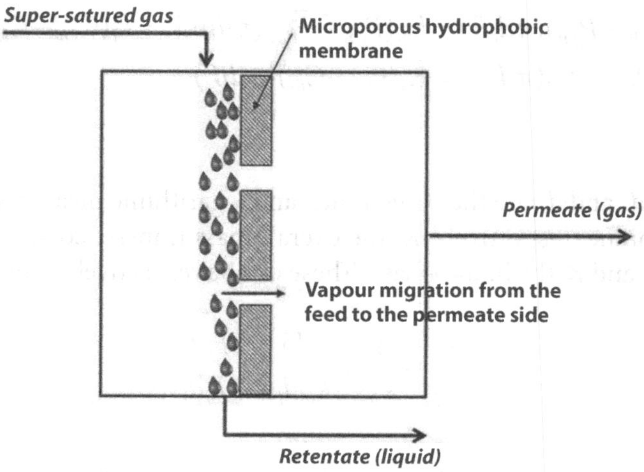

Figure 6.6 Schematic diagram of membrane condenser.

liquid water and all condensable compounds in the stream are recovered at the retentate side, whereas the other gases can be condensed on the permeate side of the membrane unit with an external condenser. In this way, the water can be purified and separated from flue gas. This process finds possible application in the pre-treatment of flue gas streams that have to be fed to another membrane unit for separation of contaminants like SO_x, NO_x, NH_3, HF, HCl CO_2, etc. [79]. Often, these compounds can cause damage to the membrane affecting the final process performance.

The water recovery is defined as the ratio between the mass of condensed water in the retentate and the total vapor water contained in the gaseous feed:

$$water\ recovery = \frac{liquid\ water\ retained\ in\ the\ retentate\ side}{total\ water\ contained\ in\ the\ feed\ stream} \times 100 \tag{6.27}$$

The vapor flux is calculated according to the reduced Knudsen-molecular diffusion transition form of the dusty gas model (DGM):

$$J_i^v = -\frac{1}{RT_{avg}}\left(\frac{K_{nw}D_{w-a}^0}{D_{w-a}^0 + P_a K_{nw}}\right)\frac{\Delta p}{\delta}M \tag{6.28}$$

where J^V is the viscous flux, K_n the Knudsen diffusion coefficient, D^0 the ordinary diffusion coefficient, and Δp the partial pressure gradient of water through both the membrane surfaces, which is generated by a temperature gradient and/or a concentration difference. The subscripts *w* and *a* are indicative of water and air, respectively.

In particular, K_n and D^0 are defined as:

$$K_{nw} = \frac{2\varepsilon r}{3\tau\sqrt{\dfrac{8RT_{avg}}{\pi M}}} \tag{6.29}$$

$$D^0_{w-a} = 4.46\times10^{-6}\left(\frac{\varepsilon}{\tau}\right)T_{avg}^{2.334} \tag{6.30}$$

where the subscript *avg* is indicative of the average value. From Henry's law, the solubility of the different gases in water has been estimated as:

$$\mathrm{kH} = \mathrm{k}_{\mathrm{H}}^{0}\exp\left(\frac{-\Delta\mathrm{solnH}}{\mathrm{R}}\left(\frac{1}{\mathrm{T}}-\frac{1}{T^{0}}\right)\right) \tag{6.31}$$

where

$$\frac{-\Delta\mathrm{solnH}}{\mathrm{R}} = \frac{-dlnk_h}{d\left(\dfrac{1}{T}\right)} \tag{6.32}$$

The values of k_H^0 and T_0 are referred to standard condition (T = 298.15 K). Furthermore, the amount of contaminants removed can be established by mass balance according to *eq. 6.33*:

$$c_{i,OUT,liquid} = \frac{moli_{i,FEED} - moli_{i,OUT,vapor}}{recovered\ water} \tag{6.33}$$

It is also relevant to observe that the inlet stream at a temperature T_i must be cooled by the condenser to a temperature T_{wr}, which is the temperature necessary to obtain the necessary recovery of water. So, the total

Table 6.3 Properties of membranes used in membrane condenser.

Membrane	ID (μm)	OD (μm)	Pore size (μm)	Porosity%	Process	Ref.
PP	n.a.	2700	0.2	70	Recovery of water and ammonia from cooling tower plume	[80]
PVDF (flat sheet)	n.a.	600	0.1	80	Recovery of pure water from flue gas with SO_x as contaminant	[81]
PVDF (fiber)	n.a	n.a	0.2	80	Recovery of evaporated vapor water from flue gas with N_2/CO_2/O_2(78:17:5, respectively)	[82]
Ethylene–Chlorotrifluoroethylene copolymer (ECTFE)	n.a	n.a	0.04	80	Water recovery from gaseous waste streams	[83]
PVDF/Hyflon AD	n.a.	1300	0.15		Particulate abatement from waste-gaseous streams	[79]

thermal energy Q required to condense the water vapor (Q^I) and to cool the exhaust gas (Q^{II}) can be calculated with the following equations:

$$Q^I = W\lambda_v \tag{6.34}$$

$$Q^{II} = fc_{pv}(T_i - T_{wr}) \tag{6.35}$$

$$Q = Q^I + Q^{II}W \tag{6.36}$$

where W is the water vapor flow rate, F the flue gas flow rate, c_{pv} the specific heat capacity, and λ_v the water latent heat of condensation.

Again, membrane pore size is a critical factor for this process. The correlation between the saturation pressure of condensable vapors and the pore size is established according to the Kelvin's Equation:

$$\ln\left(\frac{P_c}{P_o}\right) = \frac{-2\gamma V_m \cos\theta}{rRT} \tag{6.37}$$

where γ is the surface free tension, V_m the molar volume of liquid, Pc the equilibrium pressure in pores with a radius r and P_o the vapor saturation pressure on planar surface.

The membrane condenser was initially developed only for water recovery from waste gaseous streams. Other applications have been represented as pre-treatment stage for further separation units to minimize the contaminants content of various gaseous streams, as summarized in Table 6.3.

6.3 Membrane Morphology and Wetting Properties Relationships

6.3.1 Pore Size and Distribution

Membrane contactors use porous membranes as interphases between two streams. A large number of small-sized pores is always preferable to control wetting phenomena without adding resistance to transport. Small pore size can induce capillary forces that prevent direct contact of two phases touching the membrane [69]. Small pores can leave the membrane in a completely dry state under higher pressure, whereas a larger pore size suffers severe absorption causing wetting, as well-expressed by the Laplace equation (*eq. 6.19*). On the other hand, the mass transfer resistance of the membrane decreases with increasing pore size. Mass transfer resistance represents a key issue for membrane contactor technology due to a lower driving force in the process. Therefore, a large number of small pore sizes is desirable to provide a wider interfacial area and subsequently reduced resistance to mass transfer along with limited risk of wetting and mechanical instability. The pore size distribution of the membrane is well described by the most common form of the two-parameter log-normal distribution function ($f(d)$) [84]:

$$f(d) = \frac{1}{\ln(d_g) r \sqrt{2\pi}} \exp\left[-\frac{1}{2} \left(\frac{\ln\left(\frac{r}{D}*\right)}{\ln(\sigma)} \right)^2 \right] \tag{6.38}$$

where D^* and d_g are the geometric mean diameter and the geometric standard deviation that can explain the uniformity of pore size. A low value of σ_g indicates pore sizes almost close to the average pore size (narrow pore size), whereas a high value of σ_g indicates a wide variation from the average value. Moreover, a high σ_g, generally indicates easier partial wetting than membranes with narrower pore distribution. These parameters can be measured from the Levenberg–Marquardt algorithm, which correlates them with the rejection coefficient of a solute $R(a)$ and the pore size distribution parameters [85]:

$$\sum_{i=1}^{N} \{R(a_i)|_{Exp.} - R(a_i)|_{Cal.}\} 2 |_{MIN} \rightarrow (D^*, \sigma_g) \tag{6.39}$$

where $R(a_i)$ is usually calculated with concentrations of solute molecule in the feed C_f and in the permeate C_p respectively according to the following equation:

$$R(a) = \frac{C_f - C_p}{C_f} \tag{6.40}$$

Number and pore size contribute to form the overall volume of void spaces per total membrane volume, which is expressed as the overall porosity *(ε)*. This parameter describes the free pathways, which molecular components can pass through. A larger void space should allow for better mass transfer due to less resistance to transport. It can be calculated according to the following mathematical relationship:

$$\varepsilon = \left(1 - \frac{\rho_{bulk}}{\rho_{particle}}\right) * 100 = \left(\frac{V_p}{V_b}\right) * 100 \tag{6.41}$$

where ρ_{bulk} is the density of the bulk, $\rho_{particle}$ the density of the particle, which is equal to the mass of the sample divided by the volume of the sample, V_p the pore volume, and V_b the bulk volume of the sample.

A correlation of porosity with mass transfer coefficient can be expressed by the following [86, 87]:

$$k = \frac{D}{\delta \tau} \tag{6.42}$$

where K is the mass transfer coefficient and D the diffusion coefficient.

6.3.2 Surface Contact Angle

An empirical estimation for membrane wetting properties can be obtained through contact angle value measurements. The contact angle is the result of the chemistry and topography of the membrane surface. Liquid spreading suggests hydrophilic and/or oleophilic behavior, whereas liquid repellence is an indicator of a hydrophobic/oleophobic surface property. The contact angle value is regarded as the result of the balanced cohesive attraction of the liquid phase with the solid surface. For very smooth surfaces (ideal case), the chemistry of the material is a predominant factor for wetting properties. In real cases, membrane topography becomes determinant in amplifying the resistance of the membrane to liquid spreading. Pores, grooves, pillars, and asperities affect the surface profile causing increased roughness parameter (R), which amplifies the membrane resistance to liquid spreading (Figure 6.7).

The Wenzel equation well describes the dependence of contact angle on the surface roughness (R), considering the grooves filled by a liquid:

$$\cos(\theta_w) = R\left[\frac{\gamma_{SV} - \gamma_{SL}}{\gamma_{LV}}\right] \tag{6.43}$$

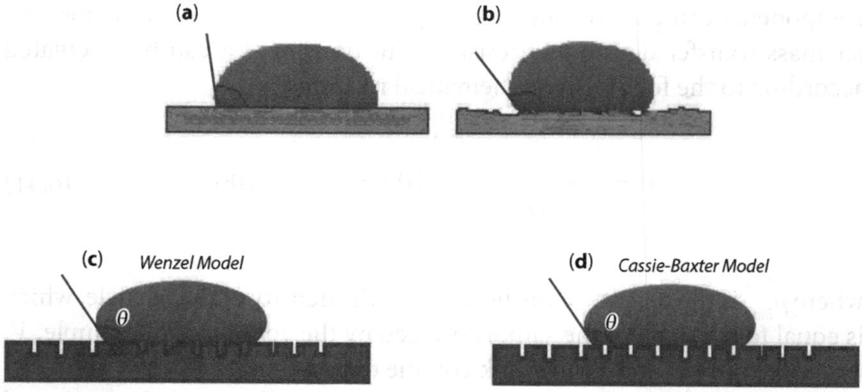

Figure 6.7 Scheme of membrane surface: (a) very flat (ideal case) membrane surface; (b) rough membrane; (c) Wenzel model and (d) Cassie-Baxter models of rough membrane surfaces.

where γ is the surface free energy at the solid vapor (*SV*), solid-liquid (*SL*), and liquid-vapor (*LV*) interfaces. Combining Wenzel and Laplace equations, the relation indicates that the liquid fills the grooves through the surfaces:

$$\cos(\theta_w) = R\cos(\theta_a) \tag{6.44}$$

when the surface roughness becomes a predominant factor, the Cassie–Baxter's model describes the relationship between the apparent contact angle (θ_w) and the fractions of the solid (f_s) and vapor (f_v) on the surface:

$$\cos(\theta_{CB}) = f_s\cos(\theta_a) + f_s\cos(\theta_v) \tag{6.45}$$

In this case, the surface grooves are filled with air, reducing the contact line between the surface and liquid, leading to a roll-to-roll effect. As explained, loss of wetting properties is a critical issue for MC technology. Therefore, the choice of materials along with the construction of singular topographies become the topic of targeted research [88].

6.4 Green Materials for More Sustainable Membrane Fabrication

Polypropylene (PP), polyethylene (PE) polyvinylidenefluoride (PVDF), poly (etheretherketone) (PEEK), polytetrafluoroethylene (PTFE), polysulfone

(PS), polyethersulfone (PES), polyimide (PI), poly(L-lactic acid) (PLLA), and many others, are among the most popular polymers used to prepare porous membranes. These materials can be used as pristine polymers or modified with other materials, i.e., polymers, oligomers, and organic/inorganic nanofillers to amplify the productivity/efficiency trade-off for each single separation [89–92]. Polymers are low-cost materials that are easy to manufacture, have good solubility in common organic solvents, exhibit satisfactory chemical and mechanical stability, and show fair fouling resistance. For every fabrication, one of the major targets is the degree of sustainability of the manufacturing process in compliance with natural resources and biodiversity management. So, a great effort has recently been

Table 6.4 List of green solvents for membrane fabrication by phase inversion.

Polymer	Green solvent	Fabrication	Pore range	Refs.
CA	Methyl acetate/ 2-methyl tetrahydrofuran/Ethyl lactate	NIPS	Nanosized	[93]
PVDF/PEEK PEI/PI/PAN	g-Butyrolactone/	NIPS/TIPS	Micro/ nanosized	[94]
CA/PVDF	Tributyl citrate/triethyl citrate/Acetyl-tributyl citrate	TIPS	Nanosized	[95]
PES/PVDF	Dihydrolevoglucosenone (Cyrene)/ Dimethylisosorbide	TIPS/VIPS	Micro/ nanosized	[96]
PVDF/PSF/ PES/CA	Methyl 5-(dimethylamino)-2-methyl-5-oxopentanoate (Polarclean®)	NIPS	Nanosized	[97–99]
PES/PVDF/ PSF/PAN/PI/ CA	Organic carbonates	NIPS	Nanosized	[100]
LDPE (Low-density polyethylen	Butyl acetate	TIPS	Nanosized	[101]
PVDF/PES	Tamisolve® NxG	NIPS	Microsized	[102]

made to identify bio-sourced solvents in place of more toxic traditional solvents, such as NMP, DMF, DMA, etc. This route is viable but requires a fine study on procedures, management, and the impact on final quality and costs of membrane manufacturing. Table 6.4 summarizes some of the most interesting polymer/solvent combinations used for the fabrication of porous membranes through well-established and scaled phase separation (PS) procedures induced by non-solvents (N), temperature (T), and vapors (V).

6.5 Manufacturing Procedures for Porous Membrane Fabrication

6.5.1 Phase Separation

Phase inversion is the most used procedure to fabricate membranes on large scale. However, the necessity to get higher structural order and defined morphology in membranes has solicited the exploration of new routes for making the membranes closest to the targets of MC technology. In any case, it is essential to know the basics of this technique to understand the potential of more sophisticated manufacturing procedures. Hereafter, a brief outline of conventional phase inversion is given, underlining related flexibility and feasibility in the production of membranes with different degrees of porosity, morphology, size, and shape [103–105]. The basic concept of this technique is to form polymeric architectures wherein pores, which are randomly variable in size, number, and geometry, can be generated according to different strategies. In all cases, the target is to induce polymer precipitation from homogenous solutions for the formation of polymeric architectures wherein pores and free gaps are frozen. The polymer precipitation can be induced by non-solvent at liquid or vapor state and temperature. Combinations of different phase separation induction are also possible. Membranes prepared according to this approach are also used for membrane contactor applications [15], because of the unavailability of membranes with suitable features on the market.

6.5.1.1 Non-Solvent Induced Phase Separation (NIPS)

A non-solvent polymer-solvent ternary mixture is used to induce phase separation (Figure 6.8). In the first stage, a homogeneous polymer solution is prepared by dissolution of the polymer in a suitable solvent. Then, the clear solution is cast as a flat film or extruded as a fiber. The precipitation of the polymer is induced through solvent-non-solvent exchange as the casting solution comes in contact with a coagulation bath [104]. During the exchange, one polymer-rich phase is generated to form the polymeric network, while the polymer lean phase leads to the porous portion of the membrane (Figure 6.9). Thermodynamics directs the phase equilibrium between components while kinetic factors control mutual diffusion between components of the mixture, yielding a large variety of morphologies. Undoubtedly, powerful tools to direct the phase separation process include the composition of the coagulation bath and polymer solution, exposure time, temperature, humidity, affinity parameters between solvent and non-solvent, etc. As an example, the Hansen solubility parameter, regarded as the result of dispersive (*d*), polar (*p*), and hydrogen bonding components (*h*), is a powerful and predictive tool of reciprocal affinity between solvent and non-solvent [106]. This parameter is able to decide the velocity of demixing [107] giving the chance to modify the architecture of the same polymer shaped in the membrane. A low solvent-non-solvent miscibility produces slow demixing leading to a sponge-like structure predominately; a fast solvent-non solvent exchange causes a denser upper layer, above of which a finger-like structure is formed as the result of slowed demixing (Figure 6.8) [108, 109].

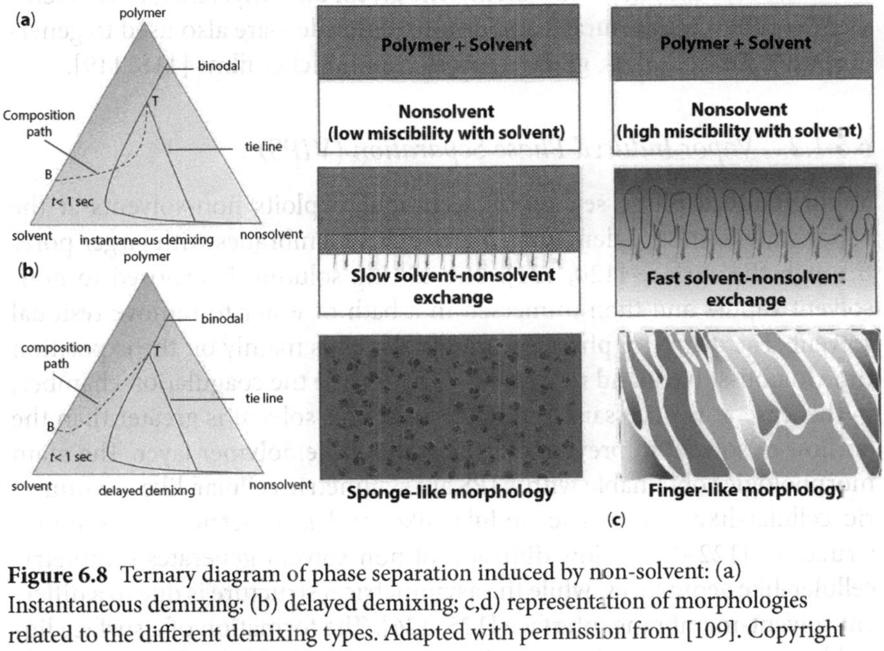

Figure 6.8 Ternary diagram of phase separation induced by non-solvent: (a) Instantaneous demixing; (b) delayed demixing; c,d) representation of morphologies related to the different demixing types. Adapted with permission from [109]. Copyright 2011 American Chemical Society.

Increased polymeric concentration allows the overall porosity to decrease, leading to a different degree of resistance to mass transfer, but also to better control of the mean pore size through the membrane surface [110]. Also, the temperature is closely related to the viscosity of the polymer solution and can have repercussions on the solvent-non-solvent exchange rate, thus providing changes in the membrane morphology. It is reported that an increase in the temperature of the coagulation bath leads to an increase in finger-like macrovoids [111, 112]. Variations in the composition of the coagulation bath can be used to reduce the solvent-non-solvent exchange rate, moving from a fast to a slow demixing state [113]. The addition of other components to the casting solution can influence the morphology of the membrane in terms of formation, structure, and interconnectivity of the pores, as well as affect the final surface properties of the membrane. The final morphology can show unconnected latex, nodules, bi-continuous structures, cellular structures, sponge-like and macrovoids, as well as particulate-like structures [110, 113]. Sometimes, pore formers can be used to increase the surface porosity of the membrane. PVP is a hydrophilic polymer frequently used as a pore former; increased PVP molecular weight in the casting solution yields membranes with thicker surface layers and denser sublayer due to a reduction of macro-voids into the bulk of the film [114–117], while hydrophobicity loss occurs. PEG and other hydrophilic additives—diethylene glycol dimethyl ether, polyoxyethylene sorbitan monolaurate, and lithium chloride—are also used to generate larger surface pores, greater porosity, and thicker films [118, 119].

6.5.1.2 Vapor Induced Phase Separation (VIPS)

Vapor-induced phase separation technique exploits non-solvents at the vapor state to induce demixing. VIPS yields membranes with larger pores through the surface [120, 121]. The casting solution is exposed to non-solvent vapors and then immersed in a bath of water to remove residual solvent. The degree of phase separation depends mainly on the exposition time to non-solvent and relative humidity inside the coagulation chamber; in this case, it is necessary that the flow of non-solvent is greater than the outflow of solvent to prevent densification of the polymer layer. The main morphologies obtainable with VIPS are symmetric cellular-like, asymmetric cellular-like, symmetric nodular-like, and symmetric bi-continuous structures [122–124]. Slow diffusion of non-solvent generates symmetric cellular-like geometries, while the asymmetrical structure is due to a different solvent-membrane interface [125, 126]. The formation of a surface liquid layer generates a solvent gradient across the system and consequently a different concentration of the polymer, which is less concentrated on the surface. This results in a slowing down of the cell growth on the surface and the formation of asymmetrical structures. The propagation of the porosity through the cross-section of the film depends on the demixing rate. This is not often enough to cause total demixing. During the washing NIPS process, asymmetry could occur, causing a denser surface layer and a finger-like sub-structure [127, 128]. Asymmetrical finger-like structures are obtained for short times of exposure to vapors, even if the VIPS process may not be sufficient to solidify the entire membrane. Solidification of the lowest layers takes place according to NIPS mechanisms. Spherical or granular structures are found in systems prepared using a strong non-solvent that generates a fast phase inversion, so that the polymer-rich phase tends to minimize its surface energy and a spherulitic structure is generated in the presence of semicrystalline polymer. Slow phase separation kinetics, determined by temperature and viscosity, lead to the formation of spongy structures with bi-continuous morphologies. Generally, the exposure time to non-solvent vapors affects porosity and the interconnection of the pores with subsequent effect on transport properties. The partial pressure of non-solvent is another critical parameter to be considered for the formation of membranes by VIPS [129, 130]. Water, expressed as relative humidity (RH), is the most common non-solvent used, so low levels of RH (10%) lead to denser structures with a random and discrete number of superficial pores and marked orange peel effects; higher RH (20%) produces more open membranes with cell-type pores or spherulitic structures, depending on the crystallinity of the polymer, while RH values of 40% cause lace-like structures (Figure 6.9).

For polysulphone (PS) membranes, an increase in RH from 70 to 100% has been observed to reduce the mean pore size, without changing the cellular-like morphology. This behavior has been ascribed to the lowest

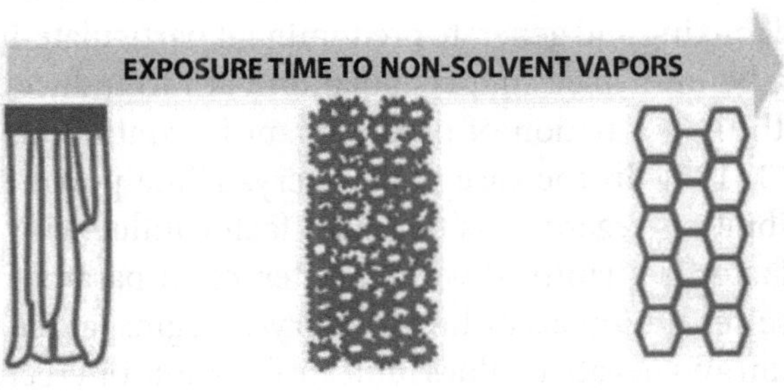

Figure 6.9 Schematic kinds of membrane morphology obtained at increasing exposure time to non-solvent vapors.

driving force of water transfer and the slowest phase separation, allowing pore growth. An increased number of cells with reduced size and anisotropy along the cross-section have been observed for PEI and cellulose acetate membranes when water vapor was used as non-solvent; a competition between droplet nucleation and growth rate occurred during condensation and diffusion through casting solutions [131]. The temperature of the chamber can affect the mass transfer rate producing dense to bi-continuous networks, and the temperature at which the polymer is dissolved can direct the formation of pores as free gaps through semicrystalline polymeric networks [2, 132]. In the case of PVDF, temperatures above 40°C cause the formation of nodules with a trend of growth in size proportional to heating temperature (1.4 to 10 μm, running from 45 to 150°C). The connection of the pores increases with decreasing temperature, and below the threshold of 32°C, while the structure becomes bi-continuous with interconnected pores. The effects of polymer concentration are comparable to those discussed for NIPS processes. A reduction in the number of pores has been observed for PVDF membranes, as the polymer concentration was increased from 10 to 20% [129, 130]. PS membranes changed from a finger-like to a cellular-like structure as the polymer concentration was increased from 10 to 20% [128]. Generally, higher polymer concentration reduces the pore size due to higher solution viscosity and subsequent lowered polymer chain mobility and solvent diffusion.

6.5.1.3 Thermally-Induced Phase Separation (TIPS)

TIPS processes use temperature as a tool to induce polymer precipitation. The polymer is dissolved in a solvent with a high boiling point, low molecular weight, and low affinity for the polymer at room temperature. The mixture is heated at an elevated temperature and then cast onto a cold support in a desired shape [133, 134]. Then, the casting solution is gradually cooled to solidify. The nascent membrane can be left in cooled air to promote liquid-solid demixing and generate predominant particulate-like morphology, or it can be immersed in a coagulation bath to induce liquid-liquid demixing with the formation of predominant bi-continuous and cellular structure [133, 135]. In the case of semi-crystalline polymers, polymer-diluent miscibility is regarded as the main factor influencing phase separation. This factor is quantified with an interaction parameter (χ), which predicts phase separation according to Flory-Huggins lattice theory [135, 136]. Minimum and negative values indicate favorable interactions whereas high values indicate poor polymer-diluent interaction. With favorable polymer-diluent interactions, the mixture substantially undergoes only solid-liquid separation, while a decrease in interaction causes instability of the mixture, and liquid-liquid separation may take place [135, 136]. Chemical composition and molecular weight of single components can affect the final morphology of the membrane. As an example, different morphologies of PE membranes have been obtained by changing the solvent and molecular weight of the polymer [137, 138]. As PE concentration increased from 10 to 40%, liquid-liquid separation occurred before crystallization, while higher concentration caused crystallization, leading to structures with connected pores. Isolated spherical pores were generated when solid-liquid demixing prevailed [139, 140]. Polymer concentration and nucleating agents can influence the number and pore size during the TIPS process, whereas post-treatments such as stretching can be applied to improve the morphological features of the formed membranes. The TIPS process takes advantage of using polymers with poor solubility in common solvents, including bio-sourced ones (Table 6.1).

6.5.2 Phase Separation and Micromolding

Micromolding exploits the concepts of hot embossing and soft lithography and can also be combined with traditional phase separation to prepare

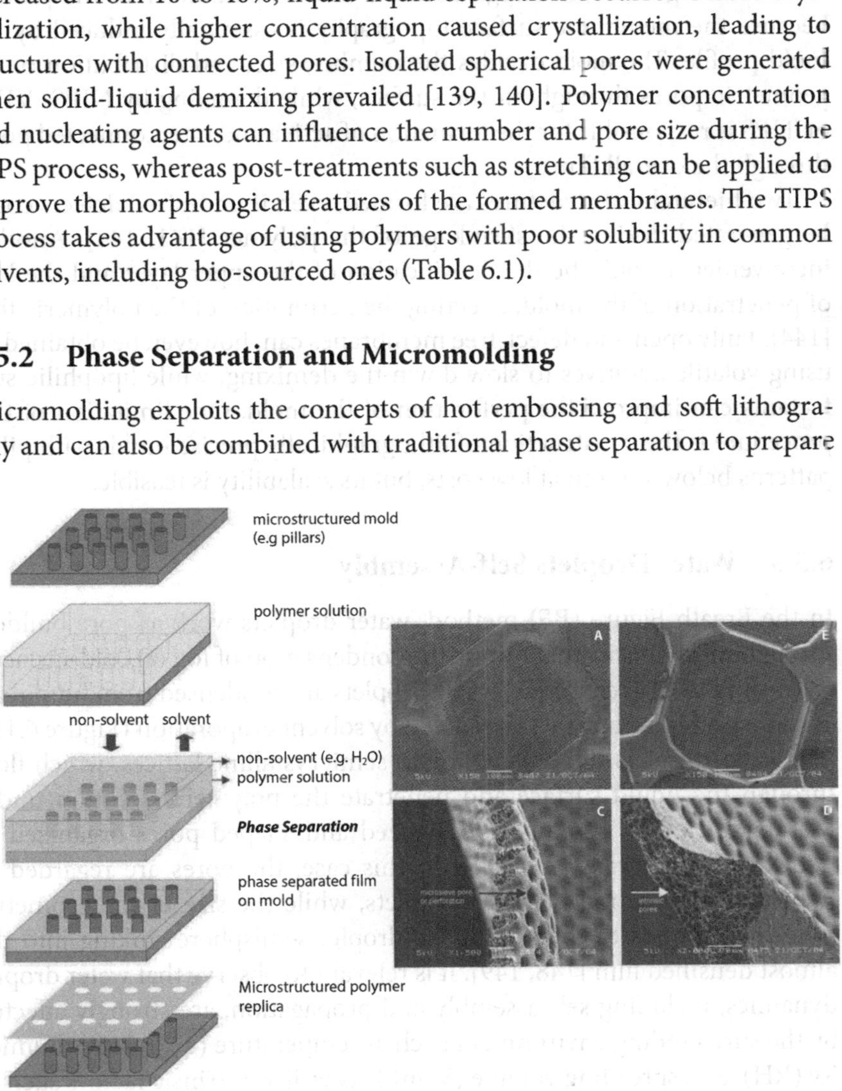

Figure 6.10 Schematic representation of the phase separation micromolding process and SEM images of a polyimide microstructure obtained with PSμM. Adapted with permission from [142]. Copyright 2006 ELSEVIER.

polymer membranes with micro and submicro scale replication [141–143]. This combined technique is an evolution of phase separation and is named phase separation micromolding (PSμM). The casting solution is usually poured or applied by knife on a mold previously prepared according to lithography procedures, and then demixing is induced according to more traditional phase separation mechanisms (Figure 6.10). The thickness of the casting solution is regarded as a critical parameter of the process because the membrane surface topography is essentially induced by the mold profile. The mold decides the number, size, and distribution of the geometric pores throughout the surface, while demixing by NIPS, VIPS, or TIPS is responsible for the formation of active and interconnected pores through the overall film.

Two hierarchical structures are formed, even if undesired defects could be generated during the shrinkage of the polymer [141, 144]. Another inconvenience could be the densification of the upper layer and the lack of penetration of the mold, affecting the perforation of the polymeric film [144]. Fully open and defect-free membranes can, however, be obtained by using volatile additives to slow down the demixing, while lipophilic surfactants can improve the perforation of the mold. One limitation of this procedure is the construction of lithographically precise molds with pillar patterns below 150 nm at low costs, but its scalability is feasible.

6.5.3 Water Droplets Self-Assembly

In the Breath Figure (BF) method, water droplets work as pore builders through mechanisms that mimic the condensation of fog on cold surfaces. According to this technique, water droplets are condensed from humid air onto casting solutions that are cooled by solvent evaporation (Figure 6.11). The water droplets self-assemble into semicrystalline lattices, which float through the liquid surface and penetrate the polymeric solution under Marangoni forces, producing well-sized and shaped pores organized in honeycomb patterns [145–147]. In this case, the pores are regarded as the imprinting action of water droplets, while the size of the geometric pore corresponds to the size of the droplet semisphere sinking into the almost densified film [148, 149]. It is relevant to observe that water droplet dynamics, including self-assembly and propagation, are strongly affected by the surrounding environment such as temperature (T), relative humidity (RH), and spreading volume (V, mL), as well as intrinsic factors such as viscosity (η), concentration (C), and surface free tension (γ) of the casting solutions and solubility parameters (δ) [150, 151].

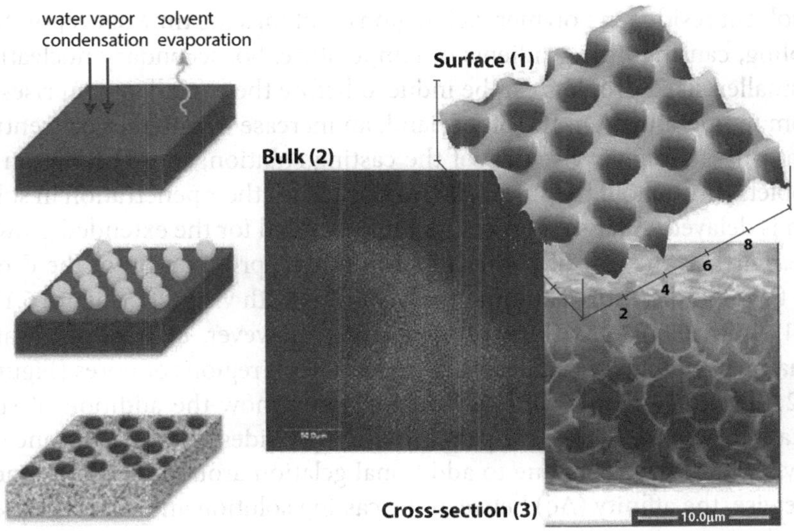

Figure 6.11 Schematic presentation of the 'breath-figure' method and micrographs of honeycomb PEEK-NO_2 membranes. Reprinted with permission from [151]. Copyright 2010 American Chemical Society.

Relative humidity plays an important role in the formation of honeycomb structures. For PS membranes, it has been observed that relative humidity lower than 46% yields surface poor in pore number. This is due to the fact that the droplet nucleation and growth are significantly limited, especially if evaporative cooling is not enough and solution viscosity is high. Differently, humidity higher than 80% leads to disordered formations of pores, wherein coalescence phenomena are dominant events [150]. In this case, massive nucleation and harsh collisions under thermocapillary forces produce very fast coverage of the surface exposed to humidity. Changes in temperature can also direct the distance between pores, determining their diameter. An increase in temperature is expected to increase the evaporative cooling, thus resulting in an enlarged driving force (ΔT). As an effect, massive nucleation and slowed growth rate of the droplets are produced. Also, the viscosity of the solution tends to reduce with temperature, thus allowing droplet submersion with a subsequent reduction of the top hemisphere of the water droplet in contact with air. However, an increase in temperature is also expected to cause irregularity in the air bubble arrays (porous patterns) due to increased droplet kinetics, which are responsible for turbulence and quick collisions. The role of polymer concentration and surface free tension is not negligible. An excess of solvent residue in polymer-rich regions prolongs the time of evaporative cooling, causing little gradients of temperature. So, secondary nucleation of smaller-sized droplets can be induced before the overall system rises to room temperature. On the other hand, an increase in polymer concentration causes a higher viscosity of the casting solution, so that flotation of droplets throughout the surface is prolonged and their penetration in solution is delayed. A larger pore size is thus obtained for the extended growth rate. Further, the risk of reduced or incomplete propagation of the droplets through the entire column of solvent rises with viscosity (Figure 6.12). [151, 152]. Controlled kinematic viscosity, however, can address water dynamics reducing the appearance of disordered regions of pores (Figures 6.12a, b). *Gugliuzza et al.* [153] demonstrated how the addition of surfactant to a PSU/DCM/2-propanol solution provides suitable assistance to prevent local disorders due to additional gelation around droplet islands. Likewise, the affinity (Δδ) between the casting solution and water droplets was demonstrated to direct the pore size as a subsequence of dissimilar flotation times for droplets through the surfaces (Figure 6.12c). The result was the formation of smaller pore sizes as a suitable affinity between the systems was provided, whereas a larger pore size was obtained reducing the degree of affinity. In both cases, the condition of immiscibility between polymer and solvent was preserved [153].

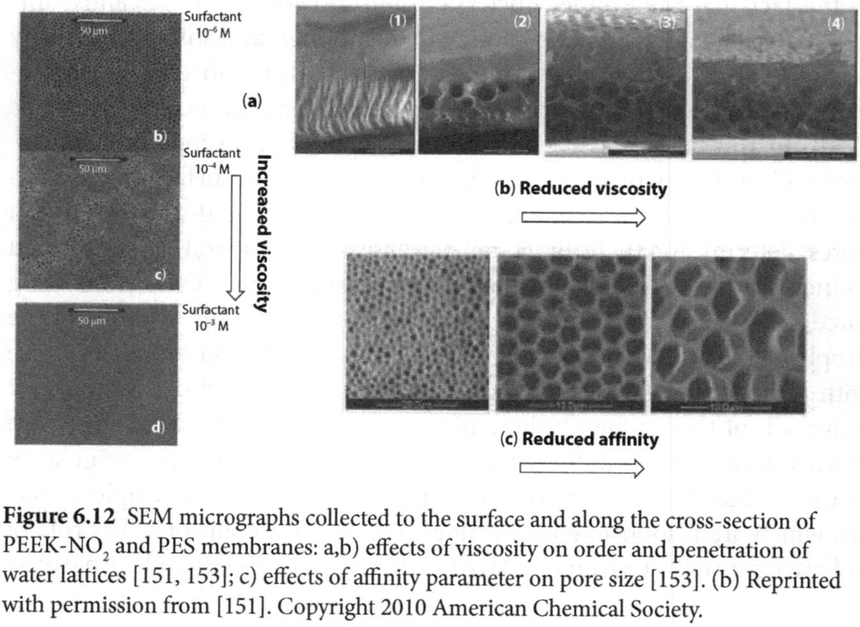

Figure 6.12 SEM micrographs collected to the surface and along the cross-section of PEEK-NO_2 and PES membranes: a,b) effects of viscosity on order and penetration of water lattices [151, 153]; c) effects of affinity parameter on pore size [153]. (b) Reprinted with permission from [151]. Copyright 2010 American Chemical Society.

In this regard, suitable solvents must have a boiling temperature lower than water and demonstrate immiscibility or partial miscibility with water. Water-immiscible solvents give multilayer films, while mono or bi-layered films are generally formed with water-miscible solvents [82]. This technique is considered green for the use of a natural template such as water and the ease of recovering the solvent in which the polymer is dissolved. However, its scalability is still limited by the difficulty of achieving controlled droplet dynamics over extensive surface areas.

6.5.4 Self-Assembly of Block Copolymers

Self-assembly of block copolymers is another popular technique to generate ordered porous membranes in the range of 5 to 100 nm [154, 155]. This manufacturing approach exploits thermodynamic incompatibility between two blocks, causing microphase separation into high-density and highly ordered pattern arrays. Microphase separation can lead to spheres (SPH), cylinders (CYL), gyroids (GYR), and lamellae (LAM) nanodomanis after thermal or vapor annealing (Figure 6.13). Final morphology and associated length scale of the structural order depend on the degree of polymerization (*N*) and the volume fraction of block A (*fA*), which changes with polymer as well as Flory–Huggins interaction parameters of two blocks (χ) related

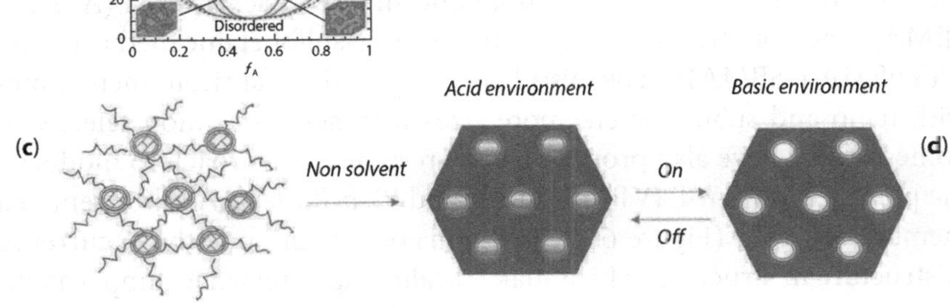

Figure 6.13 Theoretical phase diagram for A-*b*-B diblock copolymer (*SPH = body-centered-cubic spheres, CYL = hexagonally packed cylinders, GYR = bi-continuous gyroids, and LAM = lamellae*) (a) Adapted with permission from [155]. Copyright 2016 ELSEVIER; Selective etching of one block to form porous structures (b); Self-assembly of copolymers in ordered micellar structures forming switchable pores at different pH.

to the monomer pair. In a theoretical phase diagram built up for A-*b*-B diblock copolymer, it can be observed that for identical χN values, the equilibrium phases at *fA* and *1-fA* lead to the same structure, but with an inverted arrangement (Figure 6.13). Ordered self-assembled membranes can be fabricated when high χ and low N are maintained. The most used approaches are the result of combined spin-coating or solvent-coating and annealing procedures, followed by selective etching of one block to form porous structures (Figure 6.13b).

The most common block copolymers used to fabricate ordered structures via induced micro-segregation are PS-b-PMMA, PB-b-PDMS, PI-b-PLA, and PS-b-PLA. In this case, PMMA, PDMS, and PLA are etched to yield the porous component of the film. In some cases, liquid crystal molecules have been proposed to generate ordered structures with perpendicularly aligned pores to the membrane plane [156]. Under magnetic or annealing orientation, liquid crystal molecules have been assembled in cylinder phases to leave open nanosized channels after rising. However, this self-assembly-etching procedure, which takes place under thermodynamic equilibrium control, is not easy to scale and frequently does not allow getting ordered porous domains for what is the extension of the overall surface area [157]. A more practical and economical route is a procedure based on the integrated concept of self-assembly and phase inversion. In this case, di- and triblock copolymers self-assemble in ordered micellar structures in proximity to the skin layer during solvent evaporation, thereby forming open pores after coagulation in a non-solvent (Figure 6.13c). Asymmetric structures are usually obtained, and related morphologies can be rearranged according to the phase separation rules.

Zwitterionic statistical copolymers, involving a microphase separation between ionic and hydrophobic networks, which are distinctive of systems such as poly(allyl methacrylate-r-sulfobetaine methacrylate) (P(AMA-r-SBMA)) and poly(trifluoroethylmethacrylate-r-sulfobetaine methacrylate) (P(TFEMA-r-SBMA)), have also been proposed to fabricate membranes with nano and subnanometer pore sizes with superior anion selectivity. Some authors have also proposed a pH-responsive approach to modulate the pore size in PS-*b*-P4VP/PS-*b*-PAA and PI-*b*-PS-*b*-PAA self-assembled membranes [158] (Figure 6.13d). The major limitation is the occurrence of structure in structure, which makes scaling up somewhat complicated.

6.5.5 Electrospinning

Electrospinning (ES) allows the fabrication of ultra-fine fibers or fibrous structures of various polymers with diameters ranging from microns to nanometers. Membranes produced according to this technique are highly porous, with interconnected pores, exhibit low density, controllable thickness, a high surface-to-volume ratio, and exceptional mechanical strength. The most commonly used polymers are PS and PVDF [159–161] while spinning can be performed in solution or by fusion [162]. The latter has several advantages, such as reduced costs and no use of solvents. One limitation, however, is the high production temperatures and the real difficulty of producing nano-scale fibers. Therefore, the electrospinning technique in solution is the most widespread. The process involves the extrusion of the polymer mixture through a capillary tube where a strong electric field is applied. When the applied voltage exceeds the surface tension of the polymer fluid, the solution is expelled from the Taylor cone to the opposite polarity collector, part of the solvent evaporates, and solidification takes place [159, 160, 163] (Figure 6.14a). Superhydrophobic membranes with smooth and porous structures can be produced even if a lot of work is still necessary to ensure high reproducibility and controlled distribution of free gaps working as pores.

Polymer solution and process parameters can affect the final result of the manufacturing process. The solvent needs to have suitable characteristics including evaporation rate and polarity [160]. Uniform fibers require usually a dipolar aprotic solvent; however, to optimize the polarity various solvents can be mixed [164]. Despite that the mechanisms are not

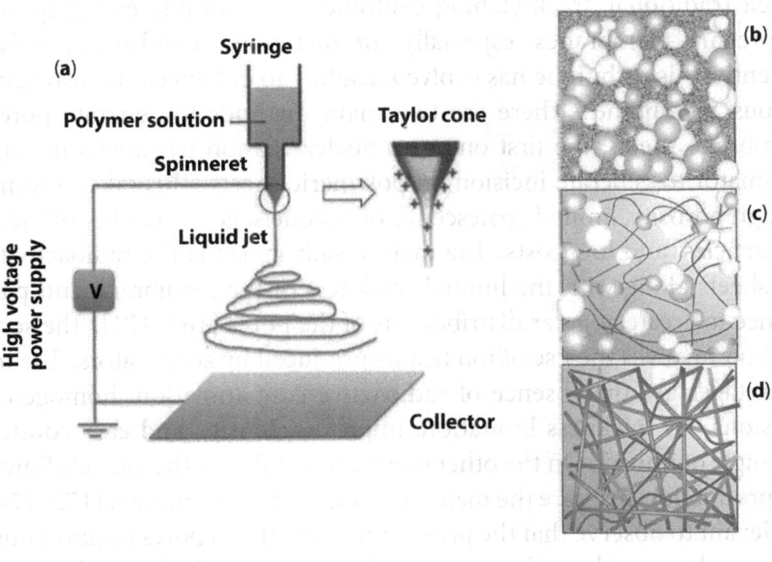

Figure 6.14 Device for electrospinning (a): possible structure obtainable with electrospinning (b) Reprinted from [163] CC-BY 2024 ELS Publishing.

being entirely clear, the surface tension of the polymeric solution could cause casting instability as well as the viscosity could produce fibers with a larger diameter or generate beads depending on the polymer concentration [164–167].

The electrical conductivity of the polymeric solution is another parameter to be considered. High conductivity decreases the diameter of the fibers, while low conductivities produce non-homogeneous fibers with the formation of beads [160]. Voltage, tip-collector distance, and flow rate also influence the morphology of the membranes. The applied voltage must exceed the surface tension, but an excessive voltage could cause instability of the Taylor cone [163]. Voltage values between 5 and 30 kV are usually considered optimal for electrospinning, producing smaller fiber diameters [165]. Variations in the diameter of PS fiber have been induced by increasing the distance between the tip and the collector due to a higher evaporation rate of the solvent. Similarly, lower electrospinning flows produce thinner and more uniform fibers, while increasing the flux fibers with bead-fiber morphology can be generated [163, 165] (Figure 6.14b). There is no limitation to the scalability of this technique, but optimized protocols are needed to control the meshes of the polymer net in order to reduce the risk of wetting.

6.5.6 Track Etching

Albeit traditional, track etching continues to be an interesting approach to prepare membranes, especially for membrane emulsification [168]. Recently, this technique has evolved, leading to enhanced nanostructured porous membranes. There are two main methods to generate pores in membrane sheets: the first one uses nuclear fission fragments through a collimator to generate incision on polymeric sheets. This takes advantage of high porosity, limited coalescence of channels, good stability of the flow of particles, and low costs. The main disadvantage is the radioactivity of the sheets along with the limited thickness of the membrane and partial chance to create angular distributions of the pores [169–171]. The second method involves the use of ion beams produced by accelerators. The main advantages are the absence of radioactive contamination, homogeneous incisions, no thickness limitation, improved density, and easy control of the angle of impact. On the other hand, the stability of the particle flow and the production costs are the main drawbacks of the technique [172–174]. It is relevant to observe that the production of uniform pores requires homogeneous damage along the trajectory (Figure 6.15). A chemical treatment allows the damaged areas to be removed and transformed into hollow

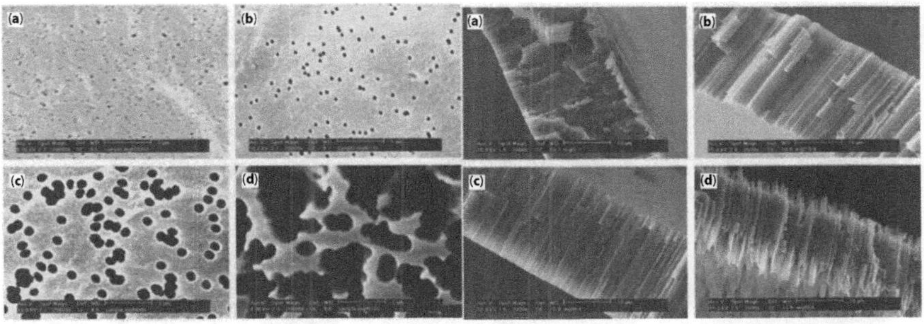

Figure 6.15 SEM image of: a) surface and b) cross section of PVDF membranes irradiated and etched with permanganate 0.25 M solution in KOH 9M, at (a) 55°C; (b) 65°C; (c) 75°C (d) 85°C. Reprinted with permission from [176]. Copyright 2005 ELSEVIER.

channels. Pore sizes greater than 1 μm can be produced depending on track and bulk etch rates. A longer track produces cylindrical pores [171], while the damaged core decides the geometry of the small pores [175]. The mass etching speed depends on materials, etching composition and temperature, while the etching speed depends on the irradiation and post irradiation conditions, material sensitivity, and etching conditions [176].

Alkali and oxidizing agents such as hydroxide solutions and permanganate are usually used to remove materials where traces have been produced by irradiation. The etching temperature can affect size and shape of the pores through the surface and across the section, leading to pore sizes of 82 to 398 nm and pore structures with straight and cylindrical geometry at a working temperature of 75°C [176]. Concerning the polymers' chemical stability and mechanical strength, various polymers allow the production of a wide range of membranes by track etching. Ceccio *et al.* [65] have recently proposed the use of non-destructive ion transmission spectroscopy (ITS) and neutron depth profiling (NDP) to analyze, monitor, and quantify the evolution of etched micron-sized pores. It has been demonstrated how to move from to a conical form (for one-sided etching) to a well-developed cylindrical geometry through a symmetrical (double-sided) etching process. LiCl and boron solutions have been used to fill and generate defined pore shape [177].

6.5.7 3D Printed Membranes

A brief introduction is also given to a recent innovative technique used to get precise control of the membrane structure. 3D printing or additive

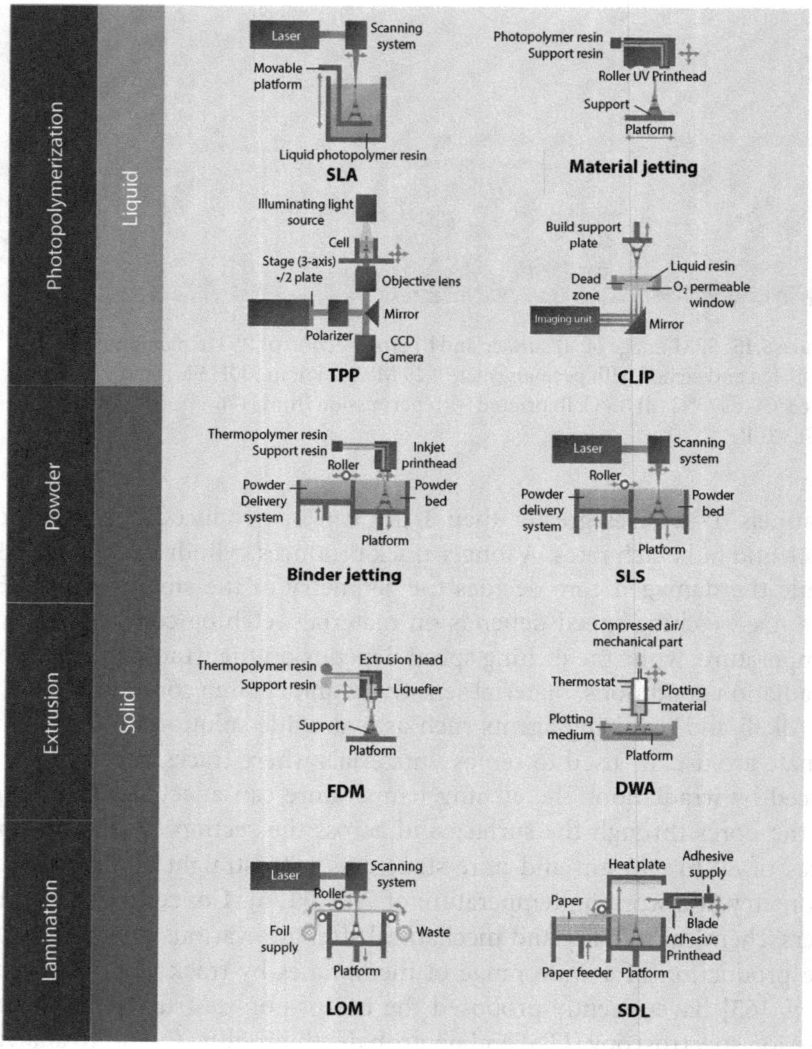

Figure 6.16 Scheme of 3D printing techniques used to generate high-defined structures. Schematic summary of AM techniques. Acronyms: stereolithography (SLA), two-photon polymerization (TPP), continuous liquid interface production (CLIP), selective laser sintering (SLS), fused deposition modeling (FDM), direct writing assembly (DWA), laminated object manufacturing (LOM), and selective deposition lamination (SDL). Reprinted from [180] CC-BY 2017 ELSEVIER.

manufacturing is under investigation for its attractive possibility to produce membranes with high-definition morphology. This new approach is regarded as a layer-by-layer (LBL) deposition of materials according to a computer-aided design model. Once the 3D file is ready, the orientation of 3D components in a predefined volume and slicing parameters is realized by using different alternative 3D printing techniques based on specific software. Based on LBL concept, these deposition techniques differ for the approach to transfer polymer layer onto each other [178–180] (Figure 6.16).

Stereolithography (SLA) exploits a photosensitive liquid resin, which is deposited through a predefined path and solidified by using an ultraviolet beam. After the first deposition and solidification, the system is raised while the laser continues to trace the cross-section of the object and a new layer of polymer sticks to the hardened part. The digital light-processing (DLP) projector replaces the UV laser to achieve resin hardening. Fused deposition modeling (FDM) consists of extrusion and deposition of a filament of polymer, while selective laser sintering (SLS) uses a laser beam to melt the polymer powder, which is deposited with high precision following the scheme of the 3D file. Successively, another layer of powder is spread onto the previous layer and melt by the laser.

The family of polymers suitable for 3DP techniques is somewhat limited with respect to traditional manufacturing approaches. Also, while it is possible to get high precision along the Z-axis, loss of resolution can occur along XY axes. This is a limiting aspect concerning control of pore size, pore distribution, density of pores, and pore thickness.

Undoubtedly, 3D printing approaches are emerging, and too few studies are reported in the literature. Preliminary investigations are encouraging, but a lot of study is still necessary to get structural features suitable for membrane processes based on membrane contactor technology [179, 181]. Also, this manufacturing approach requires longer time and is too expensive, especially for high resolution procedures, making less competitive the membranes on the market [181].

6.6 Compelling Case Studies for Water Desalination

This section examines the performance of membranes featured by structural lithographic precision through comparative analyses involving more conventional membranes. Applications in the field of water desalination are discussed. A higher degree of structural order is demonstrated to shift the productivity-efficiency trade-off when equipping MD units for the

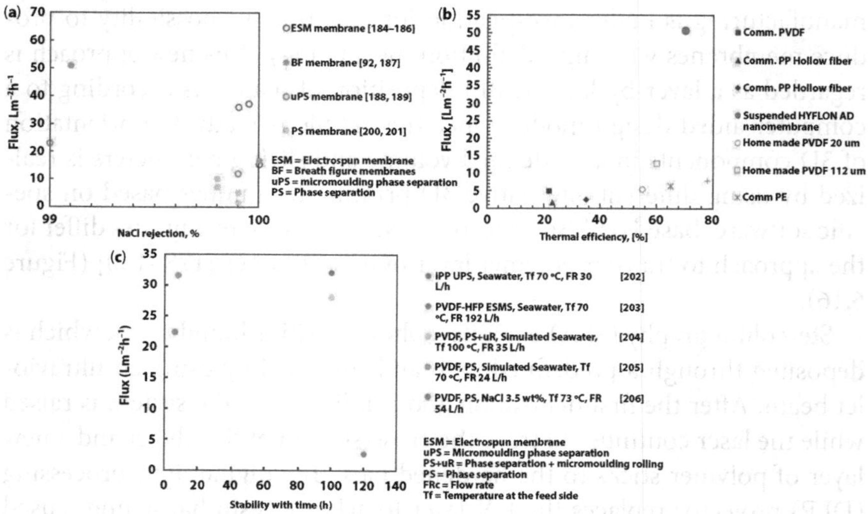

Figure 6.17 Comparative analyses between textured and conventional membranes: (a) productivity-efficiency trade-off for desalination of NaCl (35 gL^{-1}) streams; (b) productivity-thermal efficiency trade-off; (c) durability of MD operation for production of freshwater from seawater. Adapted from [163] CC-BY 2024 ELS Publishing.

treatment of synthetic seawater (NaCl 35gL^{-1}) within the range of temperature of 45–60°C (Figure 6.17a). Membranes with a finely controlled texture show better quantity-quality ratio for the production of desalted water. This can be ascribed to a more extensive L-V interface established at the entrance of pores that are more densely and uniformly distributed through the surface. Also, a larger L-V interface has beneficial effect on the flux and subsequently on improved thermal efficiency of the MD process (Figure 6.17b). The singular topography that characterizes the membranes makes the durability of the membranes longer during seawater desalination at 70°C; this is due to improved topography-related surface roughness, which prevents wetting occurrence (Figure 6.17c).

6.7 Conclusions

Membrane contactors technologies have been examined with a special focus on structure-transport relationships. Membrane morphology and related effects on mass transfer and wetting have been analyzed and discussed. The fabrication of membranes with high-defined structural features has been revised from the most traditional phase inversion to emerging manufacturing techniques, analyzing advantages and drawbacks. Top-down manufacturing approaches such as μPS, 3D printing, ESP, and track etching are easy to scale up but in some cases, the costs of fabrications are somewhat high, yielding the membranes less marketable for wide and massive applications. On the other hand, bottom-up approaches based on self-assembly concepts still require additional efforts for on scale production of membranes; this is due to the difficulty to control the dynamics of small and delicate objects such as droplets and polymer segment chains through extensive surface area. Nevertheless, lithographic precision and more broadly the structural order of the membrane have been judged to impact on the performance of specific membrane contactor operations. Some compelling case studies have been examined concerning with water desalination by membrane distillation. The productivity-efficiency trade-off and stability with time have been tuned and ascribed to the capability of the structural order of the membranes to amplify water vapor transport and contrast foulant adhesion. The explorative studies carried out so far suggest that the micromanipulation of the membrane structure is a viable tool for directing the processes outcomes towards desired targets. By considering cost-benefit balance, some of the reviewed manufacturing strategies could be regarded as forthcoming practical routes for the production of new-concept membranes for more competitive membrane separation processes.

References

1. Younas, M. and Rezakazemi, M., Introduction to Membrane Contactor Technology, in: *Membrane Contactor Technology*, Rezakazemi, M. and Younas, M. (Eds.), 2022, doi.org/10.1002/9783527831036.ch2.
2. Gugliuzza, A. and Drioli, E., PVDF and HYFLON AD membranes: Ideal interfaces for contactor applications. *J. Membr. Sci.*, 300, 51–62, 2007, doi.org/10.1016/j.memsci.2007.05.004.
3. Kukuzaki, M., Fujimoto, K., Kai, S., Ohe, K., Oshima, T., Baba, Y., Ozone mass transfer in an ozone-water contacting process with Shirasu porous glass (SPG) membranes-A comparative study of hydrophilic and hydrophobic membranes. *Sep. Purif. Technol.*, 72, 2010, 347–356, 2010, doi:10.1016/j.seppur.2010.03.004.
4. Gindl, M., Sinn, G., Gindl, W., Reiterer, A., Tschegg, S., A comparison of different methods to calculate the surface free energy of wood using contact angle measurements, Colloids Surfaces A Physicochem. *Eng. Asp.*, 181, 2001, 279–287, 2001, doi:10.1016/S0927-7757(00)00795-0.

5. Bakeri, G., Ismail, A.F., Shariaty-Niassar, M., Matsuura, T., Effect of polymer concentration on the structure and performance of polyetherimide hollow fiber membranes. *J. Memb. Sci.*, 363, 2010, 103–111, 2010, doi:10.1016/j.memsci.2010.07.018.
6. Ghasem, N., Al-Marzouqi, M., Duidar, A., Effect of PVDF concentration on the morphology and performance of hollow fiber membrane employed as gas-liquid membrane contactor for CO_2 absorption. *Sep. Purif. Technol.*, 98, 2012, 174–185, 2012, doi:10.1016/j.seppur.2012.06.036.
7. Criscuoli, A., Carnevale, M.C., Mahmoudi, H., Gaeta, S., Lentini, F., Drioli, E., Membrane contactors for the oxygen and pH control in desalination. *J. Memb. Sci.*, 376, 207–213, 2011, doi:10.1016/j.memsci.2011.04.018.
8. Abdel-Aal, E.A., Mahmoud, M.H.H., Sanad, M.M.S., Criscuoli, A., Figoli, A., Drioli, E., Membrane contactor as a novel technique for separation of iron ions from ilmenite leachant. *Int. J. Miner. Process.*, 96, 62–69, 2010, doi:10.1016/j.minpro.2010.05.002.
9. M.T., *Membrane contactors improve production yields*, vol. 4, p. 5, 2006, No Title, n.d.
10. Fakharian, H., Ganji, H., Naderifar, A., Desalination of high salinity produced water using natural gas hydrate. *J. Taiwan Inst. Chem. Eng.*, 72, 157–162, 2017, doi:10.1016/j.jtice.2017.01.025.
11. Li, B. and Sirkar, K.K., Novel membrane and device for vacuum membrane distillation-based desalination process. *J. Memb. Sci.*, 257, 60–75, 2005, doi:10.1016/j.memsci.2004.08.040.
12. Trusek-Holownia, A. and Noworyta, A., Catalytic membrane preparation for enzymatic hydrolysis reactions carried out in the membrane phase contactor. *Desalination*, 144, 427–432, 2002, doi:10.1016/S0011-9164(02)00356-9.
13. Charcosset, C., Membrane processes in biotechnology: An overview. *Biotechnol. Adv.*, 24, 482–492, 2006, doi:10.1016/j.biotechadv.2006.03.002.
14. Brunel, M.R.L., Tingry, S., Servat, K., Cretin, M., Innocent, C., Jolivalt, C., Membrane contactors for glucose/O2 biofuel cell. *Desalination*, 199, 426, 2006.
15. Gugliuzza, A. and Basile, A., Membrane contactors: fundamentals, membrane materials and key operations, in: *Handbook of Membrane Reactors: Reactor Types and Industrial Applications*, vol. 2, A. Basile (Ed.), pp. 54–106, Woodhead Publishing Series in Energy, Cambridge, UK, 2013.
16. Baghbanzadeh, M., Lan, C.Q., Rana, D., Matsuura, T., Membrane Distillation, in: *Nanostructured Polym. Membr*, pp. 419–455, John Wiley & Sons, Inc., Hoboken, NJ, USA, 2016, doi:10.1002/9781118831779.ch11.
17. Khayet, M. and Cojocaru, C., Artificial neural network modeling and optimization of desalination by air gap membrane distillation. *Sep. Purif. Technol.*, 86, 171–182, 2012, doi:10.1016/j.seppur.2011.11.001.
18. Meindersma, G.W., Guijt, C.M., de Haan, A.B., Desalination and water recycling by air gap membrane distillation. *Desalination*, 187, 291–301, 2006, doi:10.1016/j.desal.2005.04.088.

19. Kiefer, F., Spinnler, M., Sattelmayer, T., Multi-Effect Vacuum Membrane Distillation systems: Model derivation and calibration. *Desalination*, 438, 97–111, 2018, doi:10.1016/j.desal.2018.03.024.
20. Ko, C.C., Ali, A., Drioli, E., Tung, K.L., Chen, C.H., Chen, Y.R., Macedonio, F., Performance of ceramic membrane in vacuum membrane distillation and in vacuum membrane crystallization. *Desalination*, 440, 48–58, 2018, doi:10.1016/j.desal.2018.03.011.
21. Ma, Q., Ahmadi, A., Cabassud, C., Direct integration of a vacuum membrane distillation module within a solar collector for small-scale units adapted to seawater desalination in remote places: Design, modeling & evaluation of a flat-plate equipment. *J. Memb. Sci.*, 564, 617–633, 2018, doi:10.1016/j.memsci.2018.07.067.
22. Xie, Z., Duong, T., Hoang, M., Nguyen, C., Bolto, B., Ammonia removal by sweep gas membrane distillation. *Water Res.*, 43, 1693–1699, 2009, doi:10.1016/j.watres.2008.12.052.
23. Gasconsviladomat, F., Souchon, I., Athes, V., Marin, M., Membrane air-stripping of aroma compounds. *J. Membr. Sci.*, 277, 129–136, 2006, doi:10.1016/j.memsci.2005.10.023.
24. Khalifa, A.E., Performances of air gap and water gap MD desalination modules. *Water Pract. Technol.*, 13, 200–209, 2018, doi:10.2166/wpt.2018.034.
25. Winter, D., Permeate Gap Membrane Distillation (PGMD), in: *Encyclopedia of Membranes*, E. Drioli and L. Giorno (Eds.), Springer-Verlag Berlin Heidelberg, Berlin Heidelberg, 2015, DOI 10.1007/978-3-642-40872-4_447-4.
26. García-Payo, M., Rivier, C., Marison, I., von Stockar, U., Separation of binary mixtures by thermostatic sweeping gas membrane distillation. *J. Memb. Sci.*, 198, 197–210, 2002, doi:10.1016/S0376-7388(01)00649-4.
27. Memon, S., Im, B.-G., Lee, H.-S., Kim, Y.-D., Comprehensive experimental and theoretical studies on material-gap and water-gap membrane distillation using composite membranes. *J. Membr. Sci.*, 666, 121108, 2023, doi.org/10.1016/j.memsci.2022.121108.
28. Nagaraj, N., Patil, B.S., Biradar, P.M., Osmotic Membrane Distillation - A Brief Review. *Inter. J. Food Eng.*, 2, 2, 1–22, 2006, https://doi.org/10.2202/1556-3758.1095.
29. Cath, T.Y., Adams, D., Childress, A.E., Membrane contactor processes for wastewater reclamation in space: II. Combined direct osmosis, osmotic distillation, and membrane distillation for treatment of metabolic wastewater. *J. Memb. Sci.*, 257, 2005, 111–119, 2005, doi:10.1016/j.memsci.2004.07.039.
30. Savaş Bahçeci, K., Akillioılu, H.G., Gökmen, V., Osmotic and membrane distillation for the concentration of tomato juice: Effects on quality and safety characteristics. *Innov. Food Sci. Emerg. Technol.*, 31, 2015, 131–138, 2015, doi:10.1016/j.ifset.2015.07.008.
31. Gostoli, C., Thermal effects in osmotic distillation. *J. Memb. Sci.*, 162, 1999, 75–91, 1999, doi:10.1016/S0376-7388(99)00157-X.

32. Macedonio, F., Quist-Jensen, C.A., Al-Harbi, O., Alromaih, H., Al-Jlil, S.A., Al Shabouna, F., Drioli, E., Thermodynamic modeling of brine and its use in membrane crystallizer. *Desalination*, 323, 83–92, 2013, doi:10.1016/j.desal.2013.02.009.
33. Calabrò, V. and Basile, A., Economic analysis of membrane use in industrial applications. in: *In Woodhead Publishing Series in Energy, Advanced Membrane Science and Technology for Sustainable Energy and Environmental Applications*, Editor(s): Angelo Basile, Suzana Pereira Nunes, Woodhead Publishing, 2011, 90–109, 2011, doi:10.1533/9780857093790.1.90.
34. Macedonio, F. and Drioli, E., Hydrophobic membranes for salts recovery from desalination plants, *Desalin. Water Treat*, 18, 224–234, 2010, doi:10.5004/dwt.2010.1775.
35. Cui, Z., Li, X., Zhang, Y., Wang, Z., Gugliuzza, A., Militano, F., Drioli, E., Macedonio, F., Testing of three different PVDF membranes in membrane assisted-crystallization process: Influence of membrane structural-properties on process performance. *Desalination*, 440, 68–77, 2018, doi:10.1016/j.desal.2017.12.038.
36. Perrotta, M.L., Macedonio, F., Tocci, E., Giorno, L., Drioli, E., A Gugliuzza Graphene stimulates the nucleation and growth rate of NaCl crystals from hypersaline solution via membrane crystallization. *Envir. Sci. Water Res. Techn.*, 6, 6, 1723–1736, 2020. DOI: 10.1039/C9EW01124B.
37. Macedonio, F., Politano, A., Drioli, E., Gugliuzza, A., Bi2Se3-assisted membrane crystallization. *Mater. Horiz.*, 55), 912–919, 2018. DOI: 10.1039/C8MH00612A.
38. M. Frappa, Macedonio, F., Gugliuzza, A., Jin, W., Drioli, E., Performance of PVDF Based Membranes with 2D Materials for Membrane Assisted-Crystallization Process. *Membranes*, 11, 302, 2021, doi.org/10.3390/membranes11050302.
39. McPherson, A. and Gavira, J.A., Introduction to protein crystallization, Acta Crystallogr. *Sect. F Struct. Biol. Commun.*, 70, 2–20, 2014, doi:10.1107/S2053230X13033141.
40. Smatanová, I.K., Crystallization of biological macromolecules. *Mater. Struct.*, 9, 14–15, 2002.
41. Gugliuzza, A., Aceto, M.C., Drioli, E., Interactive functional poly (vinylidene fluoride) membranes with modulated lysozyme affinity: a promising class of new interfaces for contactor crystallizers. *Polym. Intern.*, 58, 12, 1452–1464, 2009, doi.org/10.1002/pi.2681.
42. Margolin, A.L. and Navia, M.A., Protein Crystals as Novel Catalytic Materials. *Angew. Chem. Int. Ed.*, 40, 2204–2222, 2001, doi:10.1002/1521-3773(20010618)40:12.
43. Falkner, J.C., Al-Somali, A.M., Jamison, J.A., Zhang, J., Adrianse, S.L., Simpson, R.L., Calabretta, M.K., Radding, W., Phillips, G.N., Colvin, V.L.,

Generation of Size-Controlled, Submicrometer Protein Crystals. *Chem. Mater.*, 17, 2679–2686, 2005, doi:10.1021/cm047924w.
44. Srisurichan, S., Jiraratananon, R., Fane, A., Mass transfer mechanisms and transport resistances in direct contact membrane distillation process. *J. Memb. Sci.*, 277, 186–194, 2006, doi:10.1016/j.memsci.2005.10.028.
45. Charcosset, C., Preparation of emulsions and particles by membrane emulsification for the food processing industry. *J. Food Eng.*, 92, 241–249, 2009, doi:10.1016/j.jfoodeng.2008.11.017.
46. Kuroiwa, T., Takada, H., Shogen, A., Saito, K., Kobayashi, I., Uemura, K., Kanazawa, A., Cross-linkable chitosan-based hydrogel microbeads with pH-responsive adsorption properties for organic dyes prepared using size-tunable microchannel emulsification technique. *Colloids Surfaces A Physicochem. Eng. Asp.* 514, 69–78, 2017, doi:10.1016/j.colsurfa.2016.11.046.
47. Niu, R., Yang, Y., Wang, S., Zhou, X., Luo, S., Zhang, C., Wang, Y., Chitosan microparticle-based immunoaffinity chromatography supports prepared by membrane emulsification technique: Characterization and application. *Int J. Biol. Macromol.*, 131, 1147–1154, 2019, doi:10.1016/j.ijbiomac.2019.04.064.
48. Piacentini, E., Yan, M., Giorno, L., Development of enzyme-loaded PVA microspheres by membrane emulsification. *J. Memb. Sci.*, 524, 79–86, 2017, doi:10.1016/j.memsci.2016.11.008.
49. Wu, J., Fan, Q., Xia, Y., Ma, G., Uniform-sized particles in biomedical field prepared by membrane emulsification technique. *Chem. Eng. Sci.*, 125, 85–97, 2015, doi:10.1016/j.ces.2014.08.016.
50. Li, D., Liu, B., Yang, F., Wang, X., Shen, H., Wu, D., Preparation of uniform starch microcapsules by premix membrane emulsion for controlled release of avermectin. *Carbohydr. Polym.*, 136, 34–349, 2016, doi:10.1016/j.carbpol.2015.09.050.
51. Charcosset, C., Limayem, I., Fessi, H., The membrane emulsification process - A review. *J. Chem. Technol. Biotechnol.*, 79, 209–218, 2004, doi:10.1002/jctb.969.
52. Joscelyne, S.M. and Trägårdh, G., Membrane emulsification - A literature review. *J. Memb. Sci.*, 169, 107–117, 2000, doi:10.1016/S0376-7388(99)00334-8.
53. Peng, S. and Williams, R.A., Controlled production of emulsions using a crossflow membrane. *Part. Part. Syst. Charact.*, 15, 3–8, 1998, doi:10.1002/(SICI)1521-4117(199802)15:1.
54. Kandori, K., Kishi, K., Ishikawa, T., Formation mechanisms of monodispersed W/O emulsions by SPG filter emulsification method. *Colloids Surf.*, 61, 269–279, 1991, doi:10.1016/0166-6622(91)80315-F.
55. Kawakatsu, T., Kikuchi, Y., Nakajima, M., Regular-sized cell creation in microchannel emulsification by visual microprocessing method, JAOCS. *J. Am. Oil Chem. Soc.*, 74, 317–321, 1997, doi:10.1007/s11746-997-0143-8.
56. Lambrich, U. and Schubert, H., Emulsification using microporous systems. *J. Memb. Sci.*, 257, 76–84, 2005, doi:10.1016/j.memsci.2004.12.040.

57. Zhang, Y. and Wang, R., Gas-liquid membrane contactors for acid gas removal: Recent advances and future challenges. *Curr. Opin. Chem. Eng.*, 2, 255–262, 2013, doi:10.1016/j.coche.2013.03.008.
58. Wang, M., Lawal, A., Stephenson, P., Sidders, J., Ramshaw, C., Post-combustion CO_2 capture with chemical absorption: A state-of-the-art review. *Chem. Eng. Res. Des.*, 89, 1609–1624, 2011, doi:10.1016/j.cherd.2010.11.005.
59. Klaassen, R., Feron, P.H.M., Jansen, A.E., Membrane contactors in industrial applications, Chem. *Eng. Res. Des.*, 83, 234–246, 2005, doi:10.1205/cherd.04196.
60. Smitha, B., Sridhar, S., Khan, A.A., Solid polymer electrolyte membranes for fuel cell applications - A review. *J. Memb. Sci.*, 259, 10–26, 2005, doi:10.1016/j.memsci.2005.01.035.
61. O'Brien, M., Baxendale, I.R., Ley, S.V., Flow ozonolysis using a semipermeable Teflon AF-2400 membrane to effect gas - Liquid contact. *Org. Lett.*, 12, 1596–1598, 2010, doi:10.1021/ol100322t.
62. Fritsch, D., Kuhr, K., Mackenzie, K., Kopinke, F.-D., Hydrodechlorination of chloroorganic compounds in ground water by palladium catalysts. *Catal. Today*, 82, 105–118, 2003, doi:10.1016/s0920-5861(03)00208-6.
63. Mavroudi, M., Kaldis, S.P., Sakellaropoulos, G.P., A study of mass transfer resistance in membrane gas-liquid contacting processes. *J. Memb. Sci.*, 272, 103–115, 2006, doi:10.1016/j.memsci.2005.07.025.
64. Simons, K., Nijmeijer, K., Wessling, M., Gas-liquid membrane contactors for CO_2 removal. *J. Memb. Sci.*, 340, 214–220, 2009, doi:10.1016/j.memsci.2009.05.035.
65. Ismail, A.F. and Lai, P.Y., Effects of phase inversion and rheological factors on formation of defect-free and ultrathin-skinned asymmetric polysulfone membranes for gas separation. *Sep. Purif. Technol.*, 33, 127–143, 2003, doi:10.1016/S1383-5866(02)00201-0.
66. Ismail, A.F., Ng, B.C., Abdul Rahman, W.A.W., Effects of shear rate and forced convection residence time on asymmetric polysulfone membranes structure and gas separation performance. *Sep. Purif. Technol.*, 33, 255–272, 2003, doi:10.1016/S1383-5866(03)00009-1.
67. Feron, P.H.M. and Jansen, A.E., CO_2 separation with polyolefin membrane contactors and dedicated absorption liquids: Performances and prospects. *Sep. Purif. Technol.*, 27, 231–242, 2002, doi:10.1016/S1383-5866(01)00207-6.
68. Bottino, A., Capannelli, G., Comite, A., Di Felice, R., Firpo, R., CO_2 removal from a gas stream by membrane contactor. *Sep. Purif. Technol.*, 59, 85–90, 2008, doi:10.1016/j.seppur.2007.05.030.
69. Atchariyawut, S., Feng, C., Wang, R., Jiraratananon, R., Liang, D.T., Effect of membrane structure on mass-transfer in the membrane gas-liquid contacting process using microporous PVDF hollow fibers. *J. Memb. Sci.*, 285, 272–281, 2006, doi:10.1016/j.memsci.2006.08.029.

70. Wang, D., Teo, W.K., Li, K., Selective removal of trace H2S from gas streams containing CO2 using hollow fibre membrane modules/contractors. *Sep. Purif. Technol.*, 35, 125–131, 2004, doi:10.1016/S1383-5866(03)00135-7.
71. deMontigny, D., Tontiwachwuthikul, P., Chakma, A., Using polypropylene and polytetrafluoroethylene membranes in a membrane contactor for CO_2 absorption. *J. Memb. Sci.*, 277, 99–107, 2006, doi:10.1016/j.memsci.2005.10.024.
72. Marzouk, S.A.M., Al-Marzouqi, M.H., El-Naas, M.H., Abdullatif, N., Ismail, Z.M., Removal of carbon dioxide from pressurized CO2-CH4 gas mixture using hollow fiber membrane contactors. *J. Memb. Sci.*, 351, 21–27, 2010, doi:10.1016/j.memsci.2010.01.023.
73. Mansourizadeh, A., Ismail, A.F., Abdullah, M.S., Ng, B.C., Preparation of polyvinylidene fluoride hollow fiber membranes for CO2 absorption using phase-inversion promoter additives. *J. Memb. Sci.*, 355, 200–207, 2010, doi:10.1016/j.memsci.2010.03.031.
74. Mansourizadeh, A. and Ismail, A.F., Effect of additives on the structure and performance of polysulfone hollow fiber membranes for CO2 absorption. *J. Memb. Sci.*, 348, 260–267, 2010, doi:10.1016/j.memsci.2009.11.010.
75. Zhang, Y., Wang, R., Yi, S., Setiawan, L., Hu, X., Fane, A.G., Novel chemical surface modification to enhance hydrophobicity of polyamide-imide (PAI) hollow fiber membranes. *J. Memb. Sci.*, 380, 241–250, 2011, doi:10.1016/j.memsci.2011.07.016.
76. Luis, P., Garea, A., Irabien, A., Zero solvent emission process for sulfur dioxide recovery using a membrane contactor and ionic liquids. *J. Memb. Sci.*, 330, 80–89, 2009, doi:10.1016/j.memsci.2008.12.046.
77. Drioli, E., Santoro, S., Simone, S., Barbieri, G., Brunetti, A., Macedonio, F., Figoli, A., ECTFE membrane preparation for recovery of humidified gas streams using membrane condenser. *React. Funct. Polym.*, 79, 1–7, 2014, doi:10.1016/j.reactfunctpolym.2014.03.003.
78. Macedonio, F., Brunetti, A., Barbieri, G., Drioli, E., Membrane condenser configurations for water recovery from waste gases. *Sep. Purif. Tech.*, 181, 60–68, 2017, doi.org/10.1016/j.seppur.2017.03.009.
79. Frappa, M., Brunetti, A., Drioli, E., Cui, Z., Pan, J., Macedonio, F., Membrane condenser for particulate abatement from waste-gaseous streams. *J. Membr. Sci. Res.*, 6, 81–89, 2020, doi:10.22079/JMSR.2019.112686.1282.
80. Macedonio, F., Frappa, M., Brunetti, A., Barbieri, G., Drioli, E., Recovery of water and contaminants from cooling tower plume. *Environ. Eng. Res.*, 25, 2019, 222–229, 2019, doi:10.4491/eer.2018.192.
81. Macedonio, F., Cersosimo, M., Brunetti, A., Barbieri, G., Drioli, E., Water recovery from humidified waste gas streams: Quality control using membrane condenser technology. *Chem. Eng. Process. Process Intensif.*, 86, 196–203, 2014, doi:10.1016/j.cep.2014.08.008.

82. Brunetti, A., Santoro, S., Macedonio, F., Figoli, A., Drioli, E., Barbieri, G., Waste Gaseous Streams: From Environmental Issue to Source of Water by Using Membrane Condensers. *CLEAN - Soil Air Water*, 42, 1145–1153, 2014, doi:10.1002/clen.201300104.
83. Macedonio, F., Brunetti, A., Barbieri, G., Drioli, E., Membrane Condenser as a New Technology for Water Recovery from Humidified "Waste" Gaseous Streams. *Ind. Eng. Chem. Res.*, 52, 1160–1167, 2013, doi:10.1021/ie203031b.
84. Boributh, S., Assabumrungrat, S., Laosiripojana, N., Jiraratananon, R., A modeling study on the effects of membrane characteristics and operating parameters on physical absorption of CO2 by hollow fiber membrane contactor. *J. Memb. Sci.*, 380, 21–33, 2011, doi:10.1016/j.memsci.2011.06.029.
85. Ren, J., Wang, R., Zhang, H.Y., Li, Z., Liang, D.T., Tay, J.H., Effect of PVDF dope rheology on the structure of hollow fiber membranes used for CO_2 capture. *J. Memb. Sci.*, 281, 334–344, 2006, doi:10.1016/j.memsci.2006.04.003.
86. Lauterböck, B., Moder, K., Germ, T., Fuchs, W., Impact of characteristic membrane parameters on the transfer rate of ammonia in membrane contactor application. *Sep. Purif. Technol.*, 116, 327–334, 2013, doi:10.1016/j.seppur.2013.06.010.
87. Iversen, S.B., Bhatia, V.K., Dam-Johansen, K., Jonsson, G., Characterization of microporous membranes for use in membrane contactors. *J. Memb. Sci.*, 130, 205–217, 1997, doi:10.1016/S0376-7388(97)00026-4.
88. Mejia Mendez, D.L., Castel, C., Lemaitre, C., Favre, E., Membrane distillation (MD) processes for water desalination applications. Can dense selfstanding membranes compete with microporous hydrophobic materials? *Chem. Eng. Sci.*, 188, 84–96, 2018, doi:10.1016/j.ces.2018.05.025.
89. Frappa, M., Del Rio Castillo, A.E., Macedonio, F., Politano, A., Drioli, E., Bonaccorso, F., Pellegrini, V., Gugliuzza, A., Few-layer graphene for advanced composite PVDF membranes dedicated to water distillation: a comparative study. *Nanoscale Adv.*, 2, 4728–4739, 2020. DOI: 10.1039/D0NA00403K.
90. Gugliuzza, A., Aceto, M., Simone, S., Curcio, E., Madonna, R., Di Profio, G., Drioli, E., Novel functional per-fluorinated membranes: suitable nucleating systems for protein crystallization. *Desalination*, 199, 200–203, 2006, doi.org/10.1016/j.desal.2006.03.047.
91. Perrotta, M.L., Saielli, G., Casella, G., Macedonio, F., Giorno, L., Drioli, E., Gugliuzza, A., An ultrathin suspended hydrophobic porous membrane for high-efficiency water desalination. *Appl. Mater. Today*, 9, 1–9, 2017, doi.org/10.1016/j.apmt.2017.04.009.
92. Frappa, M., Del Rio Castillo, A.E., Macedonio, F., Di Luca, G., Drioli, E., Gugliuzza, A., Exfoliated Bi2Te3-enabled Membranes for New Concept Water Desalination: Freshwater Production Meets New Routes. *Water Res.*, 203, 117503, 2021, doi.org/10.1016/j.watres.2021.117503.
93. Rasool, M.A., Van Goethem, C., Vankelecom, I.F., Green preparation process using methyl lactate for cellulose-acetate-based nanofiltration membranes. *Sep. Purif. Technol.*, 232, 115903, 2020.

94. Bey, S., Criscuoli, A., Simone, S., Figoli, A., Benamor, M., Drioli, E., Hydrophilic PEEK-WC Hollow Fiber Membrane Contactors for Chromium (VI) Removal. *Desalination*, 283, 16–24, 2011.
95. Cui, Z., Hassankiadeh, N.T., Lee, S.Y., Lee, J.M., Woo, K.T., Sanguineti, A., Arcella, V., Lee, Y.M., Drioli, E., Poly(vinylidene fluoride) membrane preparation with an environmental diluent via thermally induced phase separation. *J. Membr. Sci.*, 444, 223–226, 2013.
96. Marino, T., Galiano, F., Molino, A., Figoli, A., New frontiers in sustainable membrane preparation: Cyrene™ as green bioderived solvent. *J. Membr. Sci.*, 580, 224–234, 2019.
97. Wang, H.H., Jung, J.T., Kim, J.F., Kim, S., Drioli, E., Lee, Y.M., A novel green solvent alternative for polymeric membrane preparation via nonsolvent-induced phase separation (NIPS). *J. Membr. Sci.*, 574, 44–54, 2019.
98. W.Xie, A., Liu, B., Tang, P., Wang, F., Chen, S., Figoli, A., Chu, L.-Y., First Exploration on a Poly(vinylchloride) Ultrafiltration Membrane Prepared by Using the Sustainable Green Solvent PolarClean. *ACS Sustain. Chem. Eng.*, 8, 91–101, 2020.
99. Jung, J.T., Wang, H.H., Kim, J.F., Jeon, S.M., Park, S.H., Lee, W.H., Moon, S.J., Drioli, E., Lee, Y.M., Microfiber aligned hollow fiber membranes from immiscible polymer solutions by phase inversion. *J. Membr. Sci.*, 617, 118654, 2020.
100. Rasool, M.A., Pescarmona, P.P., Vankelecom, I.F., Applicability of organic carbonates as green solvents for membrane preparation. *ACS Sustain. Chem. Eng.*, 7, 3774–13785, 2019.
101. Ajari, H., Chaouachi, B., Galiano, F., Marino, T., Russo, F., Figoli A, A., A novel approach for dissolving crystalline LDPE using non-toxic solvents for membranes preparation. *Int. J. Envir. Sci. Techno.*, 16, 5375–5386, 2019.
102. Jiang, X., Yong, W.F., Gao, J., Shao, D.-D., Sun, S.-P., Understanding the role of substrates on thin film composite membranes: A green solvent approach with TamiSolve® NxG. *J. Membr. Sci.*, 635, 119530, 2021, doi.org/10.1016/j.memsci.2021.119530.
103. Van de Witte, P., Dijkstra, P.J., Van den Berg, J., Feijen, J., Phase separation processes in polymer solutions in relation to membrane formation. *J. Membr. Sci.*, 117, 1–31, 1996.
104. Wang, D.-M. and Lai, J.-Y., Recent advances in preparation and morphology control of polymeric membranes formed by nonsolvent induced phase separation, Curr. *Opin. Chem. Eng.*, 2, 229–237, 2013.
105. Reuvers, A.J. and Smolders, C.A., Formation of membrane by means of immersion precipitation Part II. The mechanism of formation of membranes prepared from the system CA/acetone/water. *J. Membr. Sci.*, 34, 67–86, 1987.
106. Hansen, C.M., The three dimensional solubility parameter. *Dan. Tech. Cph*, 14, 13–28, 1967.

107. Tan, X.M. and Rodrigue, D., A Review on Porous Polymeric Membrane Preparation. Part I: Production Techniques with Polysulfone and Poly (Vinylidene Fluoride). *Polymers*, 11, 1160, 2019.
108. Li, Q., Xu, Z.-L., Yu, L.-Y., Effects of mixed solvents and pvdf types on performances of pvdf microporous membranes. *J. Appl. Polym. Sci.*, 115, 2277–2287, 2010.
109. Guillen, G.R., Pan, Y., Li, M., Hoek, E.M.V., Preparation and characterization of membranes formed by nonsolvent induced phase separation: A review. *Ind. Eng. Chem. Res.*, 50, 3798–3817, 2011.
110. Strathmann, H., Kock, K., Amar, P., Baker, R., The formation mechanism of asymmetric membranes. *Desalin.*, 16, 179–203, 1975.
111. Tsai, H.A., Ruaan, R.C., Wang, D.M., Lai, J.Y., Effect of temperature and span series surfactant on the structure of polysulfone membranes. *J. Appl. Polym. Sci.*, 86, 166–173, 2002.
112. Zheng, Q.-Z., Wang, P., Yang, Y.-N., Cui, D.-J., The relationship between porosity and kinetics parameter of membrane formation in PSF ultrafiltration membrane. *J. Membr. Sci.*, 286, 7–11, 2006.
113. Mulder, J., *Basic Principles of Membrane Technology*, Springer Science & Business Media, New York, NY, USA, 2012.
114. Chakrabarty, B., Ghoshal, A.K., Purkait, M.K., Preparation, characterization and performance studies of polysulfone membranes using PVP as an additive. *J. Membr. Sci.*, 315, 36–47, 2008.
115. Jung, B., Yoon, J.K., Kim, B., Rhee, H.-W., Effect of molecular weight of polymeric additives on formation, permeation properties and hypochlorite treatment of asymmetric polyacrylonitrile membranes. *J. Membr. Sci.*, 243, 45–57, 2004.
116. Wienk, I., Boom, R., Beerlage, M., Bulte, A., Smolders, C., Strathmann, H., Recent advances in the formation of phase inversion membranes made from amorphous or semi-crystalline polymers. *J. Membr. Sci.*, 113, 361–371, 1996.
117. Dong, X., Al-Jumaily, A., Escobar, I., Investigation of the use of a bio-derived solvent for non-solvent-induced phase separation (NIPS) fabrication of polysulfone membranes. *Membr.*, 8, 23, 2018.
118. Kim, J.-H. and Lee, K.-H., ffect of PEG additive on membrane formation by phase inversion. *J. Membr. Sci.*, 138, 153–163, 1998.
119. Chakrabarty, B., Ghoshal, A., Purkait, M., Effect of molecular weight of PEG on membrane morphology and transport properties. *J. Membr. Sci.*, 309, 209–221, 2008.
120. Park, H.C., Kim, Y.P., Kim, H.Y., Kang, Y.S., Membrane formation by water vapor induced phase inversion. *J. Membr. Sci.*, 156, 2, 169–178, 1999.
121. Zhao, Q., Xie, R., Luo, F., Faraj, Y., Liu, Z., Ju, X.J., Wang, W., Chu, L.Y., Preparation of high strength poly (vinylidene fluoride) porous membranes with cellular structure via vapor-induced phase separation. *J. Membr. Sci.*, 549, 151–164, 2018.

122. Astakhov, E.Y., Zhironkin, S.F., Kolganov, I.M., Klinshpont, E.R., Tsarin, P.G., Study of the formation of the porous structure of membranes during phase separation of a poly(ester sulfone) solution. *Polym. Sci. Ser. A*, 53, 7, 613–620, 2011.
123. Bouyer, D., Werapun, W., Pochat-Bohatier, C., Deratani, A., Morphological properties of membranes fabricated by VIPS process using PEI/NMP/water system: SEM analysis and mass transfer modeling. *J. Membr. Sci.*, 349, 1-2, 97–112, 2010.
124. Tsai, J.T., Su, Y.S., Wang, D.M., Kuo, J.L., Lai, J.Y., Deratani, A., Retainment of pore connectivity in membranes prepared with vapor-induced phase separation. *J. Membr. Sci.*, 362, 360–373, 2010.
125. Menut, P., Su, Y.S., Chinpa, W., Pochat-Bohatier, C., Deratani, A., Wang, D.M., Huguet, P., Kuo, C.Y., Lai, J.Y., Dupuy, C., A top surface liquid layer during membrane formation using vapor-induced. *J. Membr. Sci.*, 310, 310, 278–288, 2008.
126. Caquineau, H., Menut, P., Deratani, A., Dupuy, C., Influence of the relative humidity on film formation by vapor induced phase separation. *Polym. Eng. Sci.*, 43, 4, 798–808, 2003.
127. Peng, Y., Fan, H., Dong, Y., Song, Y., Han, H., Effects of exposure time on variations in thestructure and hydrophobicity of polyvinylidene fluoride-membranes prepared via vapor-induced phase separation. *Appl. Surf. Sci.*, 258, 7872–7881, 2012.
128. Su, Y.S., Kuo, C.Y., Wang, D.M., Lai, J.Y., Deratani, A., Pochat, C., Bouyer, D., Interplay of mass transfer, phase separation, andmembrane morphology in vapor-induced phase separation. *J. Membr. Sci.*, 3338, 17–28, 2009.
129. Matsuyama, H., Teramoto, M., Nakatani, R., Maki, T., Membrane formation via phase separation induced by penetration of nonsolvent from vapor phase. I. Phase diagram and mass transfer process. *J. Appl. Polym. Sci.*, 74, 159–170, 1999.
130. Matsuyama, H., Teramoto, M., Nakatani, R., Maki, T., Membrane formation via phase separation induced by penetration of nonsolvent from vapor phase. II. Membrane morphology. *J. Appl. Polym. Sci.*, 74, 171–178, 1999.
131. Sun, H., Liu, S., Ge, B., Xing, L., Chen, H., Cellulose nitrate membrane formation via phase separation induced by penetration of nonsolvent from vapor phase. *J. Membr. Sci.*, 295, 2–10, 2007.
132. Gugliuzza, A., Ricca, F., Drioli, E., Controlled pore size, thickness and surface free energy of super-hydrophobic PVDF® and Hyflon® AD membranes. *Desalination*, 200, 26–28, 2006, doi.org/10.1016/j.desal.2006.03.229.
133. Kim, S.S. and Lloyd, D.R., Microporous membrane formation via thermally-induced phase separation. Effect of thermodynamic interactions on the structure of isotactic polypropylene membranes. *J. Membr. Sci.*, 64, 13–29, 1991.

134. Liu, Z., Cui, Z., Zhang, Y., Qin, S., Yan, F., Li, J., Fabrication of polysulfone membrane via thermally induced phase separation process. *Mater. Lett.*, 195, 190–193, 2017.
135. Lloyd, D.R., Kinzer, K.E., Tseng, H., Microporous membrane formation via thermally induced phase separation. I. Solid-liquid phase separation. *J. Membr. Sci.*, 52, 239–261, 1990.
136. Lloyd, D.R., Kim, S.S., Kinzer, K.E., Microporous membrane formation via thermally-induced phase separation. II. Liquid-Liquid phase separation. *J. Membr. Sci.*, 64, 1–11, 1991.
137. Tobo-Niño, O.M., García-Jiménez, C.D., Muvdi-Nova, C.J., Flat sheet membrane elaboration by TIPS method using palm oil as solvent and its application in membrane distillation. *Ingeniería y Competitividad*, 19, 81–90, 2017.
138. Liu, S., Zhou, C., Yu, W., Phase separation and structure control in ultra-high molecular weight polyethylene microporous membrane. *J. Membr. Sci.*, 379, 268–278, 2011.
139. Matsuyama, H., Berghmans, S., Lloyd, D.R., Formation of hydrophilic microporous membranes via thermally induced phase separation. *J. Membr. Sci.*, 142, 213–224, 1998.
140. Cha, B.J., Char, K., Kim, J.J., Kim, S.S., Kim, C.K., The effects of diluent molecular weight on the structure of thermally-induced phase separation membrane. *J. Membr. Sci.*, 108, 219–229, 1995.
141. Vogelaar, L., Barsema, J.N., Nijdam, W., van Rijn, C.J.M., Wessling, M., Phase Separation Micromolding—PSμM. *Adv. Mater.*, 15, 1385, 2003.
142. Girones, M., Akbarsyah, I.J., Nijdam, W., van Rijn, C.J.M., Jansen, H.V., Lammertink, R.G.H., Wessling, M., Polymeric microsieves produced by phase separation micromolding. *J. Membr. Sci.*, 283, 411–424, 2006.
143. Vogelaar, L., Lammertink, R.G.H., Barsema, J.N., Nijdam, W., Bolhuis-Versteeg, L.A.M., van Rijn, C.J.M., Wessling, M., Phase separation micromolding: a new generic approach for microstructuring various materials. *Small.*, 1, 645–655, 2005.
144. Jo, B.H., Van Lerberghe, L.M., Motsegood, K.M., Beebe, D.J., Three-dimensional micro-channel fabrication in polydimethylsiloxane (PDMS) elastomer. *J. Microelectromech. Syst.*, 9, 76–81, 2000.
145. Widawski, G., Rawiso, M., Francois, B., Self-organized honeycomb morphology of star-polymer polystyrene films. *Nature*, 369, 387–389, 1994.
146. Srinivasarao, M., Collings, D., Philips, A., Patel, S., Threedimensionally ordered array of air bubbles in a polymer film. *Science*, 292, 79–83, 2001.
147. Qin, S., Li, H., Yuan, W.Z., Zhang, Y., Fabrication of polymeric honeycomb microporous films: breath figures strategy and stabilization of water droplets by fluorinated diblock copolymer micelles. *J. Mater. Sci.*, 47, 6862–6871, 2012.

148. Cong, H.L., Wang, J.L., Yu, B., Tang, J., Preparation of a highly permeable ordered porous microfiltration membrane of brominated poly(phenylene oxide) on an ice substrate by the breath figure method. *Soft Matter*, 8, 8835–8839, 2012.
149. Wan, L.S., Li, J.W., Ke, B.B., Xu, Z.K., Ordered microporous membranes templated by breath figures for size-selective separation. *J. Am. Chem. Soc.*, 134, 95–98, 2012.
150. Gugliuzza, A., Aceto, M.C., Macedonio, F., Drioli, E., Water droplets as template for next-generation self-assembled poly-(etheretherketone) with cardo membranes. *J. Phys. Chem. B*, 112, 10483–10496, 2008, doi.org/10.1021/jp802130u.
151. Speranza, V., Trotta, F., Drioli, E., Gugliuzza, A., High-definition polymeric membranes: Construction of 3d lithographed channel arrays through control of natural building blocks dynamics. *ACS Appl. Mater. Interfaces.*, 2, 2, 459–466, 2010, doi.org/10.1021/am900701r.
152. Bormashenko, E., Pogreb, R., Stanevsky, O., Y., O., Gendelman, Formation of honeycomb patterns in evaporated polymer solutions: Influence of the molecular weight. *Mater. Lett.*, 59, 3553–3557, 2005.
153. Gugliuzza, A., Perrotta, M.L., Drioli, E., Controlled Bulk Properties of Composite Polymeric Solutions for Extensive Structural Order of Honeycomb Polysulfone Membranes. *Membranes*, 5, 1–12, 2016, doi:10.3390/membranes6020027.
154. Radjabian, A. and Abetz, V., Advanced porous polymer membranes from self-assembling block copolymers. *Prog. Polym. Sci.*, 102, 101219, 2020.
155. Hofman, A.H. and Brinke, K.L., Hierarchical structure formation in supramolecular comb-shaped block copolymers. *Polymer*, 107, 343–356, 2016, doi.org/10.1016/j.polymer.2016.08.021.
156. Kloos, J., Joosten, N., Schenning, A., Nijmeijer, K., Self-assembling liquid crystals as building blocks to design nanoporous membranes suitable for molecular separations. *J. Membr. Sci.*, 620, 11884, 2021.
157. Henmi, M., Nakatsuji, K., Ichikawa, T., Tomioka, H., Sakamoto, T., Yoshio, M., Kato, T., Self-organized liquid-crystalline nanostructured membranes for water treatment: selective permeation of ions. *Adv. Mater.*, 24, 2238–2241, 2012.
158. Foley, K. and Walters, K.B., Solution and Film Self-Assembly Behavior of a Block Copolymer Composed of a Poly(ionic Liquid) and a Stimuli-Responsive Weak Polyelectrolyte. *ACS Omega*, 8, 33684–3370, 2023.
159. Arribas, P., Khayet, M., García-Payo, M.C., Gil, L., Self-sustained electrospun polysulfone nano-fibrous membranes and their surface modification by interfacial polymerization for micro-and ultra-filtration. *Sep. Purif. Technol.*, 138, 118–129, 2014.
160. Bhardwaj, N. and Kundu, S.C., Electrospinning: A fascinating fiber fabrication technique. *Biotechnol. Adv.*, 28, 325–347, 2010.

161. Raghavan, P., Lim, D.-H., Ahn, J.-H., Nah, C., Sherrington, D.C., Ryu, H.-S., Ahn, H.-J., Electrospun polymer nanofibers: The booming cutting-edge technology. *React. Funct. Polym.*, 72, 915–930, 2012.
162. Demir, M.M., Yilgor, I., Yilgor, E., Erman, B., Electrospinning of polyurethane fibers. *Polymer*, 43, 3303–3309, 2002.
163. Gugliuzza, A., Review of techniques for future manufacturing of ordered and functional porous films: conceivable membranes for water desalination by membrane distillation processes. *Adv. Manuf.*, 2, 0009, 2024, doi.org/10.55092/am20240009.
164. Yang, Q., Li, Z., Hong, Y., Zhao, Y., Qiu, S., Wang, C., Wei, Y., Influence of solvents on the formation of ultrathin uniform poly(vinyl pyrrolidone) nanofibers with electrospinning. *J. Polym. Sci. Part B Polym. Phys.*, 42, 3721–3726, 2004.
165. Yuan, X. and Sheng, J., Morphology of ultrafine polysulfone fibers prepared by electrospinning. *Polym. Int.*, 53, 1704–1710, 2004.
166. Aydemir Sezer, U., Ozturk, K., Aru, B., Yanıkkaya Demirel, G., Sezer, S., A design achieved by coaxial electrospinning of polysulfone and sulfonated polysulfone as a core-shell structure to optimize mechanical strength and hemocompatibility. *Surf. Interfaces*, 10, 176–187, 2018.
167. Jalili, R., Hosseini, S.A.A., Morshed, M., The effects of operating parameters on the morphology of electrospun polyacrilonitrile nanofibers. *Iran. Polym. J.*, 14, 1074–1081, 2005.
168. Korolkov, I.V., Narmukhamedova, A.R., Melnikova, G.B., Muslimova, I.B., Yeszhanov, A.B., Zhatkanbayeva, Z.K., Chizhik, S.A., Zdorovets, M.V., Preparation of Hydrophobic PET Track-Etched Membranes for Separation of Oil–Water Emulsion. *Membranes*, 11, 637, 2021, doi.org/10.3390/membranes11080637.
169. Fleischer, R.L., Price, P.B., Symes, E.M., Novel fillter for biological materials. *Science*, 143, 249–250, 1964.
170. Spohr, R., Nuclear track irradiations at GSI. *Nucl. Tracks*, 4, 101–106, 1980.
171. Fischer, B.E. and Spohr, R., Production and use of nuclear tracks: imprinting structure on solids. *Rev. Mod. Phys.*, 55, 907–948, 1983.
172. Trautmann, C., Bouard, S., Spohr, R., Etching threshold for ion tracks in polyimide. *Nucl. Instrum. Methods Phys. Res. Sect. B Beam Interact. Mater. At.*, 116, 429–433, 1996.
173. Apel, P., P Track etching technique in membrane technology. *Radiat. Meas.*, 34, 34, 559–566, 2001.
174. Daubresse, C., Sergent-Engelen, T., Ferain, E., Schneider, Y.J., Legras, R., Characterisation of energetic heavy ion track in PVDF: Production of PVDF track-etched membrane and application. *Nucl. Instrum. Methods Phys. Res. Sect. B Beam Interact. Mater. At.*, 105, 126–129, 1995.
175. Mazzei, R., Bernaola, O.A., Saint Martin, G., Molinary de Ray, B., Submicroscopic kinetics of track formation in SSNTD. *Nucl. Instrum. Methods*, B9, 163–168, 1985.

176. Grasselli, M. and Betz, N., Making porous membranes by chemical etching of heavy-ion tracks in PVDF films. *Nucl. Inst. Methods Phys. Res. B*, 236, 501–507, 2005.
177. Ceccio, G., Vacik, J., Siegel, J., Cannavó, A., Choukourov, A., Pleskunov, P., Tosca, M., Fink, D., Etching and Doping of Pores in Polyethylene Terephthalate Analyzed by Ion Transmission Spectroscopy and Nuclear Depth Profiling. *Membranes*, 12, 1061, 2022, https://doi.org/10.3390/membranes12111061.
178. Tijing, L.D., J.Ryan, C., Ibrahim, I., Ray, A., Nisay, N., Shon, H.K., Advincula, R.C., 3D printing for membrane separation, desalination and water treatment. *Appl. Mater. Today*, 18, 100486, 2020, doi.org/10.1016/j.apmt.2019.100486.
179. Clark, M.J., Garg, T., Rankin, K.E., Bradshaw, D., Nightingale, A.M., 3D printed filtration and separation devices with integrated membranes and no post-printing assembly. *React. Chem. Eng.*, 9, 251–259, 2024, doi.org/10.1039/D3RE00245D.
180. Low, Z.X., Chua, Y.T., Ray, B.M., Mattia, D., Metcalfe, I.S., Patterson, D.A., Perspective on 3D printing of separation membranes and comparison to related unconventional fabrication techniques. *J. Membr. Sci.*, 523, 596–613, 2017, doi.org/10.1016/j.memsci.2016.10.006.
181. Yanar, N., Kallem, P., Sonc, M., Park, H., Kange, S., Choi, H., A New era of water treatment technologies: 3D printing for membranes. *J. Ind. Eng. Chem.*, 91, 1–14, 2020, doi.org/10.1016/j.jiec.2020.07.043.

PART III

RECENT ADVANCES IN MEMBRANE SEPARATIONS BASED ON POROUS MATERIALS

7

Biotech Porous Membranes

Qian Wang and Zhaoliang Cui*

State Key Laboratory of Materials-Oriented Chemical Engineering, National Engineering Research Center for Special Separation Membrane, College of Chemical Engineering, Nanjing Tech University, Nanjing, China

Abstract

A membrane bioreactor (MBR) is a sewage treatment process that combines membrane separation and biological treatment technology to replace the traditional activated sludge process (ASP). MBR takes advantage of good effluent quality, small area, and less sludge production. This technology uses porous membranes as a thin layer through which separation and purification take place according to complex mechanisms. Specifically, the MBR technology involves biological degradation and allows the total physical retention of bacterial flocs and suspended particles using membranes with a pore size within the range of 0.002 to 0.05μm. In this chapter, the MBR basics, classification, membrane fabrication, application, and related scale up are overviewed.

Keywords: Membrane bioreactors, bio-separation, purification

7.1 Introduction to MBR

7.1.1 What is an MBR?

With the continuous development of science and technology, the term "membrane" has increasingly appeared in scientific research, production, and human living. Although it is only a very thin layer, it is very useful. Of course, this does not refer to ordinary plastic films or soap bubble liquid films, but to those functional membranes with special properties. As a popular technology, membranes have received significant attention from

**Corresponding author*: zcui@njtech.edu.cn

Annarosa Gugliuzza and Wanqin Jin (eds.) Porous Membranes: Breakthroughs in Manufacturing and Applications, (175–206) © 2025 Scrivener Publishing LLC

various industries, and its development is extremely fast. So far, great progress has been made in researching membrane transport mechanisms, the development of new membrane materials, and the innovation of membrane-making processes. Membrane technology can be combined with some traditional chemical technologies to form complementary advantages [1].

Among the uses of membrane technology, since the 21st century, water shortage has become one of the restrictive factors of global sustainable development, leading to water pollution control and the reuse of reclaimed water as a historical necessity. While promulgating strict water resource protection laws and water pollution control laws, countries have also increased the research and development of new sewage treatment processes to control water pollution and recycle reclaimed water. The membrane bioreactor (MBR), as a new sewage treatment process, combines membrane separation and biological treatment technology to replace the secondary settling tank in the traditional activated sludge process (ASP) to improve the efficiency of mud-water separation. Compared with ASP, MBR has the advantages of good effluent quality, a small area, and less sludge production. Therefore, it was sought after by countries all over the world in the early 21st century. It has been regarded as the future of sewage treatment technology and has a subjective desire to replace ASP.

MBR is a common term used to describe wastewater treatment processes. The BR technology involves biological degradation techniques by activating sludge with a direct solid–liquid separation utilizing membrane filtration [2, 3]. MBR systems allow the total physical retention of bacterial flocs and practically all suspended particles within the bioreactor by employing micro- or ultrafiltration membrane technology (pore diameters within 0.002–0.05μm) [4–6].

As shown in Figure 7.1, the raw water is first pumped into the biological reaction tank. There is a large amount of sludge in the reaction tank, and microorganisms survive in it. The microorganisms digest the raw water, which represents a portion of food for them. At the same time, the effects of degrading COD, ammonia nitrogen, and total phosphorus are achieved to ensure that the water quality meets the standard. Then, through the ultrafiltration membrane, the role of filtration intercepts to achieve solid–liquid separation, so that the suspended solids and turbidity in the permeated water are close to zero. It also retains microorganisms, reducing the loss of microbial resources.

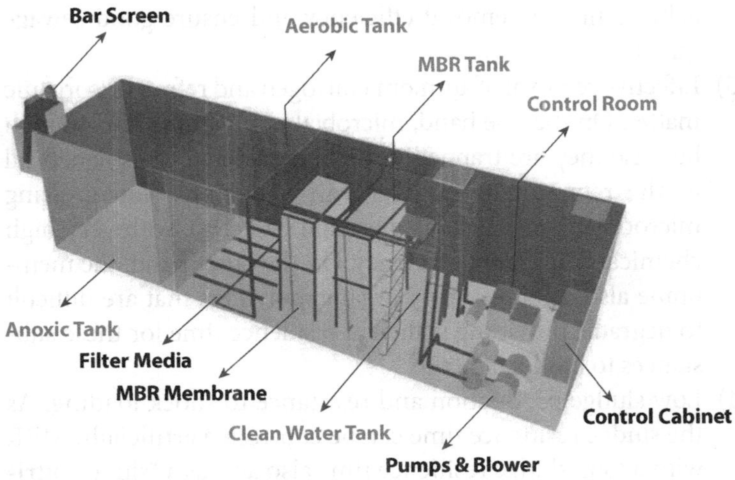

Figure 7.1 Membrane bioreactor schematic. https://spertasystems.com/product/mbr-membrane-system/.

7.1.2 MBR Features

Compared to other water treatment technologies, MBR systems have presented many unique advantages:

1) Efficient solid–liquid separation performance. Compared with traditional sedimentation tanks, the MBR process uses membrane separation technology, which can accurately filter suspended particles, macromolecule organic matter, and microorganisms in sewage, thus ensuring effluent quality. The membrane filtration precision is high, and the suspended matter and turbidity in the effluent are negligible. The filtrate can be reused directly without further purification steps.
2) Fast biochemical reaction rate. Compared with the traditional sludge activation process, MBR technology can retain activated sludge in the reactor with high efficiency. The loss of microorganisms is avoided, the concentration of microorganisms is kept at a higher level, and the cost of sludge replenishment is reduced. The biochemical reaction rate in the reactor is obviously accelerated, and the water load is greatly increased. Short hydraulic retention times can

achieve higher removal efficiency and ensure good sewage quality.

3) Effective removal of ammonia nitrogen and refractory organic matter. On the one hand, microbial populations remain high because they are trapped in the bioreactor. This is beneficial to the reproduction and growth of various slow-growing microorganisms (such as digestion bacteria), with very high chemical conversion efficiency. On the other hand, the membrane also captures many organic polymers that are difficult to degrade, providing sufficient residence time for these substances to biodegrade.
4) Low sludge production and resistance to shock loading. As the sludge residence time can be controlled artificially, MBR with a long sludge residence time also acts as a "sludge nitrification tank," thereby significantly reducing sludge production and treatment costs. At the same time, activated sludge within the MBR will not follow the effluent and loss. During operation, the sludge concentration will change in real-time with the inflow of water, and the concentration of organic matter will reach a kind of dynamic equilibrium. This feature makes the system have a higher ability to withstand shock loads and keep its stability.
5) Easy management and automatic control. Because MBR achieves complete separation of hydraulic retention time and sludge retention time, operation control is more flexible, and it is easy to implement automatic program control, making operation management more convenient.
6) Small footprint. First, MBR provides good conditions for sludge accumulation to reach the maximum concentration of the system. The bio-reactor maintains a high concentration of microbial biomass, so the sewage treatment rate is fast, and the treatment device volume load is large, saving space. Second, since the secondary sedimentation tank is replaced by membrane separation technology, its area is saved. In addition, the MBR process has a compact structure that can be divided into above-ground, semi-underground, and underground systems, further saving space. It is limited by the space of the processing place and has a wide range of applications.

Simultaneously, due to the limitations of the technology, MBR also has some shortcomings, mainly in the following aspects:

1) High membrane cost. MBR projects require a large investment, and it mainly depends on membrane materials and membrane components. Membrane materials and pollution issues affect their serviceability. Membrane components are vulnerable to contamination and need to be replaced regularly, so the depreciation cost of MBR projects is higher than that of traditional wastewater treatment processes.
2) Membrane contamination. During the long-term stable operation of MBR, the membrane is inevitably contaminated by pollutants and microorganisms. Thus, the membrane flux is affected, and the efficiency of sewage treatment is reduced. Therefore, regular cleaning is required, which brings trouble to the operation management and increases costs.
3) High energy consumption. Compared with the traditional activated sludge process, the required aeration intensity of MBR is much higher. This aspect provides a sufficiently strong membrane surface aeration and hydraulic flushing time to prevent the rapid accumulation of pollutants on the membrane surface and reduce membrane pollution. Then, energy consumption by MBR is higher than that of the traditional activated sludge process [7].

7.1.3 Classification of MBR

MBRs can be divided into aerobic membrane bioreactors and anaerobic membrane bioreactors based on whether the microorganisms need oxygen. The main goal of the aerobic membrane bioreactor is to remove organic carbon and reduce chemical oxygen demand (COD) in sewage. It is the most basic process type of MBR and also the most widely used. It is the most basic process type of MBR and also the most widely used process. It is mainly used in domestic sewage and reclaimed water. The anaerobic MBR process is a combination of membrane separation technology and traditional anaerobic process. Like traditional anaerobic processes, hydraulic retention time and sludge concentration are two important parameters of the anaerobic MBR process. However, unlike traditional anaerobic processes, due to the membrane retention, it can effectively keep the sludge concentration and anaerobic microorganisms, which can significantly improve the treatment effect of the anaerobic system. Therefore, endowing the anaerobic MBR with the complete mixing reactor, the process has many applications.

According to the combination of the membrane module and the reactor, MBR can be divided into built-in MBR and external MBR. The membrane modules used are generally hollow fiber (HF) membranes and flat sheet (FS) membranes.

The built-in (internal) MBR places the membrane module in the reactor and extracts the filtrate by negative pressure or siphon. Its advantages are that the membrane tank and the biochemical reaction tank adopt an integrated design, which does not require a mixed liquid circulation system. The energy consumption is minimal, and the footprint is less than that of the external type. However, the membrane flux of the built-in MBR is relatively low, and it is not easy to clean and replace after pollution. Therefore, an aeration device needs to be installed in the membrane module to use the shearing force generated by the bubbles during aeration and the scouring effect of the bubbles to delay and control membrane fouling. An external MBR separates the membrane module and the bioreactor. A circulating pump pressurizes the mixed solution before sending it to the membrane module's filter end. Under pressure, the clear liquid in the mixed solution passes through the membrane, while activated sludge and other solid suspensions are blocked and refluxed to the reactor with the concentrate. The main advantages of the external MBR system are their stable and reliable operation and the ease of removing, cleaning, and replacing membrane modules. The disadvantage is that the pollutants are easily deposited on the membrane surface, and the membrane cleaning cycle needs to be extended by using a large water flow. Moreover, the required energy consumption for mixed liquor recirculation is relatively high. The schematic of external and internal MBRs is reported in Figure 7.2.

In recent years, MBR technology has been developing. Configurations are broadly used in wastewater treatment in addition to the two aforementioned processes. A new MBR concept, air-lift side-stream MBRs, is also beginning to be used. The concept uses the side-flow air-lift principle and a robust and reliable side-flow configuration while combining all the benefits of a low-energy subsea system. This MBR has been applied to the treatment of sewage, landfill leachate, pharmaceutical wastewater, and municipal wastewater, as shown in Figure 7.3 [8].

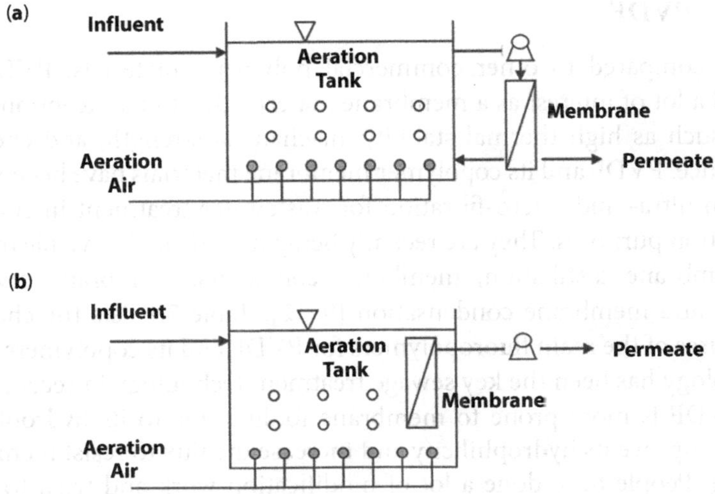

Figure 7.2 Schematic diagram of (a) external MBR and (b) internal MBR. https://www.researchgate.net/figure/Configurations-of-the-MBR-A-external-and-B-submerged_fig2_238737668.

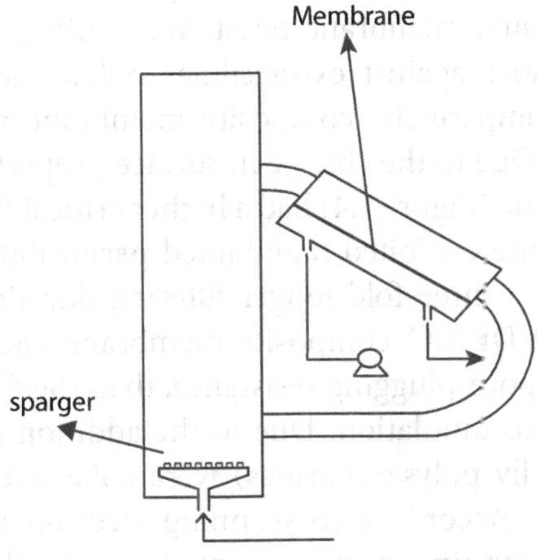

Figure 7.3 Schematic representation of the external loop airlift membrane bioreactor [3].

7.2 Membrane Materials for MBR

Since the concept of MBR was proposed, scholars have tried to use various materials to apply it. The most common ones are polyvinylidene fluoride (PVDF), polytetrafluoroethylene (PTFE), cellulose acetate (CA), polypropylene (PP), etc.

7.2.1 PVDF

When compared to other commercial polymeric materials, PVDF has gained a lot of interest as a membrane material due to its exceptional features such as high thermal stability, mechanical strength, and chemical resistance. PVDF and its copolymer membrane materials have been widely used in ultra- and micro-filtration for wastewater treatment in common separation purposes. They are recently being investigated as viable options in membrane distillation, membrane contactors, membrane crystallization, and membrane condensation [9–12]. Table 7.1 lists the chemical structures of the main fluoropolymers for PVDF and its copolymers. MBR technology has been the key sewage treatment technology in recent years, but PVDF is more prone to membrane fouling due to its hydrophobicity. To improve its hydrophilicity and increase the flux to resist membrane fouling. People have done a lot of modification work and tried to apply PVDF membrane to MBR.

Blending modification is one of the most important methods for PVDF membranes used in MBR. Zhao *et al.* [13] successfully used composite microfiltration membranes prepared by blending PVDF and hydrophilic graphene oxide (GO) nanosheets in submerged membrane bioreactor systems. The critical flux, membrane resistance, fouling behavior, and antifouling characteristics against extracellular polymeric substances (EPS) were studied to compare the composite membrane with a commercial PVDF membrane. Due to the change in surface properties, the PVDF/GO composite membrane (Figure 7.4) had a higher critical flux. The PVDF/GO composite membrane exhibited maintained permeability, reduced cleaning frequency, and a three-fold longer filtering duration than the PVDF membrane. The PVDF/GO composite membrane had lower membrane resistance, notably pore-plugging resistance, than the PVDF membrane in terms of anti-EPS accumulation. Due to the addition of GO nanosheets, fewer EPS (especially polysaccharides) were collected on the composite membrane surface. According to scanning electron microscope (SEM) and confocal laser scanning microscopy photographs, the cake cladding on the composite membrane was looser and thinner than that on the conventional PVDF membrane. As a result, the membrane surface was exposed, allowing for the long-term serviceability of high permeability [13].

In 2015, Lv *et al.* [14] synthesized graphene oxide–cellulose nanocrystal (GO-CNC) composite material. They used it as a new hydrophilic additive to be mixed into PVDF microporous membranes to improve the antifouling features of the obtained GO-CNC/PVDF membrane in the MBR. Compared to CNC/PVDF and previously reported GO/PVDF membranes,

Table 7.1 Chemical structures of PVDF and its copolymers.

Polymer		Chemical structure
PVDF homopolymer	PVDF	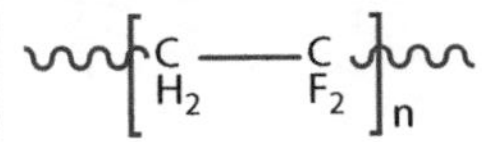
	P(VDF-co-TFE)	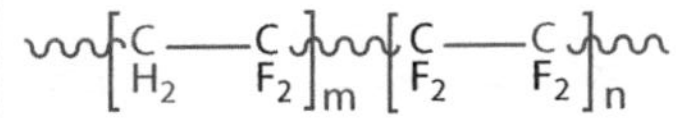
	P(VDF-co-HFP)	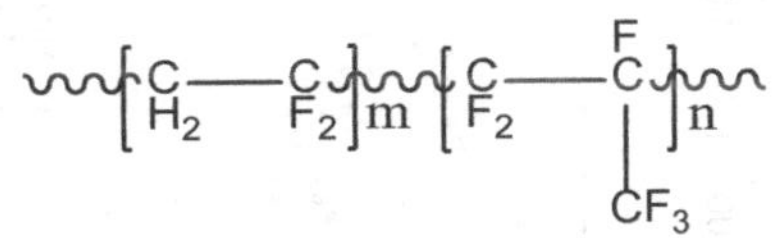
PVDF copolymer	P(VDF-co-CTFE)	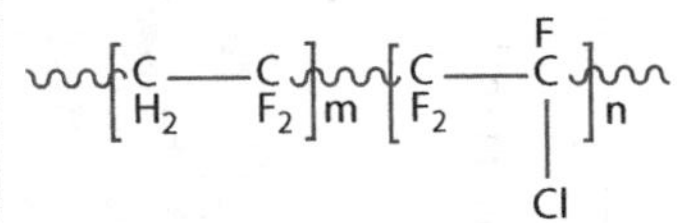
	PVDF-g-PSSA	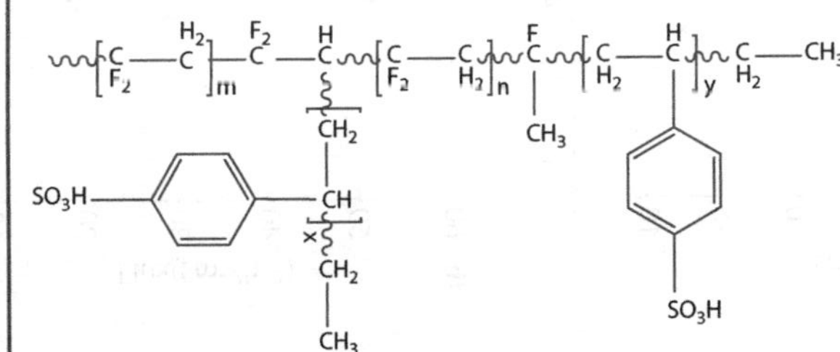
	P(VDF-TrFE)	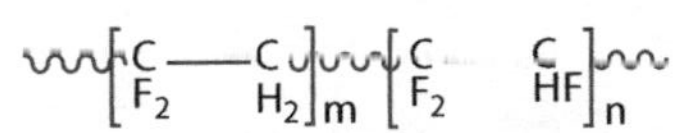

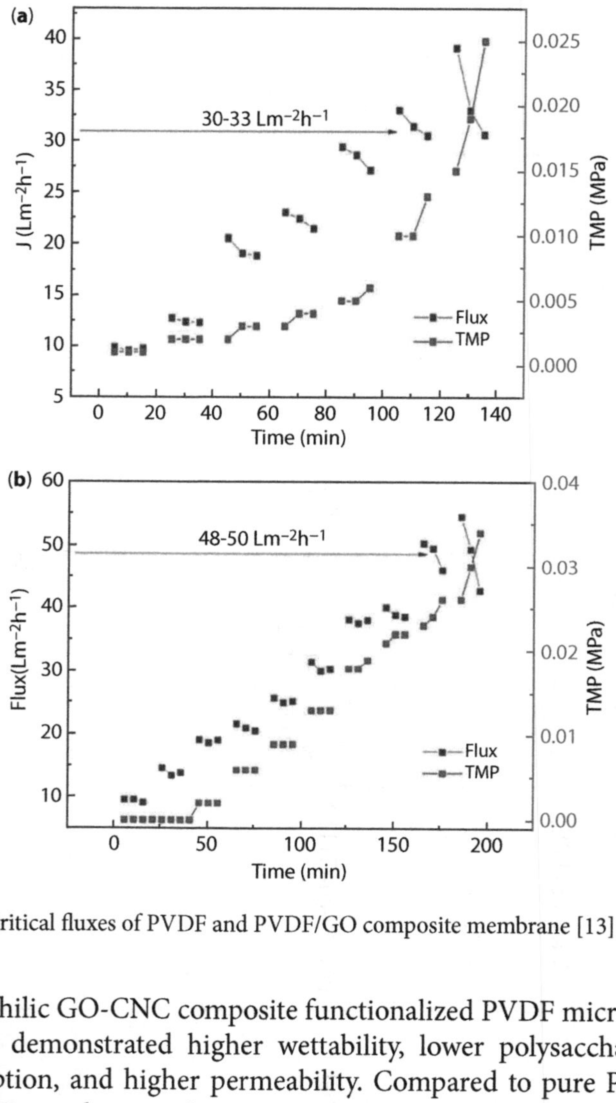

Figure 7.4 Critical fluxes of PVDF and PVDF/GO composite membrane [13].

the hydrophilic GO-CNC composite functionalized PVDF microfiltration membrane demonstrated higher wettability, lower polysaccharide/protein adsorption, and higher permeability. Compared to pure PVDF and CNC/PVDF membranes, the GO-CNC/PVDF membrane showed better antifouling properties with less EPS accumulation (Figure 7.5), lower irreversible fouling level, higher flux recovery ratio, and longer cleaning cycle during long-term operation in MBR. This was due to higher porosity and hydrophilicity, as well as more negative zeta potential. The use of GO-CNC composite functionalized membranes in MBR systems for water reclamation, and wastewater treatment is projected to grow as a result of their unique antifouling and hydrophilic properties [14].

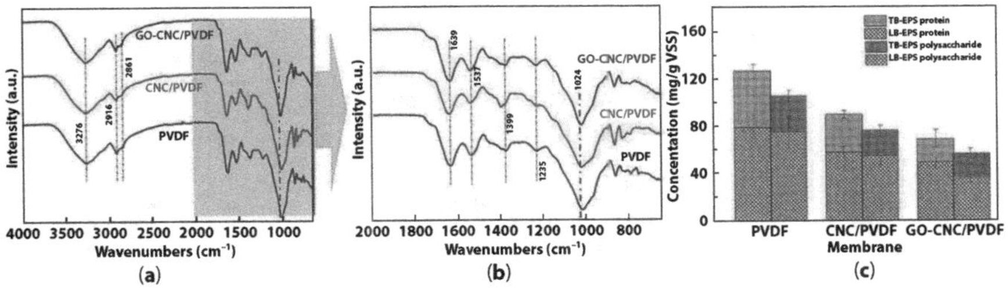

Figure 7.5 (a) The original ATR-FTIR, (b) the partially magnified ATR-FTIR, and (c) EPS results of the fouled membranes [14].

In addition to coating and blending, grafting methods have begun to attract more attention in academia. Asatekin *et al.* [15] coated commercial PVDF ultrafiltration (UF) membranes with the amphiphilic graft copolymer poly(vinylidene fluoride)-graft-poly(oxyethylene) methacrylate (PVDF-g-POEM) to produce thin-film composite (TFC) nanofiltration membranes. In all circumstances, the TFC nanofiltration (NF) membrane had much better fouling performance and effluent water quality than the pure PVDF UF membrane. Because of their antifouling properties and superior effluent quality, these NF membranes appear to be a promising potential substitute for UF membranes in MBRs with high concentrations of organic foulants [15].

In 2020, Wu *et al.* [16] assembled a polyvinylpyrrolidone (PVP)-graphene oxide (GO)/PVDF hollow fiber membrane with improved hydrophilicity and antifouling properties by chemically grafting PVP-GO nanocomposites (Figure 7.6). The PVP-GO/PVDF membrane was effectively used in an algae-membrane photo-bioreactor (MPBR) (Figure 7.7) with solids retention times (SRTs) of 30 days and hydraulic retention periods (HRTs) of 24 hours for high ammonia-nitrogen-contaminated wastewater treatment. The results showed that the PVP-GO/PVDF membrane had an excellent antifouling capability to harvest high-density microorganisms and effectively treat high ammonia-nitrogen-contaminated wastewater [16].

Although PVDF has high elongation, is not easy to break, and possesses good acid and alkali resistance, strong pollution resistance, and chemical cleaning resistance, membrane fouling is still a problem that is difficult to solve in MBR. Scholars have achieved higher critical flux, antifouling performance, and permeability through blending, grafting, coating, and combining MBR with other processes.

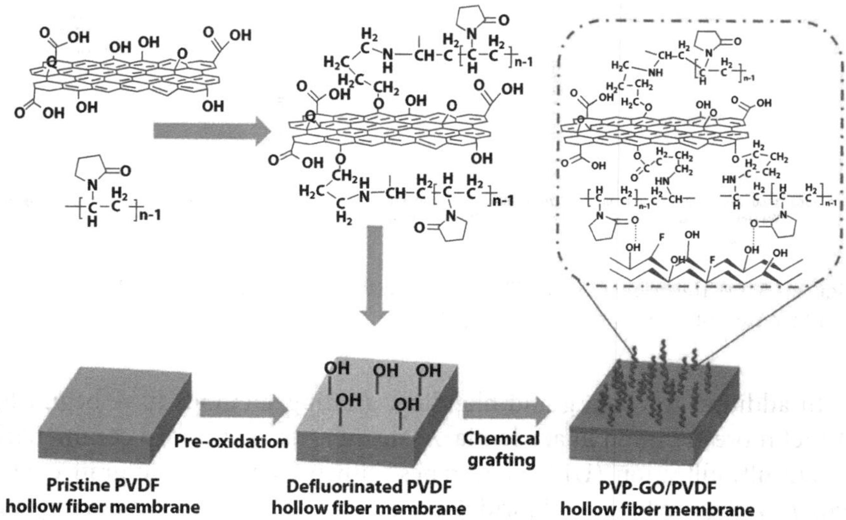

Figure 7.6 Schematic synthesis of PVP-GO nanocomposites and PVP-GO/PVDF hollow fiber membrane [16].

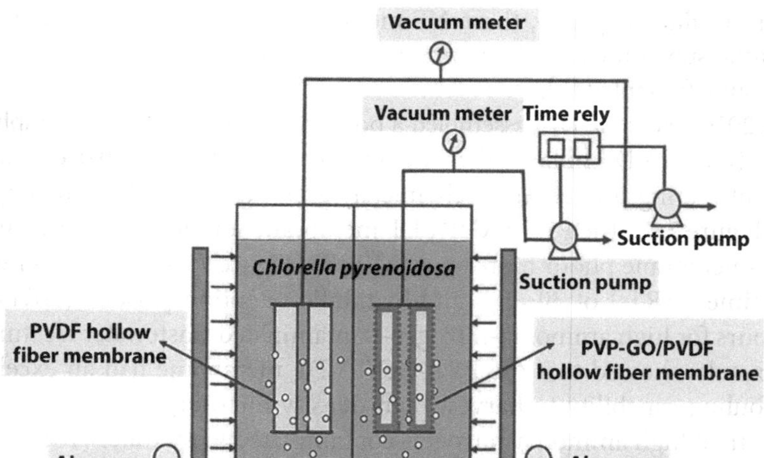

Figure 7.7 Experimental set-up of membrane photo-bioreactor (MPBR) system [16].

7.2.2 PP

Polypropylene (PP) is one of the most often utilized polymers for the fabrication of microporous membranes due to its outstanding chemical resistance, thermal stability, mechanical strength, and cheaper price [17].

Figure 7.8 shows the structure of PP. One of the biggest breakthroughs in PP membrane was using PP non-woven fabric modified membrane in an A/O-MBR (combination of anoxic and aerobic MBR compartments) process to analyze and treat polyester wastewater. Fortunately, the effluent quality of PP non-woven fabric modified membrane was better than that of commercially available PVDF membranes, but the membrane flux was limited [18]. In addition, there are studies on adding TiO_2 nanoparticles to PP membranes and embedding nano-diamonds in PP membranes, both of which improve the performance of PP membranes in MBR applications [19, 20].

The surface of a hydrophobic PP membrane was adjusted by ozone treatment followed by graft polymerization with 2-hydroxy-ethyl methacrylate (HEMA). By inserting functional groups such as carbonyl groups (C=O) and hydroxyl (-OH) groups into the membrane surface, the contact angle was lowered from 112 degrees for a pure PP membrane to 0 degrees for modified PP (MPP) membranes. The O/C ratio and membrane resistance of the MPP membrane rose as the degree of grafting (DG) increased. Membrane permeability was increased by using the MPP membrane in the crossflow operation of an anaerobic MBR [21].

Besides, different amounts of TiO_2 nanoparticles (0-1 wt.%) were added to the PP membrane. The fouling of the optimal nanocomposite membrane was significantly reduced, demonstrating great potential for its application in MBR wastewater treatment. Structural and operational analyses identified 0.75 wt% TiO_2 in the casting solution as the optimal composition. The performance and antifouling behavior of the optimized nanocomposite membrane were examined using a submerged MBR system to treat primary effluent obtained from the wastewater treatment unit of Tabriz Oil Refinery Co. (Iran). The results confirmed that the addition of TiO_2 nanoparticles improves the thermal, mechanical, and operational properties of the PP membrane [19].

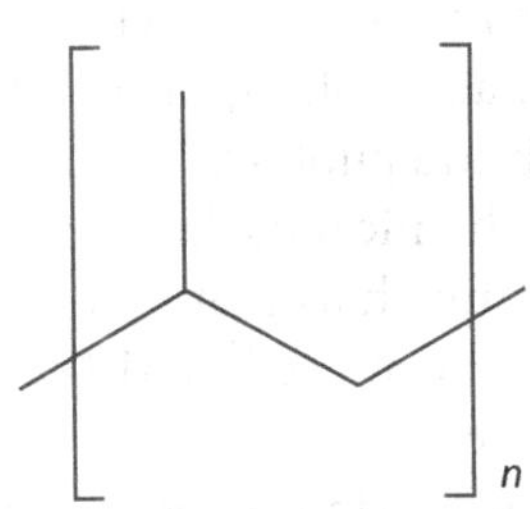

Figure 7.8 Chemical structure of PP.

The fouling behavior of the membrane surface and pores was investigated in MBR using hydrophobic and hydrophilic PP hollow fiber membranes (PPHFM). Compared to the hydrophilic membrane, the original membrane gathered substantially more protein, polysaccharides, and inorganic components. In addition, the pore size had an effect on particle distribution and accumulation, with smaller pore size membranes containing fewer pollutants and having a narrower size distribution. The transmembrane pressure (TMP) ranged narrowly between 38 and 53 KPa under a steady flow of 11.5 LMH. Furthermore, a moderately hydrophilic membrane (PP-g-AA) demonstrated reduced TMP (PP) compared to hydrophobic membranes. Indeed, the flux recovery was 30% greater than the PPHFM's original results. The results of this study add to our knowledge of membrane altering and fouling behavior in integrated MBRs [22].

7.2.3 PTFE

Polytetrafluoroethylene (PTFE; the chemical structure is shown in Table 7.2) is a perfluoropolymer material in which all hydrogen atoms in polyethylene (PE) are replaced with fluorine atoms. Its structural formula is: -[CF_2-CF_2]n-. The surface free energy of PTFE is very low, which means it hardly adheres to any substance. It also has excellent high-temperature resistance, chemical resistance, environmental resistance, electrical insulation, oxidation resistance, strong hydrophobicity, and high fracture toughness due to its strong C=C and C=F bonds and carbon skeleton, which is protected by a uniform spiral sheath formed by the electron cloud of fluorine atoms [23]. These characteristics make it suitable for a variety of applications, such as exhaust-gas treatment [24], membrane distillation (MD) [25], etc. The need for certain membrane qualities changes depending on the intended use.

Membranes made of porous PTFE are chemically inert and have been utilized in gas–solid separation, oil–water separation, and membrane distillation. Lee *et al.* [26] used a new flat membrane module based on PTFE to operate a laboratory and pilot-scale membrane system to monitor the effectiveness of chemical backwashing in membrane pollution control. It was confirmed through lab and pilot-scale experiments that using an FTFE flat sheet membrane with chemical backwashing in an MBR system could provide high flux, twice as much as that of a conventional system. This is the first time that PTFE has been used in MBR, which verifies the feasibility of PTFE used in MBR [26].

To lower energy demand during MBR operations, Miyoshi *et al.* [27] modified the PTFE hollow-fiber membrane element used in submerged

Table 7.2 Chemical structures of main fluoropolymers for PTFE membranes.

Polymer		**Chemical structure**
Poly(tetrafluoroethylene) homopolymer	PTFE	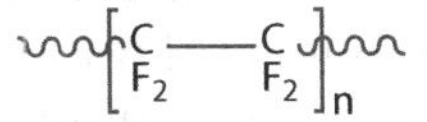
PTFE copolymer	Perfluorosulfonic acid (PFSA)	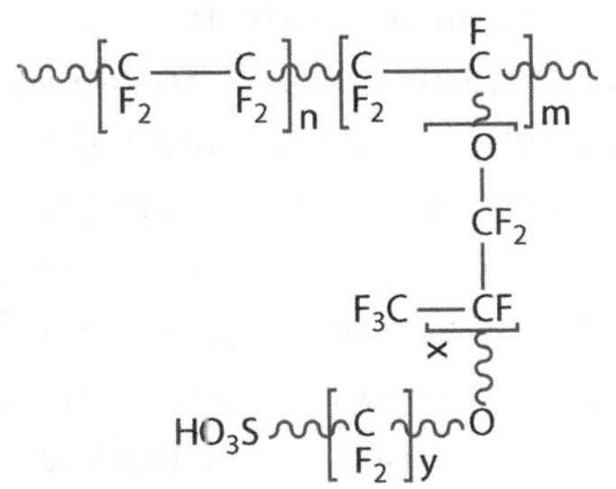
	Poly(tetrafluoroethylene-co-perfluoropropyl vinyl ether) (PFA)	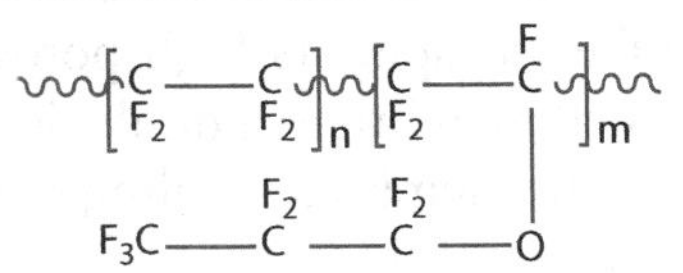
	Poly(tetrafluoroethylene-co-hexafluoropropylene) (FEP)	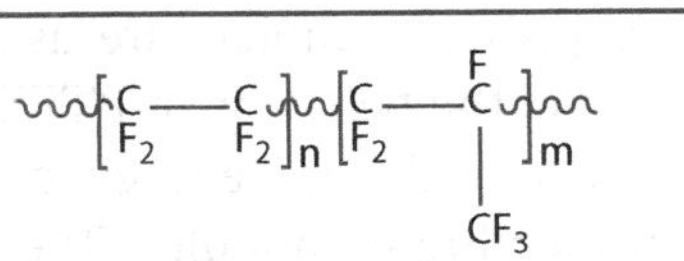
	Poly(ethylene-alt-tetrafluoroethylene) (ETFE)	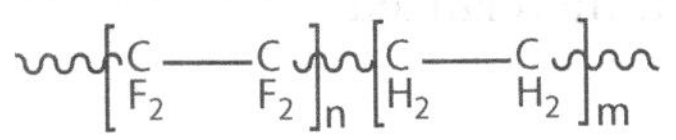

MBRs. Furthermore, by refining the membrane element arrangement, the packing density was enhanced by 20%. These changes increased the efficacy of membrane cleaning in conjunction with aeration. Operating MBRs with modified membrane components with a specific energy consumption of less than 0.4 $kWh{\cdot}m^{-3}$ is highly feasible [27].

In the same year, Li *et al.* [28] studied the performance of PTFE hollow-fiber anaerobic membrane bioreactor (AnMBR) in the treatment of organic pollutants, with varying salinity and PTFE membrane cleaning behavior. At a NaCl content of 35 $g{\cdot}L^{-1}$, the AnMBR was run for 226 days, with a total and biological COD removal effectiveness of 97.2% and 89.9%, respectively. According to the findings, the PTFE hollow-fiber AnMBR is viable for treating salt-contaminated organic wastewater [28].

With the development of technology, people began to study PTFE membranes used in MBR more intensively and in more detail. Nittami *et al.* [29]

investigated the operation of a lab-scale MBR incorporating immersed PTFE flat-sheet membrane modules with different contact angles (approximately 70 and 130 degrees) using inocula-activated sludge from the lab reactor and a municipal wastewater treatment plant (WWTP). According to the findings, the impacts of surface hydrophilicity on membrane fouling depend on the activated sludge's hydrophobicity. When the WWTP sludge was employed, the hydrophobic membrane had a substantially greater cake cladding resistance than the hydrophilic membrane. The foulants may be deterred from adhering to membrane surfaces if a dense cake layer is formed on hydrophobic PTFE membranes [29].

Moreover, to decrease membrane fouling in submerged MBR, Miyoshi *et al.* designed an intensively aerated membrane module (Figure 7.9). This module allows coarse bubbles to be injected at a higher density near the membrane fibers made of PTFE, which has a high tensile strength (90–110 N/fiber) despite its high porosity (about 80%). Applying the intensively aerated membrane module developed in this study to pilot-scale MBRs, a very low membrane fouling rate was achieved during the continuous MBR operation [30].

At present, scholars are using PTFE as a potential MBR material and studying the behavior of PTFE in MBR from the perspectives of improving flux, reducing energy consumption, and developing strong aeration membrane modules. There is still a long way to go before it is truly commercialized.

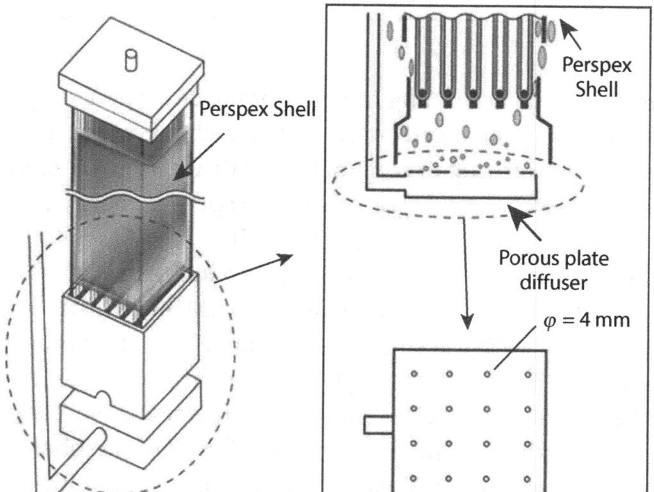

Figure 7.9 Schematic of the modified module [30].

7.2.4 CA

Cellulose acetate (CA) is the most stable membrane-forming material among all cellulosic materials, with high chlorine resistance. CA is non-toxic, odorless, and stable to light. It is a white, loose, sheet-like hydrophilic membrane material. Its chemical structure is shown in Figure 7.10 below. As seen in Figure 7.10, there is an acetyl group in the CA structure. Due to the presence of the acetyl group, the intermolecular hydrogen bond force is weakened, resulting in an increase in the intermolecular distance so that CA has good membrane formation and processability. Furthermore, CA has a wide range of sources and low prices. It is also an environmentally friendly membrane material. Therefore, CA has become one of the most widely used membrane materials [31, 32].

Around 20 years ago, people started to investigate the feasibility of CA membranes in MBR. Choi *et al.* [33] used an MBR with a CA-NF membrane to treat synthetic wastewater. The studies yielded enough water productivity to last 60 days without fouling or membrane cleaning. This increased the CA membrane's applicability to the MBR system. Since the excellent hydrophilicity of CA, it showed well antifouling performance in MBR for wastewater treatment [33].

In 2007, to test the efficacy of the NF membrane in residential wastewater treatment, a submerged NF MBR employing CA membranes was run for 240 days [34]. The results of molecular weight distribution and fractionation of dissolved organic matter in the waters showed a reduction in membrane hydrophobicity and an increase in membrane pore size with time. According to the findings, the degradation of each water-quality parameter over time could be attributed to the NF membrane's qualitative deterioration caused by CA hydrolysis, which results in an increase in pore size and porosity, as well as a decrease in surface charge and hydrophobicity of CA membranes [34].

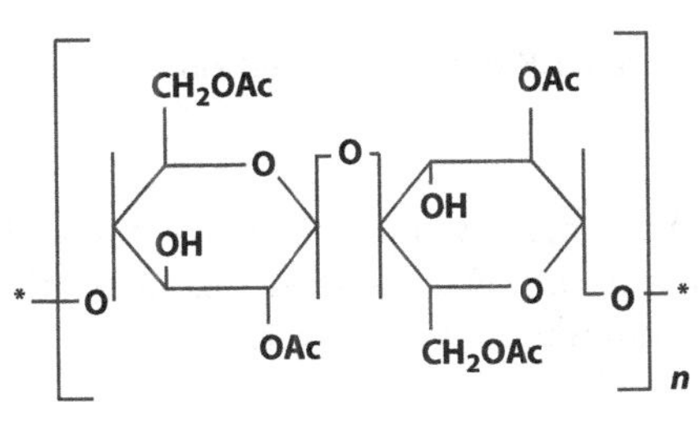

Figure 7.10 Chemical structure of CA.

The effects of membrane modification and aeration rate on the antifouling performance of CA membranes in MBR systems were investigated by Etemadi *et al.* [35] Zinc oxide (ZnO) nanoparticles were inserted in CA membrane at a concentration of 0–0.75 wt.% for membrane modification. At 0.5 wt.% injection of ZnO nanoparticles, the highest improvement in the hydrophilic nature and porosity of the CA membrane was found. Furthermore, the permeability and antifouling properties of the CA membrane were negatively affected by either a low or high aeration rate. The effect of membrane modification on the improvement of antifouling properties of the CA membrane was found to be greater than the optimal aeration rate effect [35]. CA has not been used on a large scale or pilot scale in the MBR due to its limited pH operating range (4–6) and low mechanical strength. There is a long way to go for its wide usage for water treatment, including the MBR process.

Besides PVDF, PP, PTFE, and CA, other materials are also used for the MBR process. Polyethersulfone (PES) ultrafiltration membranes fixed with Al_2O_3 have certain potential in MBR applications [36]. Adding Al_2O_3 to the PVC membrane to prepare a low-cost antifouling membrane provides ideas for developing MBR membrane materials [37]. Nanofiber polyacrylonitrile (PAN) embedded with aluminum fumarate nanoparticles in MBR can provide high flux and possess high antifouling performance [38]. With the rapid development of new materials, biotechnology, nanotechnology, and other fields, some new materials with excellent performance have been extended to the field of separation membranes. High-performance membrane materials are bound to be used in the MBR process in the future.

7.3 Commercial-Scale MBR

7.3.1 MBR Commercial Development Process

Table 7.3 lists the MBR commercial development process.

Table 7.3 Main events in MBR commercialization.

Time	Event
Late 1960s	• Dorr-Oliver has developed the first commercial MBR and applied it to ship-board sewage treatment. The system uses flat-plate ultrafiltration membranes.

(Continued)

Table 7.3 Main events in MBR commercialization. (*Continued*)

Time	Event
Early 1970s	• The Thetford system enables the commercial application of the external multi-tube Cycle-Let process in water reuse in the United States.
Early 1980s	• Commercial application of FechSep (later known as Novasep Orelis) split FS Pleiade in water reuse in Japan.
Mid 1980s	• Nitto-Denko applied for a Japanese patent on the immersed flat MBR. • The University of Tokyo has experimented with submerged hollow fiber MBR.
Early and mid-1990s	• Commercial application of immersed flat MBR in Kulbota, Japan. • Commercial application of Weir Envig's "Membralox" membrane split ADUF unit. • In North America and Europe, Zenon's vertical immersion HF "ZeeWeed" technology is commercialized. • Commercial application of the Wehrle external multi-tube Biomembrat system acquired by Thetfore in 1993. • Mitsubishi Rayon developed an immersed HF "Sterapore™" membrane MBR—horizontal device.
Early 2000s	• Commercial application of USF vertical immersion HF "MemJet®" system. • Commercial application of Htuber rotary FS MBR. • Norit X-Flow has developed a split ventilation multi-pipe system. • Puron (vertical immersion HF membrane) has been commercialized after being acquired by Koch. • Kolon and Pata (Korea) introduced a vertical immersion HF MBR. • Toray developed FS MBRs. • Mitsubishi Rayon developed a vertical immersion type HF MBR. • Asahi Kasei developed the vertical immersion type HF MBR.
Nearly 10 years	• GE commercialized LEAPmbr in 2011 and AnMBR in 2014; GE developed an MBR with a membrane accommodating carrier (MACarrier) in 2012, specifically for the treatment of severe water. • Origin Water has developed a V-MBR, which has made a huge revolutionary breakthrough in the direction of energy-saving and consumption reduction, and water quality improvement. • Kubota invented a new SP-A series immersion membrane device that is more energy-efficient.

7.3.2 Commercial MBR Technology

According to the different membrane geometries, the commercial MBR wastewater treatment technologies currently used in wastewater treatment can be divided into three types: flat sheet, hollow fiber, and multi-tube (MT).

7.3.2.1 Immersion FS Technology

7.3.2.1.1 Kubota (https://www.kubota.com/)

In the late 1980s, Kubota Company of Japan developed Kubota membrane modules. In 1990, the first Kubota pilot plant was completed and soon entered the commercial application stage. At present, more than 2,200 Kubota MBRs have been built globally, and about 10% of them are located in Europe.

The earliest FS microfiltration membrane model number is 510, consisting of a 0.5m × 1m membrane plate with a thickness of 6 mm and an effective membrane area of 0.8 m^2. The membrane material is hydrophilic chlorinated polyethylene.

In 2002, the introduction of the double-layer EK design greatly improved the membrane filtration efficiency. In this design, the effective membrane area per unit area is doubled, the cost of the membrane module is greatly

Figure 7.11 Kubota FS type submerged membrane unit. (https://www.kubota.com/products/solutions/lineup/pdf/Brochure.pdf).

reduced, and the operating cost is also reduced. Figure 7.11 shows the Kubota FS type submerged membrane unit.

7.3.2.1.2 Huber Technology (https://www.huber.cn.com/cn.html)

The Huber vacuum rotating membrane product of Huber Technology is different from other MBR systems in that the membrane module is movable, and the module rotates at a frequency of 1–2 r/min. The shearing force generated by the rotation and the scouring action of the gas generated by the central large bubble aeration on the membrane surface can prevent any part of the module from accumulating sludge, so there is no need to clean.

7.3.2.1.3 Industrial Technology Research Institute-MBR Based on Non-Woven Fabrics (https://www.itri.org.tw/english/index.aspx)

Non-woven MBR is a recent research result and has not yet been commercialized, but it has reached a certain technological scale based on a low-rejection FS membrane module made of PP non-woven fabric or non-woven felt fabric. The Taiwan Industrial Technology Research Institute took the lead in conducting the pilot-scale NWFMBR test. Compared to the MF/UF membrane module, the non-woven membrane material is very low cost and very durable. However, the membrane pore size is large so the membrane will cause serious internal pollution of the membrane pores caused by organic matter in just a few days, resulting in a significant decrease in water permeability. When there is a reasonable solution to such internal pollution problems, non-woven substrate-based MBRs may appear in the future. Figure 7.12 shows the Huber vacuum rotating membrane product.

Figure 7.12 Huber vacuum rotating membrane product. (https://www.huber.cn.com/solutions/decentralized-and-semi-centralized-wastewater-treatment/lagoon-upgrading-effluent-reuse/systems-concept-huber-pondplusr.html).

7.3.2.1.4 Toray Industries (https://www.toray.cn/)

Toray Industrial Co. Ltd. is a membrane manufacturer in Japan with a history of 25 years, mainly researching RO membranes for effluent production. In 2004, the company launched its own FS MBR MF membrane product; the membrane material is PVDF, the rated pore size is 0.08 μm, and the standard deviation is 0.03 μm. The size of the FS MBR MF membrane element is 565mm × 1608 mm, the membrane area is 1.4 m^2, and the thickness is 13.5 mm. As of December 2005, there were a total of 25 such devices in operation all over the world, with a capacity of less than 500 m^3/d, excluding three with a larger processing capacity.

7.3.2.2 Immersion HF Technology

7.3.2.2.1 Zenon Environment (https://www.copadata.com/en/product/zenon-software-platform-for-industrial-automation-energy-automation/)

Zenon Environmental is currently the world's largest MBR process technology company, with projects in 45 countries and 1,470 employees. In 1989, the company developed the first HF membrane module, the Mousic membrane module. Following the Moustic test, from 1991 to 1992, Zenon developed a reinforced HF membrane with a membrane area of 145 ft^2, so it was named ZeeWeed 145. During use, it was found that the fiber bundles surrounding the PVC frame easily caused the accumulation of solid matter, and the top of the membrane bundle was easily worn. The ZW series products designed afterward have largely solved these problems. Since

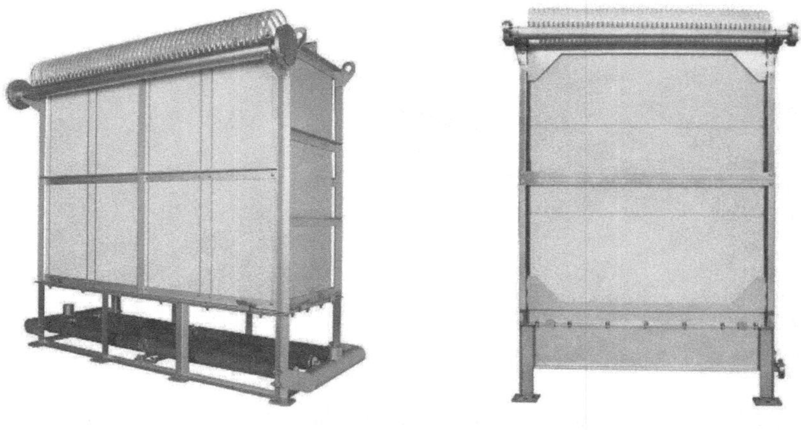

Figure 7.13 Toray submerged flat-sheet membrane bioreactor: TMR090 series(left), TMR140 series (right). (https://www.toray.cn/products/membranes/mem 0030.html).

1995, ZeeWeed membrane modules have been used for drinking water and industrial effluent treatment. Based on the early experience of using continuous aeration, from 1998 to 1999, the aeration device was improved, that is, a two-stage device was used for flushing to prevent blockage of the aeration device.

In 1997, Zenon introduced ZeeWeed 500a for sewage treatment and ZeeWeed 500b for drinking water treatment. Compared with Zeeweed 150, the membrane area of the new membrane module unit was greatly increased. The membrane area of the ZeeWeed 500a module for wastewater treatment is $500ft^2$, and the membrane area of ZeeWeed 500b for drinking water treatment is 60.4 m^2. The selectivity of the membrane material is also higher, and the nominal pore size is reduced to 0.04μm.

According to the actual operation data available, the average service life of the ZeedWeed MBR membrane is 8 years. For more than 20 years, the ZeedWeed MBR membrane has been constantly optimized, with new products launched every 3–4 years. However, the membrane material and size have never changed. These ZeedWeed MBR membrane products focus on improving membrane performance, improving membrane scrubbing, and reducing energy consumption. At present, the ZeedWeed MBR membrane products on the market are composed of ZW500d membrane assembly and 316L stainless steel membrane box (including pipes and membrane scrubbing device). Figure 7.13 shows Toray submerged flat-sheet membrane bioreactor.

7.3.2.2.2 Mitsubishi Rayon Engineering (https://www.mitsubishi.com/en/)

Mitsubishi Rayon Engineering is the world's third-largest MBR membrane manufacturer, with an installed capacity second only to Zenon and Kubota. They first introduced an MBR membrane product in 1995, and as of March 2005, the company had over 1500 of its systems installed worldwide, over 900 of these being in Japan. Its commercial profile is thus similar to that of Kubota.

7.3.2.2.3 Memcor (https://www.dupont.com/brands/memcor.html)

Memcor has been producing membrane products since 1982. It is a well-known membrane manufacturer and currently belongs to the Siemens Group. In 2002, Memcor introduced the MemJet MBR system. The system consists of vertically installed membrane modules based on PVDF HF membranes with a pore size of 0.04 μm. Bubbles entering the mixed liquid are introduced into the membrane module by a two-phase mixing system, which generates strong shear conditions, thereby minimizing the concentration polarization of the solids on the membrane surface at the ejection port.

7.3.2.2.4 Coriolis Membrane System—PURON (https://www.kochseparation.com/)

At the end of 2001, PURON was established as a subsidiary of the University of Aachen and was later acquired by Koch in 2004. Its membrane element type is a submerged HF, which is different from most other submerged HF MBR membrane modules. Its membrane material is PES. The unique feature of the PURON system is that the membrane fiber is only fixed at the bottom, while the top of the membrane filament is sealed separately. The flushing airflow is intermittently injected from the middle of the membrane filaments through the central jet pipe at the bottom of the module to reduce the blockage at the bottom of the module. The free movement of the top membrane filaments can make the coarse solids flow away, avoiding the blockage in this area. The lateral movement of the membrane filament makes it bear a certain mechanical pressure, so the inside of the membrane fiber is reinforced with fabric.

7.3.2.2.5 Asahi Kasei Chemical Company (https://www.asahi-kasei.com/)

Asahi Kasei Chemical Company is a subsidiary of Japan's Asahi Kasei Group. Since the 1970s, it has been manufacturing UF/MF hollow fiber membranes and membrane modules for various industrial purposes. In 1998, it developed Microza PVDF MF membranes for sewage treatment. Currently, the company produces 1 million square meters of MF membranes each year, mainly for drinking water treatment. Microza membrane modules are used by pallet as a standard pressurized membrane module in its industrial and urban sewage treatment systems. As of May 2005, there are more than 50 membrane filtration devices in the United States alone, providing approximately 500,000 m^3/d of drinking water. Microza membrane fibers with an outer diameter of 1.3 mm are characterized by a small pore size distribution range (rated pore size of 0.1 μm) and high resistance to oxidants.

In 1999, the company began to develop MBR components and put them on the market in June 2004. The membrane stent is composed of a series of vertical membrane elements mounted on a steel frame. Each element has a membrane area of 25 m^2, a diameter of 150 mm, and a length of 2 m. The two ends of the membrane fiber are sealed and gathered into a bundle on the top of the membrane element to allow the suspension to flow away. Using an independent aeration device, the gas enters through the diffusion plate installed at the bottom of the membrane element.

7.3.2.2.6 GE (https://www.suezwatertechnologies.com/)

As the global leader in MBR, acquired Zenon in 2006 [39]. GE has a ZeeWeed MBR water treatment capacity of over 5 million m^3/d. In 2011, GE introduced the next generation of ZeeWeed MBR technology, LEAPmbr (Low Energy Advanced Performance), based on customer challenges in water yield, footprint, simplicity, and energy efficiency. The new technology has the following characteristics: 1. High efficiency: the new generation of ZeeWeed membrane can effectively improve the water yield by 15%. 2. Low compact footprint: 20% less MBR footprint, lower construction costs. 3. Energy-saving: saves around 30% energy consumption and reduces operating costs.

In 2014, GE launched a new MBR with a membrane accommodating carrier (MACarrier), designed to help industrial companies meet stringent drainage requirements, achieve higher water reuse, and cost savings, and improve operating efficiency. This system combines a highly efficient MACarrier with GE's ZeeWeed 500D membranes. This innovation provides a system for treating wastewater with high COD, toxicity, biochemical oxygen demand (BOD), phenols, and other pollutants. The combined system draws together three technologies: the MA Carrier, which enhances the removal of COD, toxicity, phenols, and other contaminants found in wastewater; the bioreactor, which removes suspended colloidal solids and reduces organic content and concentrations of nutrients; and membrane

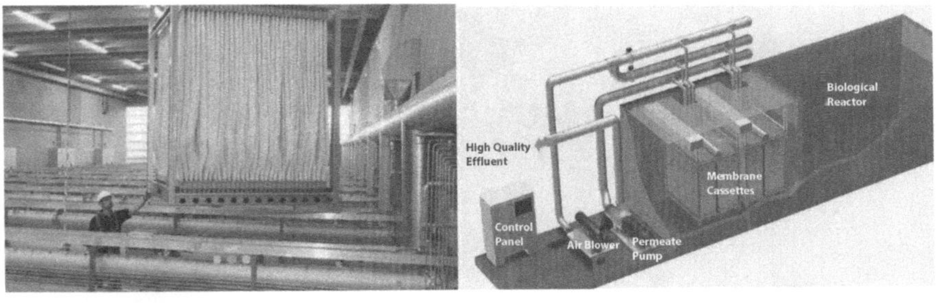

Figure 7.14 SUEZ's submerged ZeeWeed 500D membrane unit. (https://www.suezwatertechnologies.cn/products/ultrafiltration/zeeweed-500-membrane).

filtration as a physical membrane barrier to remove microorganisms found in wastewater. Figure 7.14 shows SUEZ's submerged ZeeWeed 500D membrane unit.

7.3.2.2.7 Kubota (https://www.kubota.com/)

In 2020, Kubota invented the new SP-A series immersion membrane device to make the operation of the MBR plant more energy-efficient, with more membrane area per floor area. The KUBOTA SP series (Figure 7.15) comprises SMU models optimized especially for medium to large-scale sewage treatment applications. Forty at membrane plates having 10 m^2 of membrane area and permeate collection chambers are integrated into a compact "Membrane Module". This design improves packing density and reduces scour air requirements. Multiple Membrane Modules are

Figure 7.15 Kubota SP-A series immersion membrane device. (https://www.kubota-mbr.com/en/Noticias).

assembled into a Membrane Block using simple coupling connections. The coupling connection also serves as a conduit to the permeate header. This structure simplifies the assembling procedure of the SMU during eld maintenance work. The SP series reduces required space, scour air, and assembly time during maintenance work based on its unique structure.

7.3.2.2.8 Origin Water (http://www.originwater.com/)

In 2021, Origin Water developed V-MBR, boldly innovating in terms of membrane material quality, membrane system operation, and automatic control. The use of mechanical vibration instead of membrane tank blast aeration has achieved a huge revolutionary breakthrough in the direction of energy saving, consumption reduction, and improvement of water quality. Moreover, it has the characteristics of a good denitrification effect and strong anti-pollution properties of the membrane, which has been demonstrated at the Beijing Doudian Water Reclamation Plant (Figure 7.16 and Figure 7.17).

7.3.2.3 External MBR Technology

Norit X-Flow is a major supplier of multi-tubular and capillary (CT) membrane modules. CT membrane modules are currently mainly used for

Figure 7.16 Origin water's V-MBR unit. (http://www.originwater.com/cpyjs/MF/wsclyzyhcp/6800.html).

Figure 7.17 Origin water V-MBR, launched in Beijing Doudian water reclamation plant. (http://www.originwater.com/xwzx/mtjj/6824.html).

drinking water filtration. Currently, among Norit X-Flow's products, only MT membrane modules are used for MBR. The most mature technology is based on horizontally installed pump-lifting cross-low membrane modules. Some recently installed facilities use vertically installed membrane modules and a combination of gas lift and liquid pumps (used at low flow rates). Many designers use this product in their MBR process. Most MBR facilities based on Norit X-Flow membrane modules are used for industrial wastewater treatment, but some are used for small-scale urban and household sewage treatment. Although the system has high specific energy consumption, it has good stability, simple operation, and easy control.

7.3.2.4 MABR Technology

The full English name of MABR is Membrane Aerated Biofilm Reactor. Its working principle lies in the synergy between the selective breathable membrane and the attached growth biofilm. The breathable membrane is used to transmit oxygen to the biofilm attached to the surface of the breathable membrane. At the same time, substrates such as ammonia and organic matter diffuse from the sewage to the biofilm.

Air enters the MABR membrane box, allowing oxygen to pass through the MABR membrane from the inside to the outside, and the ammonia nitrogen and organic matter in the biochemical pool are transferred to the MABR membrane. Since ammonia nitrogen has a faster mass transfer rate in the water phase, it will preferentially contact oxygen and form a biofilm dominated by nitrifying bacteria on the surface of the MABR membrane.

Compared with traditional biofilm technology, the mass transfer direction of MABR oxygen, ammonia nitrogen, and organic matter are countercurrent, which can effectively avoid the competition of nitrifying autotrophic bacteria and organic matter removal heterotrophic bacteria by oxygen. The large bubble pulse scrubbing of the exhaust gas makes the control of the thickness of the biofilm on the MABR surface simple and feasible. Therefore, MABR is a new biofilm technology more suitable for simultaneous nitrification and denitrification reactions. Figure 7.18 presents the workflow diagram of MABR.

The aeration power efficiency of MABR can reach 10 kg O_2/kWh and even 14 kg O_2/kWh under suitable conditions. This value is 3–4 times that of the traditional aeration type, and it can save up to 75% of aeration energy consumption. The traditional activated sludge method can reduce the amount of sludge by about 50%, lowering the sludge treatment cost, and decreasing the operating energy consumption of sludge return. MABR systems also have a small area, are easy to assemble, and have a large component packing density. They offer significant advantages in project upgrades, high-concentration organic wastewater treatment, distributed treatment, and other fields.

At present, the global companies with absolute leading technology in the MABR field and their related products are mainly SUEZ's Zeelung, Dupont's Oxymem, and Fluence's SUBRE, as shown in Figure 7.19.

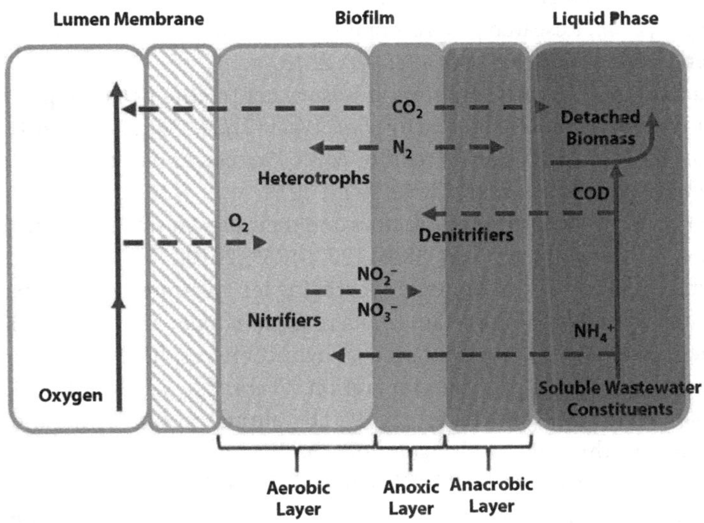

Figure 7.18 Schematic diagram of MABR.

Zeelung MABR **Oxymem** **SUBRE**

Figure 7.19 MABR mainstream products worldwide at present. Zeelung MABR: https://www.ge.com/ OxyMem: https://www.dupont.com/ SUBRE: http://www.fluencecorp.cn/.

References

1. Wang, Z.Z., *et al.*, Optimisation and performance of NaClO-assisted maintenance cleaning for fouling control in membrane bioreactors. *Water Res.*, 53, 1–11, 2014.
2. Meng, F.G., *et al.*, Recent advances in membrane bioreactors (MBRs): Membrane fouling and membrane material. *Water Res.*, 43, 6, 1489–1512, 2009.
3. Le-Clech, P., Chen, V., Fane, T.A.G., Fouling in membrane bioreactors used in wastewater treatment. *J. Membr. Sci.*, 284, 1-2, 17–53, 2006.
4. Liu, H., *et al.*, Research progress of membrane bioreactor in wastewater treatment. *Mod. Chem. Ind.*, 35, 4, 25–28, 2015.
5. Liu, H., *et al.*, Research progress in wastewater treatment techniques by using composite membrane bioreactor. *Ind. Water Treat.*, 36, 1, 7–11, 2016.
6. Hao, X., *et al.*, Status and Trend of MBR Process Application in the World. *China Water Wastewater*, 34, 20, 7–12, 2018.
7. Le-Clech, P., Membrane bioreactors and their uses in wastewater treatments. *Appl. Microbiol. Biotechnol.*, 88, 6, 1253–1260, 2010.
8. Shariati, F.P., *et al.*, Membrane bioreactor for treatment of pharmaceutical wastewater containing acetaminophen. *Desalination*, 250, 2, 798–800, 2010.
9. Gao, J.F., *et al.*, Facile preparation of hybrid microspheres for super-hydrophobic coating and oil-water separation. *Chem. Eng. J.*, 326, 443–453, 2017.
10. Miao, R., *et al.*, Understanding PVDF ultrafiltration membrane fouling behaviour through model solutions and secondary wastewater effluent. *Desalin. Water Treat.*, 52, 25–27, 5061–5067, 2014.
11. Khaisri, S., *et al.*, Comparing membrane resistance and absorption performance of three different membranes in a gas absorption membrane contactor. *Sep. Purif. Technol.*, 65, 3, 290–297, 2009.

12. Mansourizadeh, A. and Ismail, A.F., Hollow fiber gas-liquid membrane contactors for acid gas capture: A review. *J. Hazard. Mater.*, 171, 1-3, 38–53, 2009.
13. Zhao, C., *et al.*, Highly effective antifouling performance of PVDF/graphene oxide composite membrane in membrane bioreactor (MBR) system. *Desalination*, 340, 59–66, 2014.
14. Lv, J., *et al.*, Graphene oxide-cellulose nanocrystal (GO-CNC) composite functionalized PVDF membrane with improved antifouling performance in MBR: Behavior and mechanism. *Chem. Eng. J.*, 352, 765–773, 2018.
15. Asatekin, A., *et al.*, Antifouling nanofiltration membranes for membrane bioreactors from self-assembling graft copolymers. *J. Membr. Sci.*, 285, 1-2, 81–89, 2006.
16. Wu, W., *et al.*, Enhanced MPBR with polyvinylpyrrolidone-graphene oxide/PVDF hollow fiber membrane for efficient ammonia nitrogen wastewater treatment and high-density Chlorella cultivation. *Chem. Eng. J.*, 379, 122368, 2020.
17. Liang, Q.U., *et al.*, Preparation and Characterization of Polypropylene Membranes Modified with Cibacron Blue F3GA by Radiation-induced Polymerization of Glycidyl Methacrylate. *J. Chem. Eng. Chin. Univ.*, 20, 4, 538–543, 2006.
18. Fang, R., *et al.*, Non-woven modified MBR treating polyester wastewater. *Chin. J. Environ. Eng.*, 4, 9, 2005–2008, 2010.
19. Fonouni, M., *et al.*, Fouling characterization of TiO2 nanoparticle embedded polypropylene membrane in oil refinery wastewater treatment using membrane bioreactor (MBR). *Desalin. Water Treat.*, 90, 99–109, 2017.
20. Hosseinpour, S., Azimian-Kivi, M., Jafarzadeh, Y., Yegani, R., Pharmaceutical wastewater treatment using polypropylene membranes incorporated with carboxylated and PEG-grafted nanodiamond in membrane bioreactor (MBR). *Water Environ. J.*, 35, 4, 1249–1259, 2021.
21. Sainbayar, A., *et al.*, Application of surface modified polypropylene membranes to an anaerobic membrane bioreactor. *Environ. Technol.*, 22, 9, 1035–1042, 2001.
22. Guo, Y.F., Sun, P.C., Wei, J.F., New insight into the fouling behavior of hydrophobic and hydrophilic polypropylene membranes in integrated membrane bioreactors. *Environ. Technol.*, 39, 24, 3159–3168, 2018.
23. Usman, H.S., Touati, K., Rahaman, M.S., An economic evaluation of renewable energy-powered membrane distillation for desalination of brackish water. *Renew. Energy*, 169, 1294–1304, 2021.
24. Chong, M.N., *et al.*, Recent developments in photocatalytic water treatment technology: A review. *Water Res.*, 44, 10, 2997–3027, 2010.
25. Wang, T., *et al.*, Development of the separation of small molecular organic/inorganic salt mixtures by nanofiltration membrane:A review. *Chem. Ind. Eng. Prog.*, 31, 10, 2144–2151, 2012.

26. Lee, E.J., *et al.*, A study on the high-flux MBR system using PTFE flat sheet membranes with chemical backwashing. *Desalination*, 306, 35–40, 2012.
27. Miyoshi, T., *et al.*, Energy reduction of a submerged membrane bioreactor using a polytetrafluoroethylene (PTFE) hollow-fiber membrane. *Front. Environ. Sci. Eng.*, 12, 3, 1–10, 2018.
28. Li, J., *et al.*, Polytetrafluoroethylene (PTFE) hollow fiber AnMBR performance in the treatment of organic wastewater with varying salinity and membrane cleaning behavior. *Bioresour. Technol.*, 267, 363–370, 2018.
29. Nittami, T., *et al.*, Influence of surface hydrophilicity on polytetrafluoroethylene flat sheet membrane fouling in a submerged membrane bioreactor using two activated sludges with different characteristics. *J. Membr. Sci.*, 463, 183–189, 2014.
30. Miyoshi, T., *et al.*, Effect of intensive membrane aeration and membrane flux on membrane fouling in submerged membrane bioreactors: Reducing specific air demand per permeate (SAD(p)). *Sep. Purif. Technol.*, 148, 1–9, 2015.
31. Fan, Z.W., *et al.*, Structure design and performance study on braid-reinforced cellulose acetate hollow fiber membranes. *J. Membr. Sci.*, 486, 248–256, 2015.
32. Gouvea, D.M., *et al.*, Acetate cellulose film with bacteriophages for potential antimicrobial use in food packaging. *LWT Food Sci. Technol.*, 63, 1, 85–91, 2015.
33. Choi, J.H., *et al.*, A novel application of a submerged nanofiltration membrane bioreactor (NF MBR) for wastewater treatment. *Desalination*, 146, 1-3, 413–420, 2002.
34. Choi, J.-H., Fukushi, K., Yamamoto, K., A submerged nanofiltration membrane bioreactor for domestic wastewater treatment: the performance of cellulose acetate nanofiltration membranes for long-term operation. *Sep. Purif. Technol.*, 52, 3, 470–477, 2007.
35. Etemadi, H., Motamed, M., Yegani, R., Comparative study between aeration rate and membrane modification effects on antifouling properties of cellulose acetate membrane in-membrane bioreactor systems. *Desalin. Water Treat.*, 171, 67–77, 2019.
36. Maximous, N., *et al.*, Preparation, characterization and performance of Al2O3/PES membrane for wastewater filtration. *J. Membr. Sci.*, 341, 1-2, 67–75, 2009.
37. Ghazanfari, D., Bastani, D., Mousavi, S.A., Preparation and characterization of poly (vinyl chloride) (PVC) based membrane for wastewater treatment. *J. Water Process Eng.*, 16, 98–107, 2017.
38. Moradi, G., *et al.*, Fabrication of high flux and antifouling mixed matrix fumarate-alumoxane/PAN membranes via electrospinning for application in membrane bioreactors. *Appl. Surf. Sci.*, 427, 830–842, 2018.
39. Zhao, J., Progress in the Study of Membrane Bioreactor and Membrane Fouling. *China Water Wastewater*, 19, 5, 31–34, 2003.

8

Porous Imprinted Membranes for Recovering Targeted Compounds and Environmental Remediation

Laura Donato

Research Institute on Membrane Technology-National Research Council (CNR-ITM), Rende (CS), Italy

Abstract

Separations based on membrane operations are increasingly proving to be strategic approaches for the separation of various classes of compounds, in the logic of sustainable products recycled from different matrices and environmental remediation. From this viewpoint, imprinted membranes permit achieving a high separation degree of specific molecules and ions. They are inert and stable biomimetic separation tools, endowing the memory of the chemical function, shape, and size of a compound of interest, allowing for its separation from complex mixtures containing similar compounds, including structural homologs. They also permit the separation of opposite isomers and ions with similar charge and hydration radius. These smart separation membranes combine the features of membrane and imprinting technology. Similar to the traditional membranes, they are prepared in many configurations (flat sheets, hollow fibers, nanofibers) and can work in continuous operation mode. The different production strategies allow for obtaining both dense and porous imprinted membranes, depending on the fabrication method. This chapter deals with the fundamentals and the separation mechanisms of imprinted membranes. Furthermore, it devotes particular attention to the application of porous imprinted membranes in the selective separation of bioactive compounds, food contaminants, and pollutants (e.g., pharmaceuticals, pesticides, ions) from raw materials, water, and wastewaters.

Email: l.donato@itm.cnr.it

Annarosa Gugliuzza and Wanqin Jin (eds.) Porous Membranes: Breakthroughs in Manufacturing and Applications, (207–274) © 2025 Scrivener Publishing LLC

Keywords: Porous imprinted membranes, specific recognition, sustainability, resource recovery, bioactive compounds, food additives, water pollution, environmental remediation

8.1 Introduction

Humans' need to improve their living conditions has led to continuous technological progress in all fields over time. On the other hand, and also by virtue of all anthropic activities, modern society is faced with the problem of the reduction of available resources and environmental pollution. For example, the water resources of our planet are not quite enough to satisfy the increased demand, and the necessity of reducing its consumption and recycling it at industrial and agricultural levels represents an emergency that must be addressed promptly. Furthermore, natural ecosystems and human health are seriously threatened by the presence of emerging contaminants, mostly coming from industrial wastewaters. Considering all these aspects, worldwide efforts and strategic interventions need to be intensively carried out, aiming at the recycling of precious resources such as critical raw materials and valuable compounds, as well as the removal of emerging contaminants from polluted sites (water, seawater, wastewaters, lakes, biological fluids, food, air, and soil). For example, wastewaters coming from the pharmaceutical, cosmeceutical, and nutraceutical industries contain residues of valuable compounds that can be recovered and recycled, thus avoiding their accumulation in the environment and meanwhile producing clean water to be reused, with the consequent reduction of the costs of the production processes [1, 2]. Toxic compounds, such as pesticides used in agriculture, need to be removed from water and soil to avoid or limit as much as possible their presence in the food chain [3]. From these standpoints, the growing need to achieve highly efficient separations with a view to the recovery and recycling precious resources and environmental remediation finding has led to the development of highly efficient and extremely selective separation strategies toward the most disparate compounds of interest. Among them, in the framework of membrane technology, the production of advanced separation tools, such as imprinted membranes (IMs), has allowed for the realization of highly selective separation of targeted molecules or ions both in aqueous and organic environments. These intelligent membranes present specific recognition sites complementary to a targeted compound/ion called "template" and are able to separate it from a multi-component solution containing strictly related compounds, including structural homologs, opposite isomers, and similar ions. The specific interaction occurring between an IM and the template exploits the molecular fruitful recognition strategy adopted by living organisms for regulating their biological activities, which arises between specific cellular recognition sites and host molecules.

The concept of molecular recognition was pioneered in 1894 by Herman Emil Fisher to explain the mechanism of interaction taking place between an enzyme and its substrate according to the "lock and key" model [4]. Later [5], Koshland proposed the "hand in glove model", hypothesizing a certain flexibility of the active site of the enzyme during its binding with the substrate. Nowadays, numerous IMs are prepared and applied in many areas, allowing the overcoming of the problem of low stability of biological receptors (antibodies, ligands, etc.) and the high costs associated with traditional separation techniques (e.g., chromatography, liquid solvent extraction). Some examples of their application are catalysis, chiral separation, clinical diagnostic, drug delivery, food science, organic solvent nanofiltration, sensor technology, water remediation, etc. [6, 7]. The list of compounds of interest includes heavy metals, alkaline ions, anions, rare earth metals, small molecules, macromolecules, microorganisms (e.g., viruses and bacteria), cells, and plant tissues. Imprinted membranes that realize separation between molecules are called molecularly imprinted membranes (MIMs), while those separating ions are called ion imprinted membranes (IIMs). The molecular interactions occurring between the recognition sites of a MIM and the template molecules are realized *via* the formation of either covalent and non-covalent binding or their combination [6, 7]. In the case of the separation of ions, electrostatic interactions occur between the charge of the template and the opposite charge of the imprinted sites of the membrane. Imprinted membranes are produced in many configurations, like flat sheets, hollow fibers, and nanofibers by exploiting different synthetic approaches and materials [6]. Depending on the preparation method, they may have a dense or porous structure and can operate in a wide range of pH, temperature, and ionic strength. Additionally, they can work in batch and continuous operation mode, as well as alone or combined with traditional membrane-based operations and/or other separation methods in the logic of process intensification [6]. This is an interesting aspect to consider for developing sustainable processes and in the perspective of human and environmental safety, as well as for promoting the application of IMs at the industrial level, which is still not widespread despite the studies conducted in this sector demonstrating their effectiveness. Therefore, more intensive efforts are required from the scientific community to promote their diffusion. From this viewpoint, the objective of this chapter is to contribute to the spread of knowledge of the positive impact of IMs in pursuing highly efficient separations. It deals with the fundamentals and the separation mechanisms of imprinted membranes and discusses the application of the porous ones in the selective recognition of specific compounds in food science and water remediation.

8.2 Fundamentals of Molecularly Imprinted Membranes

Molecularly imprinted membranes are smart separation systems that combine the typical features of membrane and imprinting technology. They are characterized by easy preparation in a short time, high stability, reusability, and the possibility of working in continuous operation mode. Additionally, compared to traditional membranes, they are endowed with specific recognition sites that are complementary to the molecule of interest (called the template) in size, shape, and chemical functionality. This confers them an improved specific selectivity while preserving the intrinsic permeation properties. In particular, they are able to separate classes of similar compounds as well as distinguish template molecules from strictly similar compounds, thus realizing high purification and recovery degrees and overcoming the problem of separating structural homologs and opposite isomers [6, 8]. The presence of these tailored specific recognition sites on the surface or within the polymeric matrix of a membrane has traced a new path on the separation front, starting from the microscale to the nanoscale processes. Molecularly imprinted membranes can work as independent single units or combined with traditional membrane-based separations or other technologies in integrated systems, in line with the logic of process intensification to improve the separation efficiency of targeted compounds. Furthermore, they are advantageous compared to molecularly imprinted polymers (MIPs), which are characterized by restricted processability due to the high crosslinking degree and the inability to work continuously [6, 8]. The applications of MIMs include solid-phase extraction of drugs and biomacromolecules, bioactive compounds, and herbal ingredients from various sources [9–13], chiral separation, [14–16], drug delivery [17, 18], sensor technology [19], water treatment for removing pharmaceuticals, pesticides, and other toxic compounds [20–22], recovery of critical raw materials [6] and valuable compounds from food waste [23], etc. Figure 8.1 summarizes some application areas of MIMs [9].

The preparation of MIMs engages two different types of chemical imprinting methods: covalent imprinting and non-covalent imprinting.

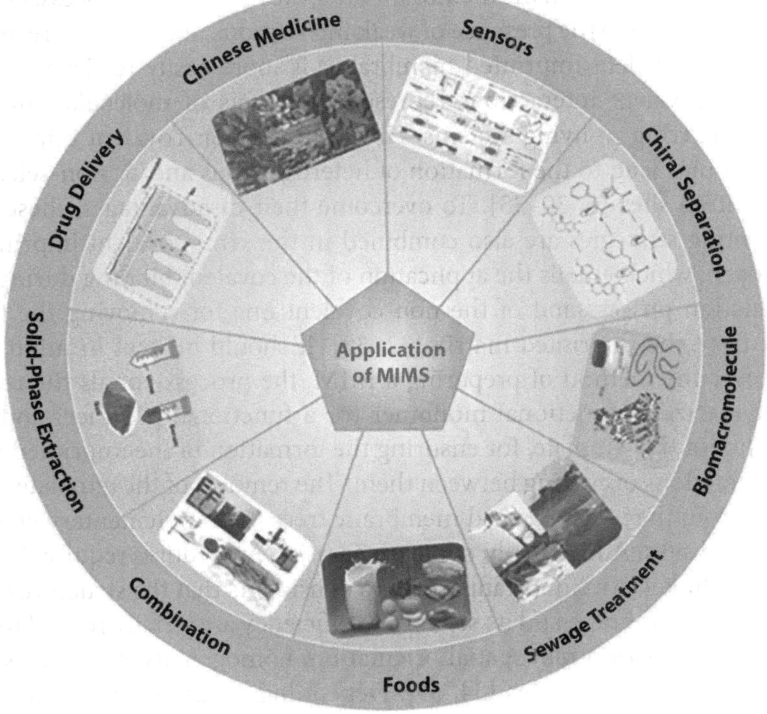

Figure 8.1 Some application fields of molecularly imprinted membranes. (Reprinted from ref. [9]: Yang, H., Liu, H.B., Tang, S.Z., *et al.*, Synthesis, performance, and application of molecularly imprinted membranes: A review. *J. Environ. Chem. Eng.*, 9, 106352, 2021. With permission of Elsevier, copyright 2021).

The first one, pioneered by Wulff and Sharan [24–26], is characterized by the formation of covalent bonds between the functional groups of the template and those of the complementary recognition sites of the membrane. It entails the occurrence of reactions of condensations using ketals/acetals, Schiff's bases, and boronated esters [24–27]. In the non-covalent imprinting approach, introduced by Mosbach and Sellergren [28, 29], non-covalent chemical bonds such as hydrogen bonds, hydrophobic interactions, ion bonds, and van der Waals forces are present. The latter method is the most applied, even if covalent imprinting allows for obtaining more homogeneous and stable recognition sites. This is because the covalently imprinted materials are characterized by a restricted number of functional moieties available for interacting with the templates. In addition, they exhibit slow binding kinetics owing to the necessity of high energy to form covalent bonds. Finally, covalent bonds require chemical reactions to achieve the template recovery from the membrane after the recognition process. On the contrary, owing to the presence of weak interactions, the template recovery from non-covalent imprinted membranes is more easily realizable. Also, the non-covalent interactions represent the basis of molecular interactions occurring in living systems. However, the non-covalent imprinting method may lead to the formation of heterogeneous and/or non-selective recognition sites [6, 30–33]. To overcome their disadvantages, these two imprinting strategies are also combined in the semi-covalent imprinting approach, which entails the application of the covalent binding during the production process and of the non-covalent one for removing the template from the imprinted matrix [32, 33]. It should be kept in mind that whatever the method of preparing a MIM, the process entails the use of a polymerizable functional monomer (or a functional polymer) and the presence of the template, for ensuring the formation of the imprinted sites by interactions occurring between them. The removal of the template molecules from the newly formed membrane frees its complementary created imprinted sites. Additionally, some preparation procedures require the use of a crosslinker. An important aspect to consider is that the structure of an imprinted membrane and its separation efficiency are strictly related to the shape, size, distribution, spatial orientation, homogeneity, and accessibility of the recognition sites [34, 35]. Depending on the preparation strategy, imprinted sites may be located on a membrane surface or distributed both on the surface and into its matrix. In the first case, they are created by the so-called two-dimensional (2D) imprinting strategy, which allows their planar distribution as a thin monolayer on the membrane surface. The second case is characteristic of the three-dimensional (3D) imprinting approach [34, 35]. The location of the recognition sites on the membrane surface favors their accessibility and interaction with the template allowing fast binding kinetics and subsequent template recovery. On the other hand, some recognition sites created *via* the 3D strategy and located in the membrane bulk are poorly accessible, with the possible consequence of a decrease in membrane performance and of an uncomplete recovery of template molecules [35, 36].

Over time, in the context of the 2D and the 3D approaches, MIMs with different configurations were produced by exploiting many preparation methods aiming to find an equilibrium between their recognition and transport properties, as well as their flexibility, selectivity, and stability. Traditionally, MIMs were prepared with organic and inorganic materials as flat sheets (thin films) and as hollow fibers (which display a spaghetti-like configuration). However, in the last decades, the development of nano imprinting technology has led to the production of fibers at the nanoscale level, with increased porosity and specific surface area, which are favorable for increasing the number of recognition sites [36]. Some polymeric materials used for preparing MIMs are acrylic copolymers [21, 37] polyamide (PA) [38, 39], polyvinylidene fluoride (PVDF), polyethersulfone (PES), polycarbonate (PC) [40], modified poly ether ether ketone (PEEK-WC) [41] polystyrene (PS) [42], polysulfone (PSf) [43], chitosan (CS) [44], cellulose (CA), cellulose acetate/sulfonated polysulfone (CA/SPSU) blends [45], sodium alginate (SA) [46], and so on. The most applied procedures for fabricating MIMs are: 1) the surface imprinting of an existing membrane; 2) the simultaneous formation of membrane structure and its imprinted sites; 3) the synthesis of an imprinted polymer (IP) and the subsequent formation of a hybrid membrane. Surface imprinting deals with the 2D approach and consists of the synthesis of a thin layer of an MIP on the surface of a membrane support *via* grafting, coating, or electropolymerization. Besides the functional monomer and the template, the presence of a crosslinker is important for stabilizing the nascent recognition sites. This approach leads to obtaining composite membranes exploiting the advantages of the mechanical strength of the porous structure of the support membrane as well as of the specific recognition properties of the upper thin imprinted polymer layer [6, 8, 47]. An example of surface imprinting is the preparation of a composite MIM by coating the surface of a commercial PVDF membrane with a thin D-tryptophan sodium alginate-based

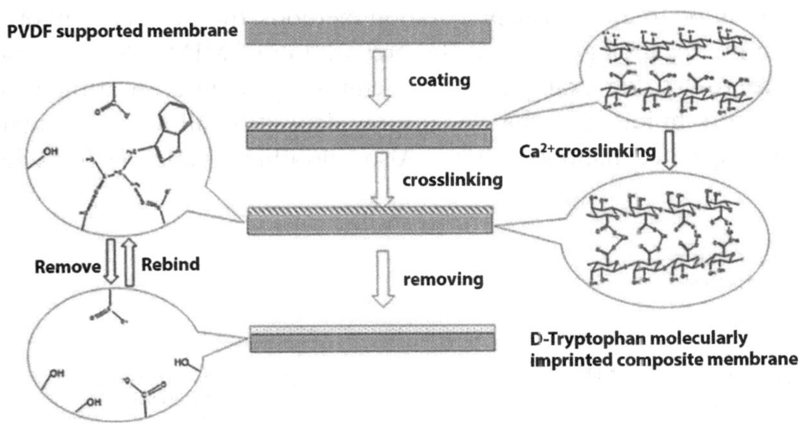

Figure 8.2 Scheme of the preparation of D-Tryptophan-molecularly imprinted composite membrane. (Reprinted from the ref. [48]: Zhou, Z., He, L., Mao, Y., *et al.*, Green preparation and selective permeation of D-tryptophan imprinted composite membrane for racemic tryptophan. *Chem. Eng. J.*, 310, 63, 2017. With the permission of Elsevier, copyright 2016).

imprinted polymer film crosslinked with calcium chloride, as illustrated in Figure 8.2 [48].

In a different work [49], the grafting method was applied for synthesizing (*via* photo-polymerization) a thin S-naproxen-imprinted polymer layer on the surface of a polypropylene support. The obtained composite MIMs exhibited permeability values typical of ultrafiltration and were capable of separating the template enantiomer from its opposite isomer [49]. The *in situ* crosslinking polymerization and the alternative molecular imprinting methods exploit the 3D synthetic approach. The first one involves the formation of a crosslinked polymeric film from a pre-polymerization solution containing the template, a functional monomer, a crosslinker, an initiator, and a plasticizer for producing a flexible membrane [6]. The alternative molecular imprinting is the most applied 3D approach and entails the preparation of a membrane *via* phase inversion using a polymer containing functional groups capable of interacting with the complementary chemical group of the template [6, 7].

The last procedure entails the application of phase inversion for preparing MIMs by hybridizing an imprinted polymer previously synthesized with a polymer forming membrane *via* either the "dry" of the "wet" phase inversion. A schematic representation of this preparation method is shown in Figure 8.3 [6].

Owing to the rapid liquid-liquid demixing that characterizes the non-solvent-induced phase separation, membranes prepared *via* the "wet" route are more porous.

In addition to these basic preparation methods, the advancement of the imprinting technology allowed the preparation of MIMs also by means of other routes. For example, the contemporary application of the phase inversion technique and the electrospray deposition allowed the obtaining of imprinted nanofibers [50]. Also, ultrathin titania gel films [51]

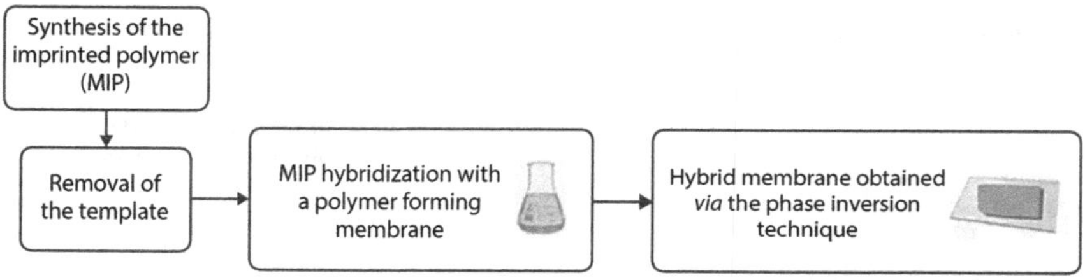

Figure 8.3 Schematic representation of the preparation of a hybrid imprinted membrane using the phase inversion technique. (Reprinted from ref. [6]: Donato, L., Drioli, E., Imprinted membranes for sustainable separation processes. *Front. Chem. Sci. Eng.*, 15, 775, 2021. With the permission of Springer, copyright 2021).

and surface-modified PSf membranes [52] were prepared *via* the sol–gel imprinting process, while multilayered imprinted composite membranes and other imprinted membranes were prepared through the click chemistry [53, 54] and the electrospinning [55–58]. These few examples represent only a small part of the numerous strategic productions of imprinted membranes. More detailed information can be found in the review papers listed in Table 8.1 and published in the last decade.

Table 8.1 Review papers about the production of imprinted membranes in the last decade.

Title	Publication year	Ref.
Recent insights into molecularly imprinted membrane technology for removal of pollutants from environmental water: from organic molecules to metal ions	2024	[59]
Advanced development of molecularly imprinted membranes for selective separation	2023	[60]
Green chemistry and molecularly imprinted membranes	2022	[8]
Molecularly imprinted affinity membrane: a review	2022	[61]
Imprinted membranes for sustainable separation processes	2021	[6]
Molecular imprinted membrane biosensor for pesticide detection: perspectives and challenges	2021	[62]
Recent advances in molecularly imprinted membranes for sample treatment and separation	2020	[63]
Imprinted composite membranes	2020	[64]
Recent advances in ion imprinted membranes: separation and detection *via* ion-selective recognition	2019	[65]
Molecularly imprinted polymer membranes and thin films for the separation and sensing of biomacromolecules	2017	[34]
Molecularly imprinted membranes: past, present, and future	2016	[36]

(Continued)

Table 8.1 Review papers about the production of imprinted membranes in the last decade. (*Continued*)

Title	Publication year	Ref.
Approaches for the assembly of molecularly imprinted electrospun nanofiber membranes and consequent use in selected target recognition	2015	[66]
Recent developments in molecularly imprinted polymer nanofibers and their applications	2015	[67]
Bio-mimetic sensors based on molecularly imprinted membranes	2014	[47]
Emerging tools for recognition and/or removal of dyes from polluted sites: molecularly imprinted membranes	2014	[68]
Separation with molecularly imprinted membranes	2014	[69]

From their studies, the reader will be able to appreciate the potential application of IMs in numerous sectors. It is important to highlight that regardless of the method adopted to prepare a MIM (or a MIP), in parallel (under the same operating conditions but in the absence of the template), the corresponding non-imprinted membrane (NIM) (or NIP) is always prepared. This is important for carrying out comparative recognition and separation studies that will establish that the recognition and selectivity properties of a MIM are due to the imprinting effect.

8.3 Separation Mechanisms and Assessment of Selective Properties of MIMs

Separations based on the employment of MIMs are characterized by the existence of a trade-off relationship between two important parameters: flux and selectivity. So far, to maintain high separation performance, membranes with a high density of recognition sites and a strong match between them and the template are required. The selective transport mechanism depend also on the distribution of the recognition sites as well as on the membrane structure and the pore-size distribution, and morphology [28, 60, 61]. More in detail, the selective separation of a given template with MIMs can be realized through two major mechanisms proposed by Ulbricht in 2004 [70]. They are distinguished as retarded permeation and facilitated permeation [47, 70]. The facilitated permeation is characterized by a quick transport of the template molecules or ions with respect to other analytes, the transport of which occurs only *via* non-specific diffusion or convection. This is due to the creation of a preferential pathway for the template owing to binding and desorption to the nearby recognition sites. It must be highlighted that the binding of the template to the recognition sites may determine membrane swelling or shrinking, with a consequent increase or reduction of permeability, respectively [60, 69–72]. Bai and co-workers [73] demonstrated that a PVDF membrane surface imprinted by the magnetic-guidance phase inversion method for the selective separation of the anti-malaria drug artemisinin exhibited a narrower pore size distribution and more efficient recognition sites with respect to those prepared *via* the conventional blending phase inversion method. Permeability studies with these membranes evidenced a relevant fast transport of artemisinin with respect to the competing compound artemether [73]. Previously, Kashani *et al.* [74] applied phase inversion for fabricating nanoporous hybrid imprinted membranes in the selective removal of the herbicide 2,4-dichlorophenoxyacetic acid (2,4-D) from water through its facilitated permeation mechanism. The diameter of the pores ranged from 300 nm to 800 nm. Permselective studies showed that the highest separation efficiency was exhibited by membranes having the smallest pore dimensions [74].

In the case of the retarded permeation, the transport of the template is retarded owing to its affinity binding with the recognition sites of the membrane until their saturation. On the contrary, other compounds in the same solution move through the membrane *via* nonspecific diffusion or convection and accumulate in the permeate stream [70–72]. An example of this separation mechanism is represented by the recovery of acetoside (an active pharmaceutical ingredient present in *Cistanche tubulosa* plant) from water with a composite PVDF/poly(4-vynilpiridine (4-VP)-based membrane that allowed the selective enrichment of acetoside with a selectivity factor acetoside/echinacoside of 2.74 [75]. Another example is the retarded permeation of gentisic acid through a ceramic alumina MIM having an ordered porous polymer network introduced in its interior through the sol–gel method with room temperature ionic liquid (RTIL) [76]. Also, hybrid porous imprinted membranes showed a retarded permeation of 4,4 methylene dianiline [77]. The phase inversion preparation procedure entailed the dispersion of pre-synthesized imprinted polymer particles into a PVDF polymer forming membrane. Experimental studies revealed an increase in membrane permeability with an increase in the polymer particles content up to 33 wt. %. All the prepared membranes showed an asymmetric morphology consisting of a dense skin layer supported by a porous structure with finger-like macrovoids ending with a sponge-like structure [77]. Figure 8.4 illustrates the two different permeation separation mechanisms characterizing MIMs [60].

Separations can also be performed by exploiting the selective adsorption mechanism. It is characterized by the presence of adsorptive interactions between the template and the complementary recognition sites located on the surface of both the membrane and its pores [71]. In this case, the template is selectively adsorbed by the membrane while the competing analytes accumulate in the treated solution. As is the case of stimuli-responsive MIMs, the separation mechanism may also be influenced by the operating conditions (e.g., temperature, pH, etc.). For example, ovalbumin-imprinted nanocomposite membranes changed their separation performance depending on temperature variation [78]. In particular, at 37 °C, they showed high binding capacity and selectivity toward the template in comparison with the competing bovine hemoglobin and lysine. When the temperature was lowered to 20 °C, the recognition sites lost their complementarity toward the template, thus showing low binding capacity and selectivity [78]. In a different work, Cheng *et al.* [79] demonstrated the presence of a retarded permeation of the template podophyllotoxin through a MIM working at a pH higher than 8.4. On the contrary, at pH below this value, the membrane exhibited a facilitated permeation separation mechanism. This different behavior was attributed to the tautomerization

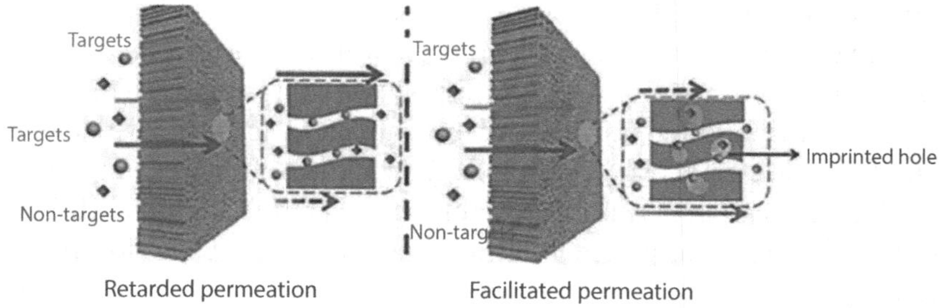

Figure 8.4 Representation of the retarded (a) and facilitated (b) permeation separation mechanism of an imprinted membrane. (Reprinted from ref. [60]: Chen, J., Wei, M., Meng, M., Advanced development of molecularly imprinted membranes for selective separation. *Molecules*, 28, 5764, 2023. Open access).

ketone/enol of the structure of the functional monomer β-diketone 1-phenyl-3-methyl-4-methacryloyl-5-pyrazolone (PMMP) when changing the operating pH value [79]. In conclusion, the separation mechanism of an imprinted membrane is determined on a case-by-case basis, depending on all the discussed aspects.

The recognition performance of an imprinted membrane is assessed by measuring three important parameters: the binding capacity (Q), the imprinting factor (IF), and the selectivity (α). The binding capacity and the imprinting factor lead to the evaluation of the capacity of the membrane to bind the template. In particular, Q represents the amount of the template bound by the membrane and is calculated be means of equation 8.1 [61, 71, 77]:

$$Q = (C_0 - C_t)V / W_m \tag{8.1}$$

Where C_0 and C_t represent the concentration of the template in the solution at the beginning and at interval time, respectively. V is the volume of the solution and W_m is referred to the membrane weight.

The difference between the binding capacity of an imprinted membrane and that of the corresponding NIM gives the specific binding capacity of the former, which represents the amount of the template bound thanks to the interaction between the imprinted sites and template molecules. Additionally, the ratio between the binding capacity of an MIM and that of its corresponding NIM makes it possible to calculate the imprinting factor.

Selectivity determines the capacity of the membrane to distinguish the template with respect to its competing similar compounds present in the treated solution. It can be determined according to equation 8.2 [61, 71, 77]:

$$\alpha = (C_{t,p} / C_{c,p}) / (C_{t,f} / C_{c,f}) \tag{8.2}$$

Where $C_{t,p}$ and $C_{t,f}$ represent the concentration value of the template in the permeate and feed streams, respectively, while $C_{c,p}$ and $C_{c,f}$ are the concentrations of the competing analyte in the same solutions.

In binding investigations, the construction of adsorption/binding isotherms (reporting the concentration of the template bound by the membrane *vs.* the concentration of the free template) is useful for measuring the maximum adsorption/binding capacity of an MIM and evaluating the binding features at equilibrium. To understand the adsorption process, these data are fitted with Langmuir and Freundlich isotherm models. In addition, the binding affinity between the template and the recognition sites, as well as the type of recognition sites are determined by the Scatchard analysis [71, 80–83]. Finally, for assessing the mechanism controlling the binding, time-dependent binding data are usually fitted with the pseudo-first-order and the pseudo-second-order kinetic models [71, 77, 80–83].

Concerning the transport separation mechanisms, in the case of facilitated permeation, the permselective factor (α) is expressed as the ratio between the transport rate of template molecules and that of the competing analytes. On the contrary, in the case of the occurrence of a retarded permeation separation mechanism, it is expressed as the ratio between the transport rate of the interferents and template molecules. In this last case, it is indicated with the symbol β.

8.4 Application of Porous Molecularly Imprinted Membranes

As it is known, the adopted industrial separation processes affect the costs and the quality of the products. Owing to their high specificity, up to date, MIMs have been applied in numerous areas, and depending on the problem to solve, they are produced as dense or porous systems, ensuring the effectiveness of the requested separation process. This chapter devotes particular attention to porous imprinted membranes. Therefore, the following paragraphs will discuss some of their applications in the selective recovery of bioactive compounds from vegetal food and herbal plants, as well as in food safety and water remediation.

8.4.1 Porous Molecularly Imprinted Membranes in Food Science

The high specific recognition properties of MIMs have rendered them useful for application in food science, thus offering an important contribution in the selective extraction of bioactive compounds from complex matrices and in the detection and removal of food additives or contaminants from food and beverages. In this context, MIMs act as affinity molecular recognition systems exhibiting mostly the retarded permeation or the specific adsorption separation mechanism. However, in some cases, a facilitated permeation of the template molecules has been observed. For example, luteolin-imprinted membranes permitted the separation of the template from the similar rutin (RT) by means of facilitated permeation of luteolin. The permselective factor was 9.1. Furthermore, in adsorption tests the maximum binding capacity toward luteolin was 29.1 mg·g^{-1}. Membranes were prepared by hybridizing PSf with the imprinted poly(vinyl triethoxysilane) in the presence of polyethylene glycol 400 (PEG400) to ensure the formation of channels into the polymeric matrix [84].

The following sections will discuss the application of porous MIMs in the selective solid-phase extraction/enrichment of bioactive compounds and in food safety.

8.4.1.1 *Selective Separation of Bioactive Compounds*

Bioactive compounds are natural substances present in vegetal food and plants. Owing to their numerous activities (e.g., antioxidant, anti-inflammatory action, antibacterial and antiviral activity, inhibition of tumor cells, regulation of detoxifying enzymes, etc.), they support the physiological functions of our body. Furthermore, they are employed at the industrial level in cosmeceutical, nutraceutical, and pharmaceutical formulations, as well as in the production of functional food and herbal medicine [23, 85, 86]. A large number of bioactive compounds exist, and their excellent properties stimulated the development of various strategies for achieving their extraction from natural or waste sources and recycling them [85–87]. Table 8.2 reports some examples of MIMs applied in the selective separation of bioactive compounds [88–104].

Table 8.2 Examples of MIMs applied for the selective extraction/enrichment of bioactive compounds.

Template	Membrane material	Sample solution	Binding capacity (mg·g^{-1})	Ref.
Rutin	Blended sodium alginate/ poly(vinyl pyrrolidone)	Ethanol/water (8:2, v/v)	4.0	[88]
10-deacetylbaccatin III	Composite cellulose/ poly(vinyl alcohol/ pethacrylic acid)	Methanol	66.90	[89]
Crysin	Composite poly(yvinyl alcohol/ ester/ methacrylic acid)	Methanol	127.5	[90]

(Continued)

Table 8.2 Examples of MIMs applied for the selective extraction/enrichment of bioactive compounds. (*Continued*)

Template	Membrane material	Sample solution	Binding capacity (mg·g−1)	Ref.
Epigallocatechin gallate	Composite regenerated cellulose/ poly(dopamine/L-cysteine)	Aqueous	26.72	[91]
Acetoside	Composite poly(vinylidene fluoride/4-vynilpyridine)	Not reported	62.83	[75]
Caffeic acid	Composite poly(vinylidene fluoride/4-vhynilpyridine)	water/methanol (3/2, v/v)	1.5	[92]
Genistein	PEG 2000/poly(acrylamide)	methanol/water (1:1, v/v)	9.0	[93]
Quercetin	Hybrid poly(vinylidenefluoride/ acrylamide)	Acetonitrile	25.62	[94]
Paclitaxel	Polysulfone	Ethanol/water (1:1, v/v)	1.63	[95]
Ferulic acid	Composite poly(vinylidenefluoride/ methacrylic acid	Aqueous	Not reported	[96]
Chlorogenic acid	Chitosan	Ethanol/water (2:8, v/v)	102.0	[97]
Kaempferol	Composite poly(phenylene sulfone/4-vynilpyridne)	Ethanol	0.5	[98]
Magnolol	Composite filter paper/ poly(methacrylamide	Ethanol	0.53	[99]
Luteolin	Polysulfone/ poly(vinyltriethoxysilane	Aqueous	29.10	[84]
Folic acid	Poly(acrylonitrile-co-acrylamide)	Aqueous	2.78	[100]
Daidzein	Composite filter paper/ poly(methacrylamide-co-acylic acid)	Ethanol	1.35	[101]
Curcumin	Poly(methacrylamide-co-methacrylic acid)	Methanol	239.68	[102]

(*Continued*)

Table 8.2 Examples of MIMs applied for the selective extraction/enrichment of bioactive compounds. (*Continued*)

Template	Membrane material	Sample solution	Binding capacity (mg·g−1)	Ref.
Naringin	Poly(acrylonitrile-co-acrylamide)	Aqueous	7.5	[103]
α-Tocopherol	Poly(methacrylate-co-acrylonitrile)	Ethanol/water	25.84	[104]

Among the different porous MIMs produced to selectively recognize bioactive compounds, an example is given by the separation of the flavonoid quercetin (QCT) by means of composite QCT-imprinted membranes prepared by synthesizing a thin porous imprinted layer on the surface of a microfiltration PVDF membrane [94]. The deposition of the imprinted layer increased the roughness of the pristine membrane, thus providing a larger surface for the recognition sites of the membrane. Previously, Nasser *et al.* [41] produced hybrid porous QCT-imprinted membranes *via* the phase inversion technique using PEEK-WC as the membrane-forming material, which was embedded with imprinted poly(acrylamide) particles previously synthesized by the precipitation polymerization method. For comparison NIMs were also fabricated; they were prepared under the same operating conditions but without the template during the polymeric synthesis. All the fabricated membranes exhibited an asymmetric structure made of a thin selective surface layer supported by a porous layer with finger-like macrovoids. According to the 3D imprinting strategy, the imprinted polymer particles were randomly distributed overall the membrane. Their presence determined a slight decrease in the membrane hydrophobicity [41]. Furthermore, the degree of the solvent uptake was very low (from 1.2% to 1.32%), indicating the existence of a weak interaction between the membrane material and the binding solvent (acetone/acetonitrile 3:1). Experimental data of the adsorption isotherms constructed for both imprinted and non-imprinted polymer fitted the Langmuir adsorption model and the adsorption kinetic was in accordance with the pseudo-second-order model [41]. The higher binding capacity (12.0 $\mu mol \cdot g^{-1}$) was shown by the membrane containing the 25 wt. % of MIP with respect to the polymer forming membrane. The corresponding NIM showed a binding capacity of 3.33 12.0 $\mu mol \cdot g^{-1}$. In permselective studies with mixed solutions of QCT and the similar compounds crysin (CRY) and naringin (NR), this MIM exhibited the retarded permeation separation mechanism and resulted selective against the competing CRY and NR. The selectivity factor was calculated as the ratio between the binding capacity toward the template and those toward similar compounds. For the couple QCT/CRY, it was 1.8, while that for QCT/NR was 4.5. The difference between the two couples was attributed to the more hydrophobic character of CRY, which competed much more with QCT for interacting with the recognition sites of the membrane. In fact, being more hydrophilic, NR was less bound by the membrane. Figure 8.5 shows the chemical structure of the functional monomer acrylamide, the template QCT, and the similar CRY and NR [41].

Composite MIMs made of PVDF and a thin surface layer (synthesized *via* UV irradiation) of PMAA were also applied for the specific recognition of ferulic acid in an aqueous solution and separating it from the structural analog cinnamic acid through a retarded permeation separation mechanism [96]. The selectivity factor was 2.59. The addition of SiO_2 nanoparticles to the pre polymerization mixture increased the hydrophilicity of the imprinted membrane and promoted the formation of a microporous

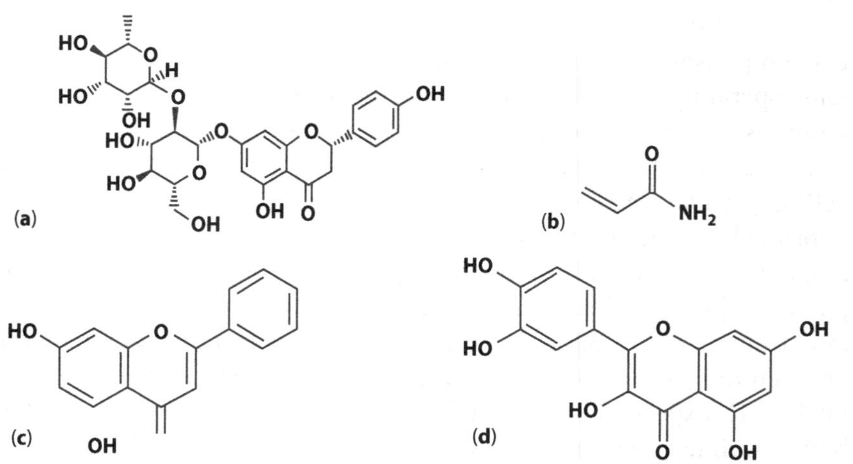

Figure 8.5 The chemical structure of naringin (a), acrylamide (b), crysin (c), and quercetin (d). (Reprinted from ref. [41]: Nasser, I.I, Algieri, C., Garofalo, A., *et al.*, Hybrid imprinted membranes for selective recognition of quercetin. *Sep. Purif. Technol.*, 163, 331, 2016. With the permission of Elsevier, copyright 2016).

structure. The value of the average porosity of MIM containing the 0.25 wt. % of 15 nm SiO_2 nanoparticles was 72.10 %. The irradiation time and the water permeate flux were 140 minutes and 2687.6 L m^{-2} h^{-1}, respectively. In the case of the membrane prepared in the same conditions but without nanoparticles, the value of porosity was 62.80 %, while the water flux was 245.6 L m^{-2} h^{-1} [96]. In a different study, the PVDF was blended with TiO_2 imprinted nanoparticles of 18 nm to produce high-flux hybrid dual-template membranes *via* phase inversion. Cinnamic acid and ferulic acid were the print molecules [105]. Also, the flavonoid kaempferol (KP) was selectively separated from RT in a mixed ethanol solution with poly(phenylene) sulfone (PPSU) phase inversion-based imprinted membranes [98]. This bioactive compound is present in numerous edible and medical plants [98, 106]. The addition of the inorganic additive LiCl or $ZnCl_2$ to the dope solution increased the amount and dimensions of the pores, which were typical of macroporous membranes. Additionally, a decrease in the membrane contact angle was observed, indicating an increase in the membrane's hydrophilicity. Figure 8.6 shows the scanning electron micrograph (SEM) image of the cross-section of the imprinted membrane without any inorganic additive (Figure 8.6a) and the imprinted membrane containing the 25 wt. % of LiCl (Figure 8.6b). The contact angle as a function of the additive concentration (Figure 8.6c) is also shown [98].

The optimized dope solution composition was: 14.0 wt. % of PPSU, 0.3 wt. % of kaempferol-imprinted poly(4-VP) particles and 2.5 wt. % of LiCl. In selectivity studies, these membranes exhibited a separation factor KP/RT of 4.12 while, for membranes prepared without additive the value of the separation factor was 3.3 [98]. In 2024, Ouarda *et al.* [88] demonstrated the possibility of recovery RT from food matrices by means of novel MIMs based on sodium alginate and poly(vinyl pyrrolidone, PVP) as the functional polymer and the pore-forming agent, respectively [88]. Isothermal adsorption studies evidenced a higher adsorption of the template molecules on MIM with respect NIM. At an initial feed concentration of 50 mg·L^{-1}, their adsorption capacity was about 35 mg·g^{-1} and 15 mg·g^{-1}, respectively. Results confirmed the occurrence of specific recognition in MIM and therefore the efficacy of the imprinting procedure. For both membranes, the equilibrium was reached after 10 minutes. The adsorption behavior was in agreement with the Freundlich adsorption model and with the pseudo-second-order kinetic model [71, 88]. The removal of the adsorbed template molecules was rapidly achieved *via* complexation with

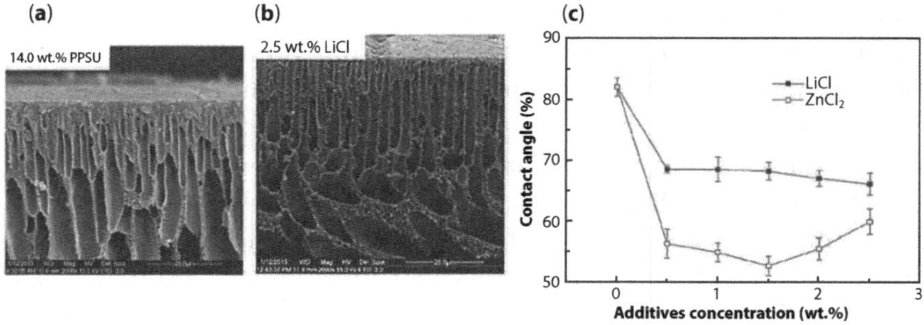

Figure 8.6 SEM image of the cross-section of the kaempferol-imprinted membrane without additives (a), the kaempferol-imprinted membrane containing the 25 wt.% of LiCl (b), and the contact angle as a function of the additive concentration into the imprinted membrane (c). (Reprinted from ref. [98]: Huang, Z., Zhang, P., Yun, Y., Preparing molecularly imprinted membranes by phase inversion to separate kaempferol. *Polym. Adv. Technol.*, 28, 373, 2017. With the permission of John Wiley & Sons, copyright 2016).

calcium. Differently [107], the integration of membrane crystallization with MIMs allowed to recovery oleanolic acid (OA) by inducing its crystallization (OA) in supercritical CO_2 ($ScCO_2$) [107]. More deeply, the chosen functional monomer MAA was copolymerized with the surface of a porous ultrafiltration PSf membrane support *via* UV irradiation. Afterwards, the template molecules were recovered from the novel membrane by means of extractions with methanol/acetic mixture leading to the availability of the created imprinted sites for subsequent specific recognition process. Figure 8.7 shows the membrane preparation procedure.

Results of this study demonstrated the feasibility of this preparation strategy for obtaining good membrane performance in separating bioactive compounds from natural matrices through a rapid and efficient route. The crystallization and the purity were of 45.3% and 98.3%, respectively. Figure 8.8 shows the chromatogram of the purified crystals of OA.

The separation of Ebracteolata compound B from *Euphorbia fischeriana* was also done by means of a porous reticular molecular imprinted membrane (having a pore diameter in the range of 10–50 nm) synthesized on stainless steel wire mesh as support [13]. An interesting case is the selective separation of emodin with antibacterial, high-flux, and 3D porous molecularly imprinted nanocomposite sponge membranes [108]. Emodin is an antibacterial, antitussive, and antitumor agent abounding in aloe, buckthorn, and rhubarb plants together with its analogs. Owing to the numerous benefits, its phytochemical purification assumed great

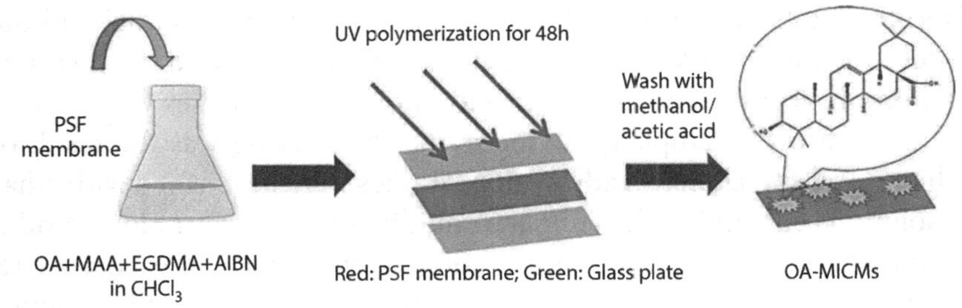

Figure 8.7 Preparation procedure of composite oleanolic acid-imprinted membrane. (Reprinted from ref. [107]: Zhang, W., Zhang, Q., Zhang, X., *et al.*, Preparation and evaluation of molecularly imprinted composite membranes for inducing crystallization of oleanolic acid in supercritical CO_2. *Anal. Methods.*, 8, 5651, 2016. With the permission of the Royal Society of Chemistry, copyright 2016).

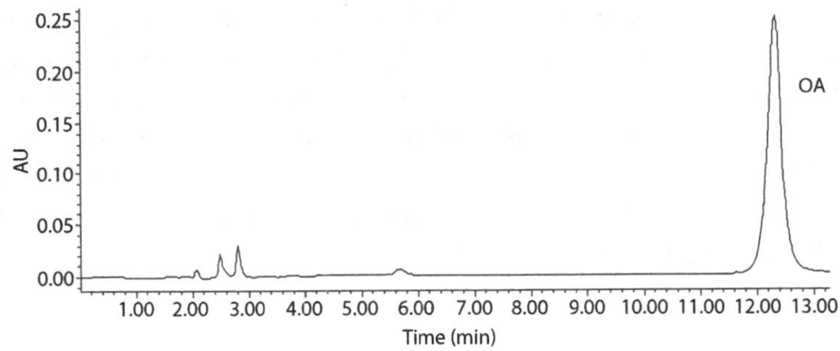

Figure 8.8 HPLC-UV chromatogram of purified oleanolic acid crystals *via* inducing crystallization in supercritical CO_2. (Reprinted from ref. [107]: Zhang, W., Zhang, Q., Zhang, X., *et al.*, Preparation and evaluation of molecularly imprinted composite membranes for inducing crystallization of oleanolic acid in supercritical CO_2. *Anal. Methods.*, 8, 5651, 2016. With the permission of the Royal Society of Chemistry, copyright 2016).

importance both at research and industrial level [108, 109]. For membrane preparation, a commercial sponge with a three-dimensional (3D) porous structure and large pore diameters was modified with graphene oxide (GO), silver nanoparticles and 3-(trimethoxysilyl) propyl methacrylate (KH570) in sequence. Graphene oxide (GO) ensured the activation of the sponge, promoting the subsequent modification step with KH570 that created a secondary active platform warranting the adhesion of an imprinted layer of poly(acrylamide). The addition of silver nanoparticles conferred antibacterial properties to the membrane to prevent the accumulation of microorganisms on its surface and interior structure [108]. Data from static adsorption tests fitted well with the Langmuir isothermal model and the pseudo-second-order kinetic model. These results evidenced the presence of a homogeneous monolayer distribution of imprinted sites with similar affinity toward the template. Furthermore, the binding was the result of both physical and chemical adsorption. Studies were performed with ethanol solutions containing the template emodin and its similar aloe-emodin and physcion. At an initial concentration of 100 mg·L^{-1} of each analyte, the emodin-imprinted membrane (Emodin-MIS) and the pristine sponge bound 24.32 mg·g^{-1} and 7.42 mg·g^{-1} of emodin, respectively. On the other hand, the binding capacity toward aloe-emodin and physcion was much lower. The IF was 3.8, while the adsorption selectivity occurring at the membrane surface emodin/aloe-emodin and emodin/physcion was 1.61 and 1.70, respectively. After 10 adsorption/desorption cycles including an interval of 15 days between the fifth and sixth cycle the adsorption capacity of Emodin-MIS was still the 85 % of the initial value [108]. Finally, crossflow filtrations revealed a retarded permeation of emodin through the Emodin-MIS, while the competing analogs permeated faster. Additionally, all the investigated compounds were poorly bound by the pristine sponge. Figure 8.9a shows the separation mechanism with the Emodin-MIS, while Figure 8.9b shows the cross-flow transport rate of the three compounds through the same membrane [108].

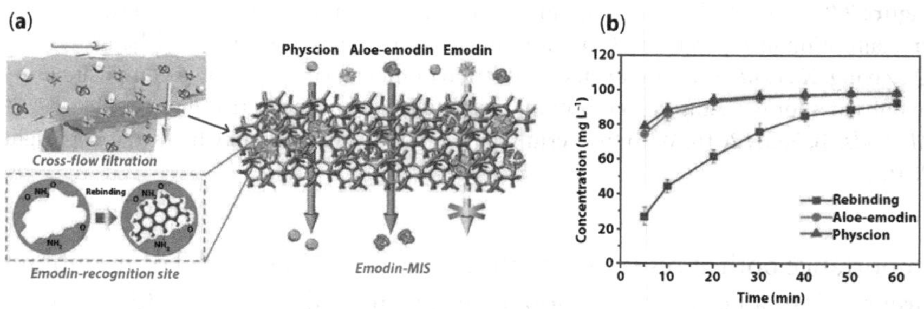

Figure 8.9 Retarded permeation separation mechanism of emodin through an emodin-imprinted membrane (Emodin-MIS) (a) and crossflow transport of permeants emodin, aloe-emodin and physcion through the same imprinted membrane (b). Reprinted from ref. [108]: Lu, J., Qin, Y., Zhang, Q., *et al.*, Antibacterial, high-flux and 3D porous molecularly imprinted nanocomposite sponge membranes for cross-flow filtration of emodin from analogs. *Chem. Eng. J.*, 360, 483, 2019. With the permission of Elsevier, copyright 201.

The highest permselective factor for emodin/aloe-emodin and emodin/physcion was 7.66 and 10.00, respectively. These values result from the contribution of the specific adsorption on the membrane surface and that occurring at the internal membrane's pores and recognition sites [108].

Other cases of application of MIMs in the selective separation of bioactive compounds can be found in Tables 8.1 and 8.2.

8.4.1.2 *Food Safety*

It is widely assessed that many types of contaminants can pollute food and beverages, resulting in health problems. Some pollutants are pesticides, drugs, secondary metabolites, and compounds from the chemical industry or human activity [3, 8]. For example, the continuous low levels of antibiotics in food, beverages, and water can lead to antibiotic resistance, negatively affecting the environment and human health. Additionally, residues of pesticides used in agriculture can be present in water and food chain, thus promoting the insurgence of serious diseases in plants, animals, and human people [3, 8, 110]. From this perspective, the development of remediation strategies to remove various pollutants from food matrices to ensure food safety is extremely important. In the context of the different applied approaches (e.g., adsorption, bioremediation, electrochemical remediation, advanced oxidation technology, and membrane technology), MIMs offer the possibility of achieving a high level of selective removal of the interested contaminant. This paragraph deals with the employment of porous imprinted membranes in food control and remediation. The first example is the removal of the antibiotic cloxacillin from shrimp samples with composite MIMs based on a PVDF support covered with a thin layer of imprinted PMAA [111]. This β-lactam antibiotic is used as a growth-stimulating agent for preventing animal infections and therefore increase feed utilization. However, its excessive illegal use in food-producing animals can allow the insurgence of problems in aquatic fauna and humans. Therefore, in view to reduce its presence, it is important to develop strategies for monitoring illegal abuse in livestock farming and aquaculture [111, 112]. Owing to its excellent thermal stability, chemical inertness, and low toxicity, PVDF is largely used in the production of IMs. The fabricated membranes allowed for the detection and removal of cloxacillin from 20 real shrimp samples from different markets. Results of this investigation revealed the presence of the antibiotic only in three cases, thus demonstrating effective control of its use in local food-producing animals. For these cases, the extraction recovery rate ranged from 76.0 % to 84.3% [111]. Scatchard's analysis showed the existence of high-affinity and low-affinity binding sites in MIMs, while in NIMs only low-affinity binding sites were present. Furthermore, the cloxacillin adsorption on MIMs and NIMs fitted the Freundlich isotherm, while the adsorption kinetics agreed with the pseudo-second-order model [111]. More recently, nanoporous composite membranes were applied for the selective detection and separation of the macrolide antibiotics erythromycin (ERY) and spiramycin (SPI) from cow milk [113]. They are used in veterinary medicine and sometimes as feed additives in the role of growth promoting agents. Their presence in milk can cause allergic reactions in some hypersensitive subjects. To achieve the fast removal of these two antibiotics from commercial doped semi-skimmed cow milk samples, composite MIMs were fabricated employing the dual template imprinting strategy *via* the surface imprinting of fiberglass microfiltration membrane through UV photo grafting. The functional monomer was MAA, the crosslinker was ethylene glycol dimethacrylate (EGDMA) and the initiator was sodium-diethyl dithiocarbamate (Na-DTC). Additionally, to avoid the saturation of the recognition sites of the membranes, an online solid phase extraction procedure with suitable solvents was carried out [113]. The optimal elution solvent was methanol, which was highly efficient in destroying the chemical bonds between the templates and the MIM. The recovery rate of the two antibiotics was calculated as the percentage of the amount extracted with methanol with respect to the amount retained by the membrane through a specific surface adsorption. The recovery rate of ERY reached the value of 83.43 $\pm$ 0.65 % while that of SPI was 24.57 $\pm$ 2.82 %. The remarkable difference was attributed to possible competition between the two template molecules (as can occur in the multi-template imprinting approach) [114, 115]. In a different case, the exploitation of click chemistry allowed the production of antifouling and hydrophilic lincomycin-imprinted membranes for removing this antibiotic in a milk pre-treatment step [115]. The successful removal of the mycotoxin citrinin (CIT) from rice was also done by Lee *et al.* [116] through the employment of MIMs produced by means of the dummy template imprinting strategy. In this context, the authors used the non-toxic l-naphthol (NA) as a dummy template for the first time. Membranes were prepared according to the hybrid imprinting method. The procedure entailed the synthesis of core-shell-naphthol imprinted polymer particles that were subsequently hybridized with the PES scaffold polymer by the phase inversion process. During the synthetic step of MIP, methacryloyl chloride and divinyl benzene were used as the functional monomer and the crosslinker, respectively. Figure 8.10 shows the chemical structure of both CIT and NA, as well as the representation of the formation of the imprinted sites in the presence of NA as the dummy template molecule [116].

High recovery of citrinin in spiked rice samples (89.7–94.2%) suggests the promising application of this strategy for the removal of CIT from rice and other food-polluted sources [116]. Additionally, this work clearly demonstrated the efficacy of the dummy template imprinting as a green strategy in avoiding the use of the toxic CT during the membrane preparation procedure [116].

Regarding pesticides, polypropylene-based hollow fiber membranes (of 0.2 μm pore size) with packed thiabendazole (TBZ) imprinted microspheres (MIP) in their lumen were successfully developed for the selective removal of this pesticide from orange and lemon samples [117]. Binding studies with MIP and NIP in acetonitrile solutions of TBZ evidenced that the adsorption capacity was strictly dependent on time. The produced membranes were tested in extractive orbital stirring of vials containing them and chopped bulk of fruits immersed in acetonitrile. Hollow fibers created a protective environment for the microspheres, which being imprinted with thiabendazole, interacted with it *via* hydrogen bonding. Interestingly, the porous nature of the fibers promoted the transport of the template from the shell to the membrane lumen containing the imprinted

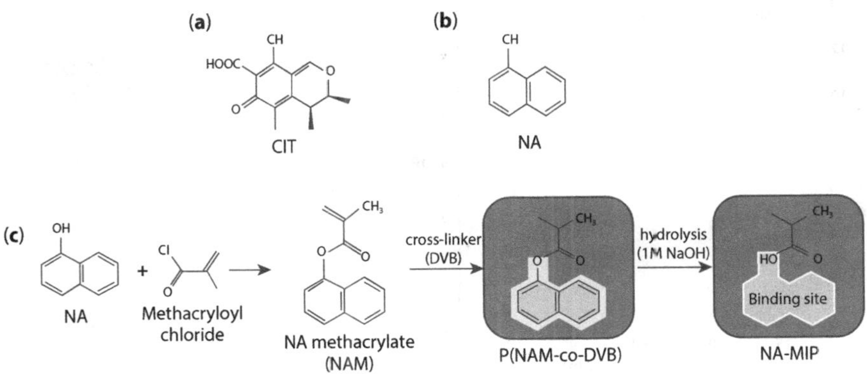

Figure 8.10 Chemical structure of CIT (a) and NA (b). Formation of recognition sites in an imprinted membrane prepared using NA as dummy template. (Reprinted from ref. [116]: Lee, T.P., Saad, B., Nakajima, L., *et al.*, Preparation and characterization of hybrid molecularly imprinted polymer membranes for the determination of citrinin in rice. *Sain. Malays*, 48, 1661, 2019. Open Access).

polymer, thereby promoting the interaction with its complementary imprinted sites. Parallel experiments with the empty fibers confirmed the absence of interactions between them and the template molecules during their migration from the solution to the lumen side [117]. Figure 8.11 shows the adsorption trend for MIP and NIP, as well as the vial before and after the extraction of TBZ from real citrus samples with the imprinted membrane.

More in detail, Figure 8.11a shows that the binding capacity of MIP increased up to 90 minutes and then decreased owing to the saturation of imprinted sites. On the other hand, the binding capacity of NIP increased throughout the investigated time but was much lower than that of MIP. The highest IF was obtained at 60 minutes of loading time. Figure 8.11b clearly demonstrates a change in color in the vial B at the end of the extraction process with a real sample (vial A), thus confirming the successful application of these TBZ-MIMs in the removal of the pesticide. The recovery rate from the real samples ranged from 5.1% to 6.1%; the initial concentration of thiabendazole in spiked samples was $830 \cdot 10^{-3}\ mg \cdot g^{-1}$. The extraction time was 60 minutes [117]. Similar membranes were produced for removing indole-3-butyric acid (IBA) from mung bean sprouts in aqueous solution [118]. Figure 8.12 shows the photos and the SEM images of the empty

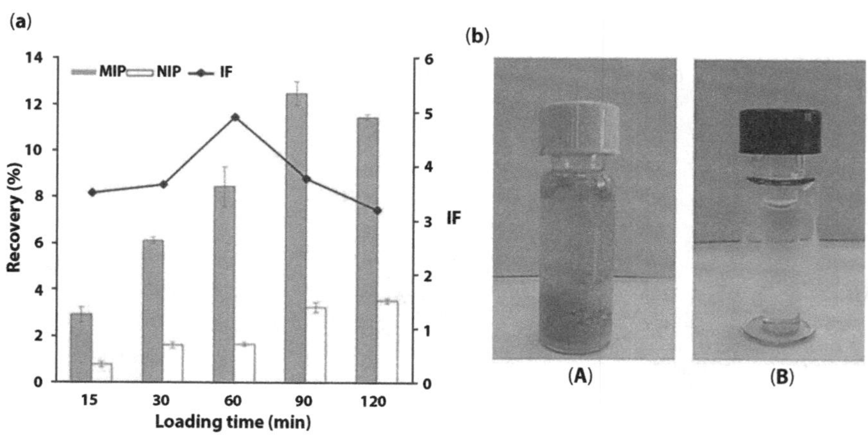

Figure 8.11 Recoveries (%) obtained in the extraction of TBZ in both MIP and NIP at different loading times (a), vial ready for the extractive process with thiabendazole-imprinted hollow fiber membranes (bA) and final extract (bB). Reprinted from ref. [117]: Díaz-Álvarez, M., Martín-Esteban, A., Hollow fiber membrane-protected molecularly imprinted microspheres for micro solid-phase extraction and clean-up of thiabendazole in citrus samples. *J. Chromatogr. A.*, 39, 1531, 2018. With the permission of Elsevier, copyright 2017.

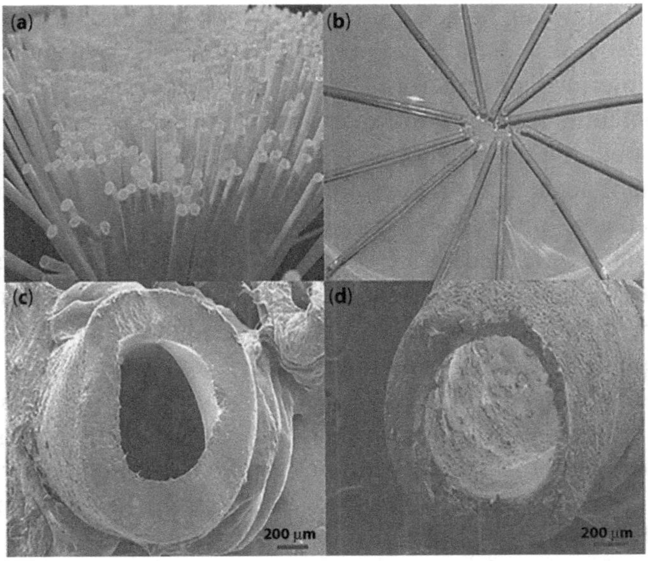

Figure 8.12 Photo of hollow fibers: empty (a) and packed (b); cross-section SEM images: empty (c) and packed (d) hollow fibers. Reprinted from ref. [118]: Aihebaier, S., Muhammad, T., Wei, A., *et al.*, Membrane-protected molecularly imprinted polymer for the microextraction of indole-3-butyric acid in mung bean sprouts. *A.C.S. Omega.*, 4, 16789, 2019. Open Access.

fibers and those packed with poly(vinyl imidazole) imprinted polymer particles having diameter in the range 38–75 μm.

The extraction of IBA was carried out in 2.0 $mg{\cdot}L^{-1}$ of mung bean samples with phosphate buffer. The binding capacity of the imprinted and the corresponding non-imprinted membrane were 1.13 $mg{\cdot}g^{-1}$ and 0.87 $mg{\cdot}g^{-1}$, respectively [118]. In another work, the selective and simultaneous recognition of traces of bisphenol A and tebuconazole in vegetable and juice samples was also done [119]. Other MIMs allowed for the removal of riboflavin and rhodamine, which are dyes used in food industries [120, 121].

8.4.2 Water Remediation

Water pollution by various contaminants has long been a problem for the entire world population due to its repercussions on environment, flora, fauna, and humans. Firstly, the excessive use of pesticides in agriculture, in addition to being harmful to the health of aquatic flora and fauna, also has negative repercussions on the health of the soil and humans through the food chain. Another source of pollution is represented by uncontrolled industrial wastewater. In fact, in addition to causing air pollution in numerous cases, it leads to the presence of pharmaceutical agents, chemical synthesis intermediates, microorganisms, metal ions, dyes, and other recalcitrant contaminants in groundwater, lakes, rivers, seawater, and drinking water. Last but not least, pollution from sanitary and domestic wastewater due to pharmacological residues and metabolic intermediates present in biological liquids must also be considered [59, 62, 122]. All this triggered an emergency, which led to the enactment of laws aimed at reducing the release into the environment of numerous substances with proven toxicity. At the same time, this problem has led the world of research and industry to develop effective sustainable intervention strategies aimed at carrying out the remediation of polluted sites and the recovery of recyclable compounds, in the logic of preserving the equilibrium of the ecosystem and in agreement with the circular economy concept. Among the various proposed technologies, membrane-based processes, due to their undisputed and widely demonstrated features (no requirement of additives or phase change, low energy and time consumption, easy realization and scalability, mild operating conditions, environmental friendliness, high performance) are competitive to the traditional techniques [59, 122, 123]. In this context, a special role is played by the use of porous MIMs thanks to their high specificity and selectivity [6, 8, 9, 59–62]. This paragraph discusses some examples of applications of porous MIMs in the selective recovery of targeted compounds and remediation strategies.

8.4.2.1 Removal of Pharmaceuticals

The increased consumption of pharmaceutical treatments in human and veterinary medicine, as well as the large use of antibiotics, hormones, and other growth factors in farms and aquacultures, has also increased the risk of contamination of water. In fact, different classes of active pharmaceutical compounds and their byproduct residues can reach environmental and drinking waters. Some of them are antibiotics, anticancer drugs, antiepileptics, beta-blockers, lipid regulators, non-steroidal anti-inflammatory drugs (NSAIDs), vasodilators, and so on [59, 122–124]. In the framework of water remediation strategies based on membrane processes, the literature reports good results obtained through the use of MIMs. In particular, the employment of porous membranes allowed high removal efficiency of specific contaminants from tap water, river water, industrial wastewater, and much more. For example, various MIMs were produced to remove residual antibiotics in water [59, 122–124]. Their use is increasing resulting in a great widespread problem owing to the insurgence of antibiotic resistance related to their persistent presence in polluted sites. Therefore, they are included in the list of emerging contaminants of water. Statistical information tells us that 105,600 tons of veterinary medications are going to be used per year by 2030 [59, 122–126]. An example is ciprofloxacin (CP), a highly water-soluble antibiotic present in wastewater of pharmaceutical industries and hospitals. Various MIMs were produced for its recovery. In 2017 [127], a porous ciprofloxacin imprinted nanocomposite membrane was fabricated by synthesizing a ciprofloxacin imprinted polymer layer (*via* atom transfer radical polymerization) on a polydopamine (PDA) layer covering the surface of a microfiltration PVDF-based membrane activated with SiO_2 nanoparticles. Membranes showed rapid adsorption of template molecules. The specific adsorption capacity of MIM was 64.54 $mg{\cdot}g^{-1}$, while that of the corresponding NIM was 12.89 $mg{\cdot}g^{-1}$. After 10 regeneration cycles, the recognition performance remained almost constant [127]. Permselective studies performed with a mix solution of CP and interfering analytes using an apparatus consisting of a donor and a receiving chamber (containing the mix feeding solution and water, respectively), indicated that owing to the existence of a concentration gradient between them, all the compounds permeated through the porous structure of the membranes until reaching equilibrium. However, for MIM, the transport of template molecules was lower due to their binding or binding and desorption to the imprinted sites of the membrane. On the contrary, the competing ciprofloxacin lactate (CPFL), and moxifloxacin (MOFX) did not interact with the imprinted sites and permeated the membrane faster. After 60 minutes, the permeation factor CPFL/CP was 2.27, while that of the couple MOFX/CP was 2.33. When using NIM, no difference in permeation rate between all the tested analytes was observed [127]. More recently [128], the same separation mechanism was exhibited by imprinted membranes fabricated through an innovative approach entailing the construction of a porous sleeve with PVDF as membrane-forming material in the presence of β-cyclodextrin (β-CD) *via* the *in situ* polymerization [128]. The presence of PDA thermally embedded into the membrane allowed the creation of a chemical bridge between the membrane and a ciprofloxacin imprinted polymer layer, synthesized *via* click chemistry using MAA as the functional monomer and pentaerythritol tetra(3-mercaptopropionate) (PT3M) as the crosslinker. The adoption of this strategy, rather than polymerizing PDA on the membrane surface *via* grafting, enhanced the interpenetrating bicontinuous morphology of membrane pores and the number of recognition sites of the membrane microchannels [128]. Adsorption isotherms and kinetics studies evidenced a homogeneous monolayer distribution of the surface imprinted sites and that the binding was promoted by chemiosorption and physical diffusion. The higher imprinting factor (3.40) was achieved at pH 7 for the membrane produced at 4 h of polymerization time. At this pH and an initial concentration of 10 mg·L^{-1}, the rebinding capacity of MIM was 121.12 mg·g^{-1}. Furthermore, in selective rebinding studies with mixed solutions of ciprofloxacin, enrofloxacin (ENR), norfloxacin (NOR), and ofloxacin (OFL) the separation factors CIP/ENR, CIP/NOR, and CIP/OFL was 5.70, 2.89, and 8.28, respectively.

Permeation studies demonstrated the occurrence of a retarded permeation of template molecules owing to their interaction with the recognition sites of the membrane pores [128]. Figure 8.13a shows the contact angle of a MIM prepared *via* the thermal induced PDA embedding and that of a MIM prepared *via* the traditional surface grafting. Figure 8.13b shows the selective separation mechanism [128].

From Figure 8.13a, it is evident that the thermally induced embedding membranes completely adsorbed the water drop in less than 1s, while the contact angle of the membrane prepared *via* the traditional surface grafting was still 85.9 [128]. Many other papers deal with the selective recognition of CIP [129–131]. Another antibiotic quite present in water is tetracycline, which is largely used for its activity against a wide spectrum of bacteria. Trotta *et al.* [132] fabricated a porous tetracycline hydrochloride

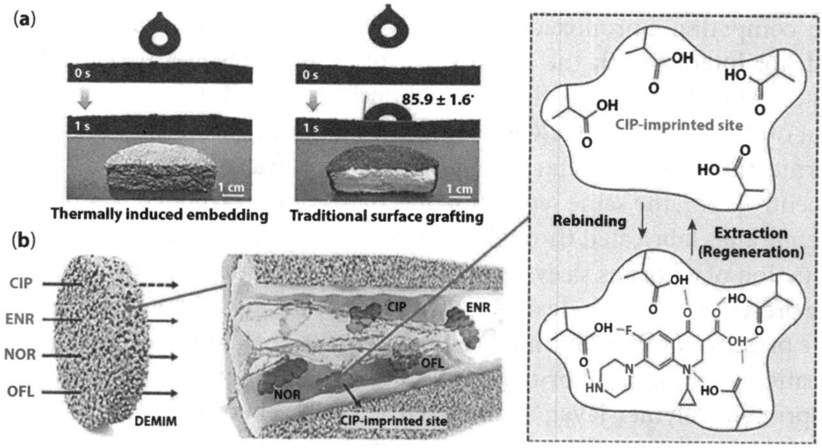

Figure 8.13 Contact angle of a MIM prepared *via* the thermal induced PDA embedding and that of a MIM prepared *via* the traditional surface grafting a); selective separation mechanism of ciprofloxacin-imprinted membranes with embedded PDA. Readapted from ref. [128]: Wang, X, Liu, Z., Lu, J., *et al.*, Highly selective membrane for efficient separation of environmental determinants: enhanced molecular imprinting in polydopamine embedded porous sleeve. *Chem. Eng. J.*, 449, 137825, 2022. With the permission of Elsevier, copyright 2022.

(TCH)-imprinted membrane *via* the phase-inversion technique using the pre-synthesized poly(acrylonitrile-co-acrylic acid) (P(AN-co-AA) as functional polymer. The MIM had a typical asymmetric porous structure with permeation properties in the range of ultrafiltration and was selective toward the template with respect to the competing chloramphenicol. The binding capacity toward TCH and CHP was 0.14 mg·g^{-1} and 5.17·10^{-2} mg·g^{-1}, respectively. On the contrary, the PAN-based membrane and the non-imprinted one only exhibited low unselective binding. Results indicated that the carboxylic groups of the functional polymer exercised an important role in the selective recognition process owing to their interaction with the aminic group of the antibiotic [132]. Over time, other tetracycline-based MIMs were produced *via* free-radical and *in situ* polymerization, as well as *via* the sol–gel imprinting and other strategies [133–141] for extracting tetracycline from environmental water. For example, the research group of Wu [136] first fabricated well-designed multilevel/hierarchical 3DW tetracycline-imprinted membranes (3DW-MIMs). The procedure entailed a first step of basswood surface modification with TiO_2@PDA-based layers, followed by the synthesis of a TC imprinted polymer layer employing 3-aminopropyltriethoxysilane (APTES) and tetraethyl orthosilicate (TEOS) as the functional monomer and the crosslinker, respectively. Finally, a sandwich-like mesoporous membrane (3DW-MIM) was obtained. Figure 8.14 shows the selective recognition behavior of 3DW-MIMs toward the template and the similar antibiotics cefalexin (CEX), sulfamethazine (SMZ), and oxytetracycline (OTC) after isothermal static rebinding (a) and permeation studies (b).

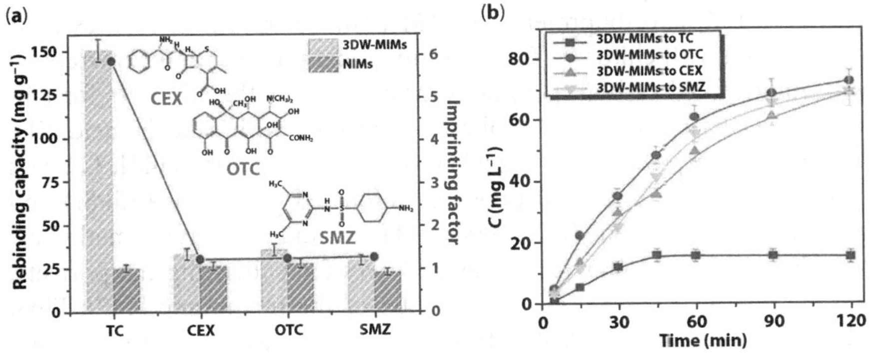

Figure 8.14 Rebinding capacity (a) and transport rate (b) of TC and its similar compounds with TC-imprinted and non-imprinted membranes produced by Wu *et al.*, (Reprinted from ref. [136]: Wu, Y., Xing, W., Ma, F., *et al.*, Three-dimensional basswood-based membrane with well-designed multilevel/hierarchical imprinting surface: a high-efficiency selective separation system. *Chem. Eng. J.*, 398, 125636, 2020. With the permission of Elsevier, copyright 2020).

In Figure 8.14a is possible to see that the imprinted membrane (the initial concentration of analytes in solution was 60 mg·L^{-1}) exhibited high binding of TC than to its structural homologs, while the non-imprinted one showed a similar and poor binding toward all the investigated analytes. Furthermore, from Figure 8.14b is evident as the template molecules permeated slowly the membrane while the interfering analytes permeated faster [136]. Other developed MIMs for water remediation deal with the selective removal of norfloxacin [142], enoxacin [143] and pyrimethamine [144]. A last example is the very recent exploitation of the wet phase inversion for producing sulfamethoxazole-cellulose-based porous imprinted membranes [145]. CA was used as the functional polymer forming membrane and polyethylene glycol as the pore forming agent. The SEM analysis evidenced as both MIMs and NIMs had a uniform microporous structure with enhanced porosity in comparison with a membrane prepared only with CA, which appeared dense [145]. The produced MIMs were successfully combined with a smartphone-based sensor for selectively detect and remove sulfamethoxazole from real river waters and wastewater through a specific selective adsorption. The recovery rate was in the range 91–103.6 %. The adsorption occurred on a monolayered homogeneous surface and was based on a chemical adsorption [145]. Sulfamethoxazole is a member of the widespread family of sulfonamides antibiotics. The United States Environmental Protection Agency (EPA) fixed at 0.1 μg mL^{-1} its maximum tolerable concentration in water, and similarly to other antibiotics it needs to be monitored for avoiding environmental pollution [145]. Imprinted membranes were also produced for removing other type of water contaminants, as for example the non-steroidal anti-inflammatory drugs salicylic acid [146-148] and ibuprofen [149–151], and so on. In this context, CA was blended with chitosan (CS) as a functional polymer and CS wrapped with multiwalled carbon nanotubes (MWCNTs) as the additives for fabricating a phase inversion membrane imprinted with salicylic acid. The binding capacity was 22.25 mg·g^{-1} [146]. In another case [147], CA was blended with the functional polymer sulfonated polysulfone (SPSf) and polyethylene glycol 4000 (PEG 4000) and ionic liquid 1-butyl-3-methylimidazolium chloride ([BMIM]Cl) as additives for enhancing the porosity, the hydrophilicity and the permeate flux of the membrane [147]. Final MIMs exhibited a porous structure containing macrovoids, low contact angle (50.1°), robust adsorption capacity, as well as high permeate flux (1166 L·m^{-2}·h^{-1}). The permeation separation mechanism was the retarded transport of the template [147]. The adsorption equilibrium was reached faster (15 min), owing to the creation of a great number of recognition sites with high affinity toward SA [147]. Salicylic acid-based MIMs were also prepared *via* the phase inversion using CA as polymeric matrix and ethylene imine as the functional monomer. Afterwards, for enhancing the antifouling and hydrophilic properties of the new borne MIMs they were deep coated with PVA and TiO_2 nanoparticles and successfully used for removing this drug from wastewater [148]. The binding capacity of MIMs reached the value of 24.3 $mg{\cdot}g^{-1}$, while the selectivity salicylic acid/(p-hydroxybenzoic acid) and salicylic acid/methyl salicylate was 3.87 and 3.55, respectively [148]. Differently, porous PVDF membrane and PDA modified TiO_2 (pDA@TiO_2) microspheres were used for developing an ibuprofen imprinted membrane [149]. Porous basswood-based MIMs were also developed using the betablocker propranolol as the targeted drug [152]. Membranes were prepared through different steps, which entailed the modification of the basswood surfaces with PDA, silver and TiO_2 nanoparticles and the subsequent synthesis of a propranolol imprinted layer *via* the sol-gel process (see Figure 8.15).

Membranes showed excellent specific binding (118.6 $mg{\cdot}g^{-1}$) and permselective properties. The separation mechanism was the retarded permeation of propranolol with respect to the competing atenolol, bisoprolol, and celiprolol [152]. These few cases that have been discussed represent a part of the existing works on this topic. More information is present in the already cited literature.

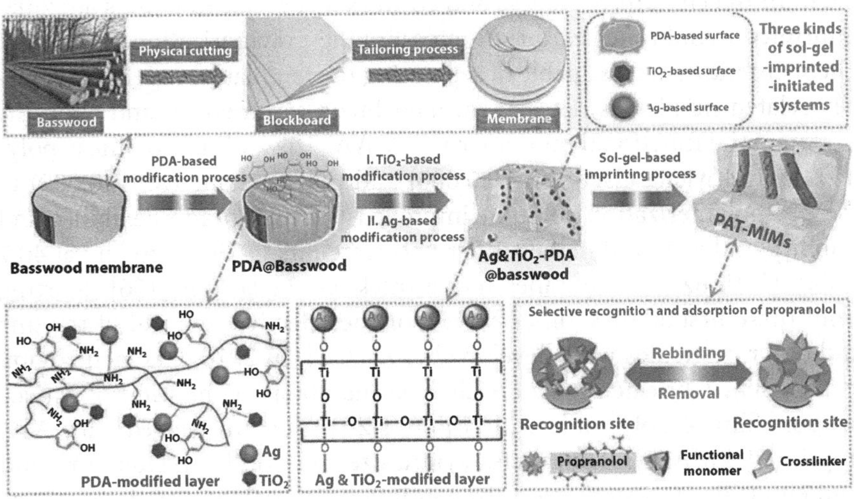

Figure 8.15 Scheme of the synthetic process of wood-based multi-layered porous propranolol-imprinted membranes. (Reprinted from ref. [152]: Yan, J., Sui, Y., Yan, M., *et al.*, Wood-based multi-layered porous imprinted membranes with dual metal composite nanostructures: synergistic promotion of rebinding capacity, selectivity and permeability. *J. Environ. Chem. Eng.*, 10, 107774, 2022. With the permission of Elsevier, copyright 2022).

8.4.2.2 Removal of Pesticides and Other Recalcitrant Contaminants

Other water-polluting agents are represented by pesticides and other organic recalcitrant. Similar to pharmaceuticals, their excessive presence in water results in danger and therefore requires continuous control and eventual removal actions. For example, the widespread use of pesticides in agriculture has resulted in their persistence in soil, air, and water, and, as already discussed, their unwanted presence in our food crosses the food chain. They enter our body not only *via* ingestion but also through inhalation and cutaneous contact and are responsible for the insurgence of many pathologies, like autoimmune and neurological diseases, cancer, liver damage, neonatal defects, reproductive problems, etc. [3, 153]. Owing to their high toxicity, pesticides are subject to strict regulations limiting their concentration. In the field of membrane technology, various imprinted membranes for removing pesticides from different water sources were produced. One of the most representative pesticides is atrazine (ATZ), a member of the family of triazines, which are heterocyclic aromatic compounds used as herbicides. Atrazine is present in environmental water. It is an organochlorine herbicide that is extremely dangerous for our endocrine system. Furthermore, it can accumulate in organisms promoting oxidative stress and cell injury. It is present in the list of possible carcinogens by the Environmental Protection Agency (EPA) [154]. Already at the beginning of the third millennium, Sergeyeva and co-workers developed porous atrazine-imprinted membranes *via in situ* crosslinking polymerization [155, 156]. The membranes were capable of separating the template from the structurally related herbicides desmetryn, metribuzin, prometryn, and simazine in water. The functional monomer was MAA. The addition of linear polymers polyethylene glycol (Mw 20,000) and polyurethane (Mw 40,000) to the pre-polymerization solution improved the membrane flexibility and stability, and achieved higher water permeate flux (15305 $L{\cdot}m^{-2}{\cdot}h^{-1}$ at 40.7 MPa). Optimized membranes had a thickness of 60 μm [156]. Starting from these first studies, other MIMs for the selective removal of triazine and other pesticides were developed [157–162]. In 2016, Gkementzoglou *et al.* [158], fabricated a sandwich-type membrane consisting of a two-support film of polyamide with a pore size of 0.1 μm packed with ATZ imprinted PMAA nanoparticles synthesized *via* both precipitation and miniemulsion polymerization. In continuous crossflow filtrations, membranes proved highly efficient in removing ATZ present in water samples at very low concentrations (< 1ppb) [158]. More recently, Xing's group [162] developed porous hierarchically non-imprinted (HP-NINMs) and atrazine-imprinted membranes (HP-MINMs) containing metal-organic frameworks (MOFs)-based nanoparticles surface imprinted with a thin layer of PMAA *via* reversible addition fragmentation chain transfer (RAFT) polymerization. The prepared membranes had nanopores distributed both on the surface and in their interior matrix. In addition, they were characterized by the presence of multiple interpenetrated transfer channels originating from the interconnections between the pores of a nanofiber and the imprinted MOF nanoparticles, as well as with those of adjacent nanofibers. The occurrence of these channels facilitated access to the imprinted sites and enhanced the permeation flux by providing an extra pathway for the permeating solution [162]. The porosity of HP-MINMs reached the value of 90%. Figure 8.16 shows the results of batch isothermal (a) and kinetic (b) adsorption studies, as well as the adsorption behavior toward template molecules and the competing ametryn (AMT) and terbuthylazine (TBL) depending on concentration (c) and rebinding time (d).

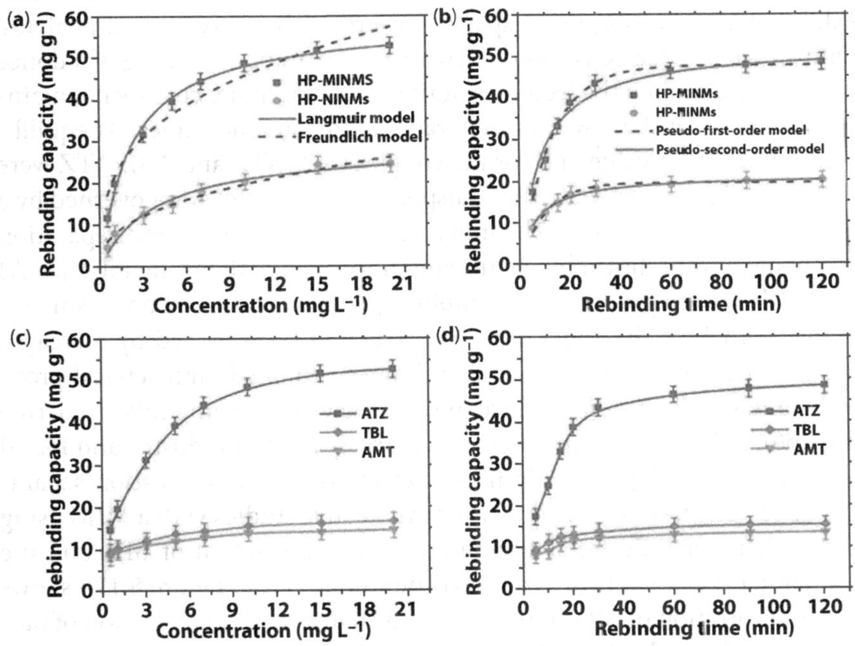

Figure 8.16 Isothermal (a) and kinetic (b) models fitting of rebinding capacity of HP-MINMs and HP-NINMs; selective rebinding performance of HP-MINMs at different concentrations (c) and rebinding times (d). Reprinted from ref. [162]: Xing, W., Ma, Z., Wang, C., *et al.*, Hierarchically porous molecularly imprinted membranes with multiple transfer channels for micropollutants selective separation. *Desalination*, 547, 116226, 2023. With the permission of Elsevier, copyright 2022.

In Figure 8.16, it is clearly evident that as the initial feed concentration (a) and the adsorption time (b) increase, the amount of ATZ adsorbed by both HP-MINMs and HP-NINMs also increases. However, the adsorption capacity of the imprinted membrane is much higher due to the specific interactions between the template and the imprinted sites. For optimized MIMs, the imprinting factor was 2.66. All the membranes were characterized by monolayer adsorption typical of the Langmuir model. Furthermore, the adsorption kinetics data fitted well with the pseudo-second-order model [162]. In selective adsorption investigations, HP-MINMs were selective with respect to the competing analytes AMTZ and TBL, whose adsorption remained almost constant and at a low level even when both initial feed concentration (c) and adsorption time (d) increased. This was because these compounds were not able to interact with the imprinted sites of the membrane *via* specific binding through chemical interactions and were adsorbed only through non-specific interactions. This last type of binding also occurred when using HP-NINMs. In fact, they were not selective and exhibited a low adsorption capacity toward all the investigated analytes. Diffusional permselective studies (with an apparatus consisting of a donor chamber containing the feeding solution and a receiving chamber containing water) showed a low transport rate of template molecules. At equilibrium, the highest value of β for the couple AMT/ATZ and TBL/ATZ were 3.52 and 3.22, respectively. The transport of the analytes was governed by a concentration gradient [162]. Similar recognition behavior and separation mechanism were shown by membranes imprinted with bisphenol A (BPA) [163]. They were produced by combining hierarchical microporous membranes with MIPs having a dot array structure (synthesized by mercapto alkenes click chemistry). These membranes exhibited antibacterial properties and excellent performance in selectively recognizing this endocrine disruptor, which is a phenol derivative largely used in industry and found in natural water. Figure 8.17a illustrates the retarded permeation separation mechanism occurring during permselective studies with a MIM using a two-chamber diffusional system and a mixed solution of BPA and the similar 4,4′-bisphenol (BIP) and hydroquinone (HQ). Figure 8.17b shows the concentration of BPA in the receiving chamber during filtration of natural water through BPA-MIMs and corresponding NIMs. The concentration of BPA in the initial feed solution was 75 $mg \cdot L^{-1}$ [163].

From Figure 8.17b, it is possible to see that in the case of NIMs, the high concentration trend of BPA confirmed their low nonspecific and weaker rebinding capacity toward template molecules, resulting in a short time to reach the equilibrium. On the contrary, for BPA-MIMs after 120 minutes, the concentration of BPA was still distant from the saturation. The

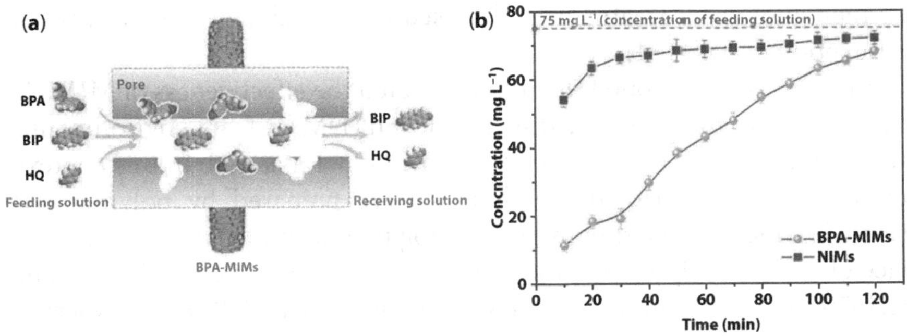

Figure 8.17 Schematic diagram of the selective separation of BPA by BPA-MIMs (a); Concentration of BPA in semi-practical samples after filtration by BPA-MIMs and NIMs (b). Reprinted from ref. [163]: Lu, J., Qin, Y., Li, C., *et al.*, Irregular dot array nanocomposite molecularly imprinted membranes with enhanced antibacterial property: Synergistic promotion of selectivity, rebinding capacity and flux. *Chem. Eng. J.*, 405, 126716, 2021. With the permission of Elsevier, copyright 2020.

permselective coefficient BIP/BPA was 2.78 while that of the couple HQ/BPA was 7.57 [163].

Another emergent water contaminant is the herbicide 2,4-D, a member of the phenoxy acetic acids largely used in agriculture. Owing to its toxicity for humans and animals, its acceptable limit was fixed at 70 $\mu g \cdot L^{-1}$ by the World Health Organization (WHO) [59, 164]. In 2020, 2,4-D-MIMs (prepared by γ-initiated polymerization using N-vinyl imidazole monomer and ethylene glycol dimethacrylate (EGDMA) as the functional monomer and crosslinker, respectively) permitted the removal of this herbicide from real tap water and spring water samples [164]. The higher binding capacity (23.0 $\mu mol \cdot g^{-1}$) was observed at an initial concentration of 5 $mg \cdot L^{-1}$ of ultrapure water aqueous solution at pH 7.0 for 100 minutes of testing time. In the case of real samples, the binding capacity was lower (18 $\mu mol \cdot g^{-1}$ and 15 $\mu mol \cdot g^{-1}$ for spring and tap water, respectively) owing to the presence of other interfering species (ions and other molecules) and their possible interactions with the membrane [164]. In recognition tests with multicomponent solutions, 2,4-D-MIMs proved to be selective toward the 2,4-D with respect to the analogs 2-(2,4-dichlorophenoxy) propionic acid (2,4-DP) and 4-chlorophenoxyacetic acid (4-CPA), while NIMs showed low nonspecific adsorption toward all analytes. Scanning electron micrography analysis of MIMs surfaces revealed a porous structure similar to a cauliflower, while the surface of NIMs appeared smoother [164]. The adsorption behavior was in agreement with the pseudo-second-order kinetic model. After each binding step, membranes were regenerated by methanol/water treatment and reused for a total of seven adsorption/desorption cycles [164].

In 2023, Wang's group [165] demonstrated the effectiveness of MIMs in the removal of the fungicide dimethomorph (DMM) from environmental water and complex *Panax notoginseng* medicinal material. In this context, dimethomorph-imprinted membranes were prepared by polymerizing (*via* UV irradiation) a thin imprinted polymer layer of PMAA on the surface of a PVDF microporous support membrane modified with PDA and KH570 [165]. The modification steps with PDA and KH570 preserved the surface and internal porous structure of the native PVDF, while the addition of the imprinted layer determined a little decrease in the surface porosity. Figure 8.18 shows the photos (a) and the SEM images of the pristine PVDF membrane and the different modification steps at 2k (b) and 20k (c) magnification.

The newly developed membranes had a high hydrophilic character; the contact angle was 36.53°, while that of the pristine PVDF-based membrane was 128.93°. Kinetic and thermodynamic adsorption studies in batch operating mode showed that the binding capacity of both DMM-IMs and NIMs increased over time until equilibrium was reached. The adsorption behavior was in agreement with both the Langmuir model and

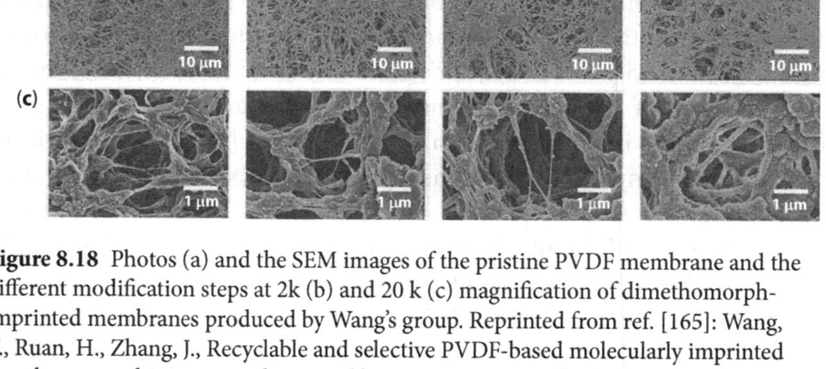

Figure 8.18 Photos (a) and the SEM images of the pristine PVDF membrane and the different modification steps at 2k (b) and 20 k (c) magnification of dimethomorph-imprinted membranes produced by Wang's group. Reprinted from ref. [165]: Wang, Y., Ruan, H., Zhang, J., Recyclable and selective PVDF-based molecularly imprinted membrane combining mussel-inspired biomimetic strategy for dimethomorph elimination. *Chem. Eng. J.*, 478, 147322, 2023. With the permission of Elsevier, copyright 2023.

the pseudo-second-order kinetic model. However, DMM-IMs exhibited higher adsorption capacity with respect to NIMs, and the equilibrium was reached after 120 minutes, while in the case of NIMs, the equilibrium was reached after 60 minutes. The maximum adsorption capacity of DMM-IMs (314 $\mu g \cdot cm^{-2}$) was obtained at 0.05 $mmol \cdot L^{-1}$ of the initial template concentration. Under these conditions, the adsorption capacity of NIMs was 195 $\mu g \cdot cm^{-2}$ [165]. This difference was due to the presence of recognition sites in DMM-MIMs, which were characterized by a specific adsorption recognition mechanism primarily based on chemical interactions between the template and the recognition sites and low nonspecific chemical adsorption. In this context, template molecules moved from the solution to the membrane by diffusive transport. In the initial step, the specific adsorption arose mostly on the membrane surface, while subsequently, it occurred at the level of internal recognition sites. On the contrary, NIMs exhibited only nonspecific adsorption. The application of the imprinted membrane in the adsorptive removal of DMM from real tap water samples containing it at residual concentrations of 0.02 $mmol \cdot L^{-1}$, 0.05 $mmol \cdot L^{-1}$, and 1.0 $mmol \cdot L^{-1}$ resulted in removal rates of 74.43%, 82.65%, and 86.12%, respectively [165]. Differently, the exploitation of the phase inversion technique allowed for the production of MIMs applied in the removal of the notorious polychlorinated biphenyls (PCBs) 2-2,3,3′,4,4′-pentachlorobiphenyl, 3,3′,4,4′,5-pentachlorobiphenyl, 2,4,5,2′,4′,5′-hexa-chlorobiphenyl, and 1,1′-biphenyl, 2,2′,3,4,4′,5′,6-heptachlorobiphenyl from environmental waters, where they are often present [166]. In this context, MIMs were prepared *via* wet phase inversion, hybridizing PSf with a previously synthesized MIP using MAA as the functional monomer and 4-phenylphenol (BP-OH) as the dummy template [166]. Morphological analysis evidenced the presence of small pores on the membrane surface and a porous internal structure consisting of finger-like macrovoids and pore channels. NIMs had a similar but less porous structure. Water flux was typical of ultrafiltration; the highest value (247 $L \cdot m^{-2} \cdot h^{-1}$) was shown by BP-OH-MIMs containing the 18 wt. % of PSf and the 20 wt. % of MIP with respect to PSf. In dead-end filtrations of river water samples containing the PCBs and the structural analogs anthracene and dichlorodiphenyltrichloroethane, a facilitated mass transport of PCBs through BP-OH-MIMs was observed [166]. Phase inversion porous PSf-based IMs were also developed for the removal of the polycyclic aromatic naphthalene from petrochemical wastewater effluents [167].

Another class of water pollutants is represented by dyes, which are largely used in textile industries and other sectors such as paper production, cosmetics, food industries, etc. [168]. Traditional approaches exploited for removing them from polluted water have been disadvantageous due to their high operating costs, as well as oxidation and stability problems [169]. Although literature reports only a few papers dealing with dye-based IMs to date, their application has shown promising potential for real-world applications [68, 170]. As a representative example, methylene blue (MB) photocatalytic imprinted membranes, fabricated by blending PSf-based UF membrane with imprinted TiO_2 nanoparticles, allowed for the removal of this dye from water in an integrated photodegradation-ultrafiltration system [171]. The process entailed the photodegradation of the dye through UV irradiation during a crossflow filtration (under a pressure gradient) through the imprinted membrane and the accumulation of the degradation product into the permeate stream. Filtrations carried out with water samples containing MB and methyl orange (MO) revealed that the presence of the imprinted photocatalytic nanoparticles into the membrane promoted the photodegradation of template molecules with respect to the competing dye. The selectivity value of MB removal with respect to MO was 2.83 [171]. The molecular recognition occurred at the surface of the photocatalyst. The mechanism was based on attractive electrostatic interactions between the positive charge of MB and the negative charge of PMAA polymerized on the TiO_2 nanoparticles' surface (see Figure 8.19).

Conversely, being negatively charged, MO was not able of to interact with the photocatalyst. Finally, the presence of the surface imprinted

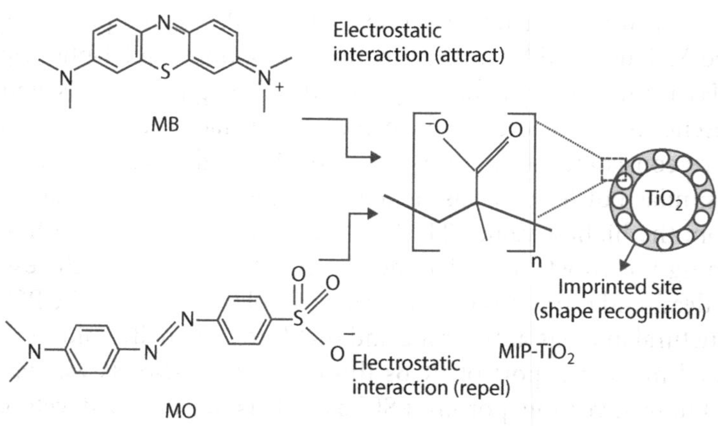

Figure 8.19 Mechanism of selective removal of MB and MO of MIP-TiO_2 after dissociation in solution. Reprinted form ref. [171]: Melvin, Ng, H.K., Leo, C.P., Abdullah, A.Z., Selective removal of dyes by molecular imprinted TiO_2 nanoparticles in polysulfone ultrafiltration membrane. *J. Environ. Chem. Eng.*, 5, 3991, 2017. With the permission of Elsevier, copyright 2017.

TiO_2 nanoparticles increased membrane porosity and flux. Sericin/PVA electrospun fibers and CA/PSf-based MIMs were also produced for the selective removal of MB from wastewater [172, 173]. Additionally, surface imprinted membranes were applied in the removal of rhodamine from wastewater in batch adsorption/desorption cycles [174], while porous mixed matrix membranes based on PVDF blend and hydrophilic molecularly imprinted MIL-101 (Cr) as filler were successfully applied in the removal of the red dye erythrosin B (EB), through a specific adsorption recognition mechanism [175].

8.5 Ion Imprinted Membranes and Removal of Ions

The contamination of numerous matrices by ions is another important problem that should not be underestimated. Their use at an industrial level has led to their continuous presence in the environment, resulting in the contamination of soil, the atmosphere, surface and groundwater, as well as water resources accessible from runoff and those used in agriculture. This also represents a health problem for living organisms as they can be absorbed by plants and aquatic fauna and end up on our table through the food chain causing harmful results. For example, continuous exposure to heavy metals (HMs) *via* contact, inhalation, or ingestion, can lead to their accumulation in fauna, the human body, and aqueous matrices, contributing to the insurgence of numerous pathological states and the alteration of the equilibrium of the aquatic ecosystem [176, 177]. Figure 8.20 illustrates the sources and the distribution of HMs in the aquatic environment [176].

Among them, the divalent lead may accumulate in blood, bones, and various body organs with deleterious effects on the skeletal, neurological, and reproductive systems [176, 178]. Trivalent arsenic is responsible for DNA damage, cancer, cardiovascular and neuronal diseases, gastrointestinal problems, skin lesions, and more [179]. Other dangerous metals are aluminum (Al), barium (Ba), beryllium (Be), cadmium (Cd), cobalt (Co), chromium (Cr), copper (Cu), manganese (Mn), mercury (Hg), nickel (Ni), ruthenium (Rt), thallium (Tl), vanadium (Va), zinc (Zn), etc. [176, 177]. Another ion frequently present in water is the alkaline monovalent cation lithium (Li), which is largely used in industry. It is not dangerous at low concentrations, but at high amounts, it can be toxic to aquatic and terrestrial environments [180, 181].

Due to the greater general persistence of contaminating ions, strategies and regulations have been adopted to minimize their concentration in industrial wastewaters before their discharge into the environment.

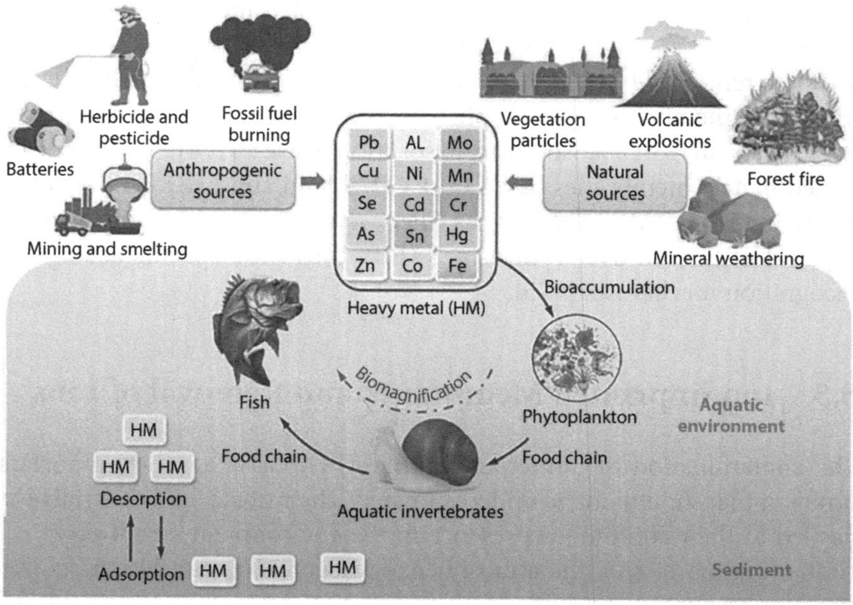

Figure 8.20 Sources of heavy metals and their distribution in the aquatic environment. Reprinted from ref. [176]: Elnabi M.K.A., Elkaliny, N.E., Elyazied, M.M., *et al.*, Toxicity of heavy metals and recent advances in their removal: a review. *Toxics*, 11, 580, 2023. Open Access.

Furthermore, technological progress has allowed the development of advanced processes to reduce their presence in contaminated sites and at the same time be able to reuse them when possible, in the logic of a circular economy. From this viewpoint, once again, membrane separation processes have proven advantageous compared to classical water treatment methods (e.g., coagulation, flocculation, and biological treatment) [182, 183]. A prodigious contribution has also come from the application of ion imprinted membranes (IIMs), which exploit the characteristics of membrane-based separations and imprinting technology. Ion imprinted membranes are useful for highly selective recognition and separation of ions. They are applicable on various matrices and permit the separation of ions having similar charge and radius [184]. The separation processes involving the use of IIMs are mostly based on the occurrence of electrostatic interactions, which are compatible with aqueous environments. Overall, the preparation procedures of IIMs exploit the same strategies applied for preparing MIMs. However, it must be considered that even if ions have similar chemical and physical properties to organic molecules, they have minor functional groups and small dimensions. Therefore, the synthesis of IIMs requires the use of ligands having appropriate coordination groups able to interact with the ion of interest, promoting the formation of ion–ligand complexes usually stabilized by the aid of a crosslinker, which is chosen based on the functional monomer used. The process is directed by the charge, size, and shape of the template ion. Its removal at the end of the synthetic process leads to the formation of the imprinted sites [184–186]. Some examples of ligands are carboxylic acids, calixarene derivatives, carboxylic acids, cyclodextrins, crown ethers, and β-diketones. Their coordination groups contain either electro-donor or electro-receiving atoms (e.g., azo-derivatives, oxygen, nitrogen, sulfur, vinylated groups, etc.) depending on the ion to be imprinted. The most employed functional monomers are acrylamide, methacrylic acid, 4-vinylpyridine, 1-vinylimidazole, propyl acetamide, tetraethoxysilane, and so on. Sometimes, ligands also act as functional monomers [184–186]. The parameters that govern the recognition process are pH, temperature, extraction time, template-monomer, and template-ligand ratio. Figure 8.21 illustrates a general representation of the ion imprinting technology [65].

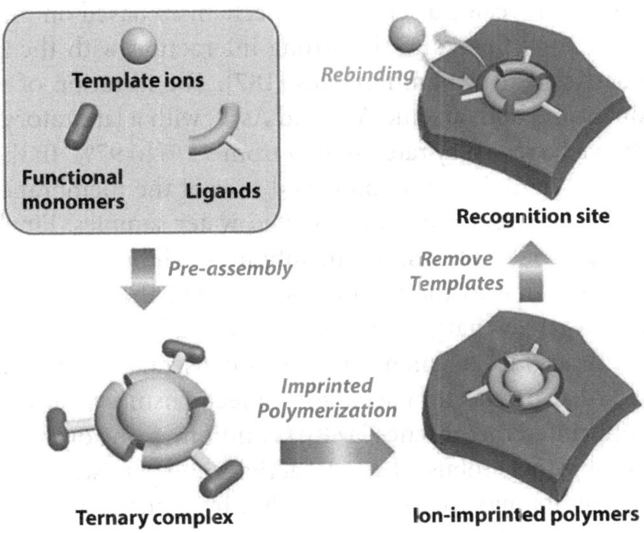

Figure 8.21 Schematic representation of the ion imprinting technology. Reprinted from ref. [65]: Lu, J., Qin, Y., Wu, Y., *et al.*, Recent advances in ion imprinted membranes: separation and detection *via* ion-selective recognition. *Environ. Sci. Wat. Res. & Technol.*, 5, 1626, 2019. With the permission of the Royal Society of Chemistry, copyright 2019.

Similarly to MIMs, the separation mechanisms characterizing IIMs are the selective adsorption of the template, as well as its facilitated or retarded permeation [184–186]. Until now, the production of IIMs has been lower with respect to that of MIMS even if it has increased in the last decade. Regarding their application in water remediation, some papers deal with the removal of heavy metals, toxic anions, monovalent cations, as well as alkaline and rare earth metals [182–186]. Heavy metals are transition elements that are very prone to forming coordination complexes with suitable ligands thanks to their large hydration radius. In 2016, Roy *et al.* [187] developed cysteine@ZnS:TiO_2 nanoparticles modified IIMs for removing arsenic and arsenate (As^{3+} and As^{5+}) from water samples and drinking water samples collected from different areas of India. Membranes were prepared through the "grafting from" method using carboxy methyl cellulose (CMC) and (PVA) as polymers, the monomer acrylamide, and vinyl cysteine/zinc sulfide/titanium oxide nanoparticles to ensure the interaction with the template ion. The crosslinker was N,N'-methylene bisacrylamide. In batch rebinding studies with aqueous solutions containing As^{3+} and As^{5+}, membranes exhibited a very high specific adsorption capacity toward them with respect to the non-imprinted ones [187]. Furthermore, permeation experiments carried out with arsenic As^{3+} and various interfering ions including sodium (Na^{+}), copper (Cu^{2+}), zinc (Zn^{2+}), cadmium (Cd^{2+}), chloride, fluoride, and nitrate revealed a separation mechanism based on the retarded permeation of As^{3+}, thanks to the strong interaction with the imprinted TiO_2 nanoparticles of recognition sites [187]. The filtration of real water samples containing both arsenic As^{3+} and As^{35+}, with a laboratory filtration system, achieved a recovery rate ranging from 93 % to 97%. In the adsorption of both As^{3+} and As^{5+}, the imprinted sites of the membrane played a central role in their removal from polluted water samples. Finally, in the presence of bacterial cultures, the membranes exhibited an antibacterial activity. This was due to the interaction between the positive charge of zinc and the negative charge of the bacterial cell, leading to the destruction of the intracellular elements and, as a consequence, bacterial death [187]. A retarded permeation separation mechanism was also shown by cadmium-imprinted membranes [188]. Cadmium is one of the most toxic heavy metals due to its robust chemical activity and persistence. The WHO and the EPA set its maximum consumption by humans at 1 $\mu g{\cdot}kg^{-1}$ day^{-1} [188–190]. Membranes were produced through the surface modification of a PVDF membrane, creating specific host-guest binding sites along its pores through a high-temperature crosslinking process [188]. The ligand was 3-allylrhodanine, while EGDMA was the crosslinker. Batch rebinding studies showed that Cd^{2+} bound by the IIM was 1.84 times higher than that of the corresponding NIM. The investigation of mass transport revealed a permeation amount of Cd^{2+} (1.05 μg·cm^{-2}·h^{-1}) through IIM of 60.95% lower than that of NIM. It was also lower than that of Ag^{-} and Mg^{2+}, which have hydration and Stokes radii closer to Cd^{2+}. The recognition mechanism was based on the occurrence of electrostatic interactions between the positive charge of cadmium and the negative charge of the sulfur present in the 3-allylrhodanine. The binding rate was characterized by three steps: at the beginning, rapid binding of Cd^{2+} occurred thanks to the presence of the high number of recognition sites. In the second step, the binding rate decreased, and the template ion reached the binding site owing to a concentration gradient. In the final step, the binding rate was very slow owing to the exhaustion of the imprinted sites [188]. More recently, Hong and co-workers [189] fabricated porous composite Cd-IIMs by the surface imprinting of a polytetrafluoroethylene (PTFE) membrane support using the new N,N'-pyridine-2,6-bis(2-methacrylamide) (PMA) as the functional monomer and EGDMA as the crosslinker. When used for removing cadmium from water by soaking, these Cd-IIMs exhibited excellent adsorption capacity. Adsorption kinetics studies evidenced that experimental data were in agreement with the pseudo-second-order model. The performance was affected by the pH: in the interval 3–8 of pH the adsorption capacity of both IIM and NIM increased, while above pH 8 it decreased owing to the scarcity of H^{+} ions in the solution, which influenced the membrane stability and allowed a variation of the structure of imprinted sites. After 6 minutes of soaking, the maximum adsorption capacity of the optimized IIM (named Cd(II)-IICM$_{17}$) was 509.93 mg·g^{-1}. For this membrane, the molar ratio template/functional monomer/cross-linker was 1:4:50. The initial Cd(II) concentration was 25.6 mg·mL^{-1}, while the adsorptive IF was 2.15 [189]. Osmotic selectivity studies performed on real water samples collected from the nearby areas revealed a facilitated permeation of the template ion in comparison with the interfering copper and nickel (see Table 8.3).

This was because a privileged pathway for the template was created *via* binding and dissociating from strictly closed recognition sites [49, 72].

Other IIMs were focused on the selective removal of chromium III (Cr(III)) and chromium VI (Cr(VI)) [191–193]. Among them, an interesting work deals with the production of Cr(VI)-imprinted membranes (Cr(VI)-IICMs) the newly synthesized pH-responsive functional monomer 2-allyl-1,3-diphenyl-1,3-propanedione (ADPD) for modifying the surface of nylon membranes in the presence of the template ions [191]. Brunauer-Emmett-Teller (BET) adsorption/desorption analysis assessed that the pristine and the novel membranes (both imprinted and non-imprinted)

Table 8.3 Experimental results on the selectivity of $Cd_{(II)}$-$IICM_{17}$ in real water sample. Reprinted from ref [189]: Hong, Y., Wang, X., Fu, D., Synthesis of cadmium(II) ion imprinted composite membrane with a pyridine functional monomer and characterization of its adsorption performance. *e-Polymers*, 23, 2022810, 2023. Open access.

Ion	**J (mg·cm^{-2}·s^{-1})**	**P(cm^2·s^{-1})**	***βCd/Co,Ni,Cu***
Cd(II)	1.019 x 10^{-6}	1.567 x 10^{-10}	---
Ni(II)	3.395 x 10^{-7}	4.597 x 10^{-11}	3.41
Cu(II)	8.642 x 10^{-7}	1.174 x 10^{-10}	1.34

were mesoporous, with an average pore size of 2–50 nm. Figure 8.22 shows the effect of temperature and pH on the adsorption performance of the different prepared membranes.

From Figure 8.22a, it is evident that an increase in temperature above 293 kelvin (K) resulted in a general decrease in adsorption capacity and a reduction in the difference of the adsorption between the Cr(VI)-IICMs and the corresponding non-imprinted composite membrane (NICMs). This behavior was attributed to the fact that the high temperatures increased the kinetic motion of Cr(VI), reducing the possibility of its interaction with the imprinted sites. Furthermore, it allowed the conversion of the enolic form of ADPD into its ketonic form, which was not suitable for promoting the adsorption of Cr(VI) [191, 194]. Regarding the pH

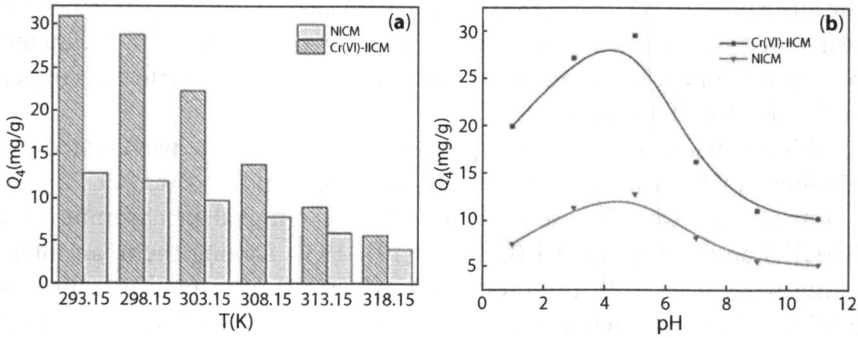

Figure 8.22 Effect of (a) temperature and (b) pH on the Cr(VI) adsorption on Cr(VI)-IICMs and corresponding NICMs. (Reprinted from ref. [191]: Li, P., Wang, X., Wang, G., *et al.*, Synthesis and evaluation of ion-imprinted composite membranes of Cr(VI) based on β-diketone functional monomers. *RSC Adv.*, 11, 38915, 2021. Open access).

effect shown in Figure 8.22b, within the pH range 1–5, the performance of Cr(VI)-IICM increased, while at pH values above 5, a rapid decrease in adsorption capacity occurred. This was due to the presence of ADPD enol anions (C=C–O–) under basic operating conditions, which hindered the adsorption process at the imprinted sites. In fact, at acidic pH, the presence of the cationic enol form of ADPD (C–C=OH) promoted the adsorption of $CrO4^{2-}$ (see Figure 8.23).

Nevertheless, under extreme acidic conditions (pH < 3), the occurrence of the ketonic structures (C=O) of ADPD led to the reduction in the amount of the enolic form with the consequent reduction of the adsorption capacity of the membrane [191, 195].

In the framework of alkaline, alkaline–earth, and transition metals IIMs were successfully applied in the recovery of lithium (Li^{1+}) [65, 196–198], calcium (Ca^{2+}) [199], magnesium (Mg^{2+}) [200], and silver (Ag^{2+}) [201].

Owing to its large industrial use, the monovalent cation Li^{1+} is one of the most focused metals. Lithium is a raw material primarily found in clay deposits, salt-lake brines, seawater, and hard rock. It is used in the production of batteries, ceramics, glass, mobile phones, pharmaceuticals, polymers, portable computers, and so on [65, 198, 202, 203] and represents an emerging contaminant. The presence of lithium in industrial wastewaters and the increased demand in comparison with the production has led to the

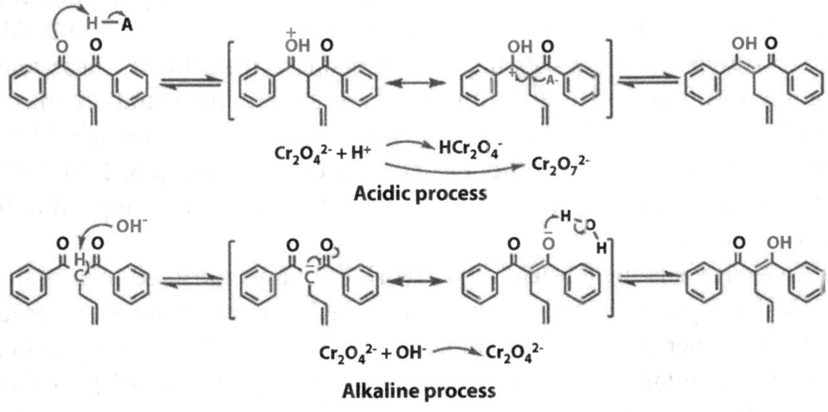

Figure 8.23 Mechanism of the structural variations of ADPD under acid or basic conditions. Reprinted from ref. [191]: Li, P., Wang, X., Wang, G., *et al.*, Synthesis and evaluation of ion-imprinted composite membranes of Cr(VI) based on β-diketone functional monomers. *RSC Adv.*, 11, 38915, 2021. Open access.

development of many strategies for its extraction, not only from primary sources but also for its recovery and recycling from secondary resources. This aims to prevent its future scarcity and environmental accumulation, as well as address the economic point of view. Considering that the recycling of lithium is around 1% and the consumption has reached 16% against a production of 12%, these are important aspects that should not be underestimated [198, 202, 203]. Up to now, lithium has been recovered and recycled from ion batteries of the metallurgical industry, aquifer brines, seawater, and wastewaters. Traditional strategies include crystallization, evaporation, chemical precipitation, solvent extraction, chromatography, and membrane technology [186, 198, 202, 203]. Despite their agreement with the concept of circular economy, most of these methods are expensive and harmful, owing to the high water and energy consumption, CO_2 emissions, the use of high volume of organic solvents, etc. [198, 202, 203]. The employment of traditional membrane-based separations proved more advantageous and sustainable with respect to other processes, but their use is still limited owing to the poor selectivity toward monovalent ions [202, 203]. This limit is overcome by the application of IIMs, which exhibit high selective binding (adsorption) properties toward lithium [186, 196–198]. Many papers deal with the development and application of Li-based IMs. In 2024, Zavahir and co-workers critically reviewed the application of various IIMs in lithium recovery from brine and seawater [196]. A useful strategy for producing Li-IMs is the surface modification of microporous membrane support through different routes. It must be highlighted that the Li^{1+} forms highly stable complexes with the 12-crown-4-ether (12CAE) and calixarenes (CALIXs) owing to their richness in electrons. These compounds are often used both as functional monomers and ligands, thanks to the close matching of the 12CAE ring and the cup-like structure of CALIXs with Li^{1+} rather than other monovalent and divalent ions [65, 196]. Table 8.4 lists some examples of Li-surface imprinted membranes applied in its selective recovery.

An example is the fabrication of hybrid lithium-imprinted membranes (LIIMs) *via* phase inversion, embedding TiO_2 nanoparticles in a porous PVDF polymer matrix. The nanoparticles were imprinted in a previous step, which entailed the immobilization of the ligand (12CAE) on their surface by hydrolytic polymerization. The functional monomer was the 3-(aminopropyl)tryethoxilane), while the crosslinker was ethyl orthosilicate. Finally, during the membrane preparation, PVP was added to the dope solution as a pore-forming agent [206]. All the membranes exhibited a spongy porous structure, together with improved hydrophilicity and

Table 8.4 Some examples of surface imprinted membranes applied in the selective recovery of lithium.

Membrane support material	Functional monomer	Ligand	Separation mechanism	Selectivity	Ref.
Polyvinylidene fluoride	Methacrylic acid	12-crown-4-ether	Selective adsorption	$\alpha_{Li}{}^{1+}/{}_{k}{}^{1+} = 4.36$ $\alpha_{Li}{}^{1+}/{}_{Na}{}^{1+} = 5.03$ $\alpha_{Li}{}^{1+}/{}_{Mg}{}^{2+} = 3.55$	[204]
Polyvinylidene fluoride	Methacrylic acid	2- methylol-12-crown-4 ether	Retarded permeation	$\beta_{K}{}^{1+}{}_{/Li}{}^{1+} = 8.23$ $\beta_{Na}{}^{1+}{}_{/Li}{}^{1+} = 8.30$ $\beta_{Mg}{}^{2+}{}_{/Li}{}^{1+} = 8.58$	[205]
Polyvinylidene fluoride	3-(Aminopropyl) tryethoxilane)	12-crown-4-ether	Selective adsorption	$\alpha_{Li}{}^{1+}/{}_{k}{}^{1+} = 17.00$ $\alpha_{Li}{}^{1+}/{}_{Na}{}^{1+} = 24.60$ $\alpha_{Li}{}^{1+}/{}_{Ca}{}^{2+} = 21.30$ $\alpha_{Li}{}^{1+}/{}_{Mg}{}^{2+} = 6.80$	[206]
Polyvinylidene fluoride	12-crown-4-ether	-crown 4-ether	Selecttive adsorption	$\alpha_{Li}{}^{1+}/{}_{Mn}{}^{2+} = 6.71$ $\alpha_{Li}{}^{+}/{}_{Co}{}^{2+} = 5.84$ $\alpha_{Li}{}^{2+}/{}_{Ni}{}^{2+} = 3.03$	[207]
Polydimethylsiloxane	4-tert-Butylcalix [4] arene	4-tert-Butylcalix [4] arene	Retarded permeation	$\beta_{Na}{}^{+}/{}_{Li}{}^{+} = 1.71$ $\beta_{K}{}^{+}/{}_{Li}{}^{+} = 4.56$ $\beta_{Rb}{}^{2+}{}_{/Li}{}^{1+} = 3.80$	[208]

(Continued)

Table 8.4 Some examples of surface imprinted membranes applied in the selective recovery of lithium. (*Continued*)

Membrane support material	Functional monomer	Ligand	Separation mechanism	Selectivity	Ref.
Polypyrrole	2-hydroxymethyl-12-crown-4 ether	2-hydroxymethyl-12-crown-4 ether	Selective adsorption	$\alpha_{Li^{1+}/k^{1+}} = 4.11$ $\alpha_{Li^{1+}/Na^{1+}} = 4.20$ $\alpha_{Li^{1+}/Mg^{2+}} = 4.13$ $\alpha_{Li^{1+}/Al^{3+}} = 4.30$ $\alpha_{Li^{1+}/Fe^{3+}} = 4.28$	[209]
Polyvinylidene fluoride	12-crown-4-ether	12-crown 4-ether (12C4E)	Retarded permeation	$\beta_{K^{+}/Li^{1+}} = 5.38$ $\beta_{Ca^{2+}/Li^{1+}} = 21.94$ $\beta_{Mg^{2+}/Li^{1+}} = 15.56$	[210]
Polyvinylidene fluoride	2-(allyloxy) methyl-12-crown-4	2-(allyloxy)methyl-12-crown-4	Retarded permeation	$\beta_{Mg^{2+}/Li^{1+}} = 8.94$ $\beta_{Mg^{2+}/Li^{1+}} = 5.50$	[211] [212]
Polyethersulfone	Methacrylic acid	12-4-crown-ether	Retarded permeation	$\beta_{Na^{+}/Li^{+}} = 7.39$ $\beta_{K^{+}/Li^{+}} = 9.86$	[213]

roughness compared to the simple PVDF-based membrane. Additionally, they showed an optimal specific adsorption capacity characterized by homogeneous monolayer adsorption. Adsorption behavior fitted well with the pseudo-second-order kinetic model, indicating the occurrence of a chemical adsorption mechanism. More in detail, the maximum adsorption capacity calculated with the kinetic model was 131.90 mg·g^{-1}, while the experimental value was 132.00 mg·g^{-1}. Osmotic filtration experiments confirmed the presence of a retarded permeation separation mechanism owing to the presence of an adsorption/desorption process of template ions at the imprinted sites of the membrane. Meanwhile, the competing ion species Na^{1+}, K^{1+}, Mg^{2+}, and Ca^{2+} freely passed through the membrane pores. Figure 8.24 shows the permeability behavior over time of the investigated ions (a), the osmotic flux of Li^{1+} through LIIMs and NIMs over time (b), and the schematic diagram of the permeation separation mechanism (c) [206].

In a different case, imprinted microfibers with an ordinated porous structure allowed a scalable recovery of L^{1+} from coin cells [197]. The

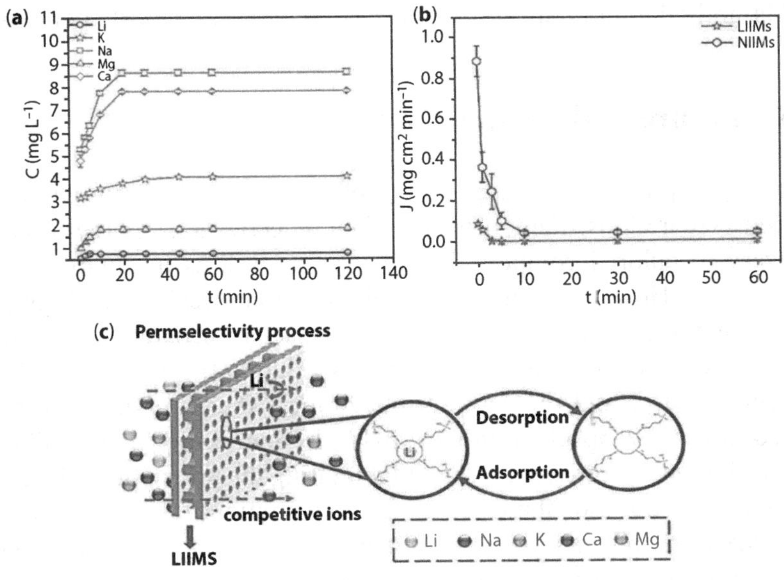

Figure 8.24 Permeation ability of LIIMs (a). Osmotic flux curves of LIIMs and NIIMs to Li^{1+} over time (b), and schematic diagram of permeation mechanism (c) (100 mg·L^{-1} of L^{1+}, Na^{1+}, K^{1+}, Mg^{2+}, and Ca^{2+} mixed solution). Reprinted from ref. [206]: Yang, J., Qu, G., Liu, C., *et al.*, An effective lithium ion-imprinted membrane containing 12-crown ether-4 for selective recovery of lithium. *Chem. Eng. Res. Des.*, 184, 639, 2022. With the permission of Elsevier, copyright 2022.

diameter of the fibers was 176.5±3.44 μm, that of the pores was 2.29±0.31 μm and the ion exchange capacity was 8.45 meq·g^{-1}. Differently, flat sheet membranes were useful for their selective recovery from aqueous environments [211]. Other interesting applications deal with the selective recovery of rare earth metals (REMs) such as dysprosium (Dy^{3+}) [214, 215], europium (Eu^{3+}) [216, 217], gadolinium (Gd^{3+}) [218, 219], neodymium (Nd^{3+}) [220], and more. REMs are 17 elements of the periodic table that attracted the attention of the industrial community. Thanks to their excellent magnetic, optical, and electric properties, they are employed in the production of clean energy, petrochemical engineering, high-tech productions, etc. However, their exacerbated use caused an increase in their cost and environmental emissions, as well as the risk of their future availability. Today, they are viewed as critical raw materials to be recovered from environmentally polluted sites and recycled as much as possible [6, 65, 221].

To date, various IIMs have been developed for the selective removal of other ions both from water and food matrices. The recognition and separation mechanisms characterizing them were similar to the already discussed examples and were assessed on a case-by-case basis. For more detailed information regarding the whole production of IIMs toward lithium, REMs, and other ionic species, readers can refer to the reported literature.

8.6 Future and Perspectives

This chapter presented the applications of imprinted membranes in highly efficient separations, demonstrating their effectiveness in the selective recovery of tailored compounds and ions, as well as in environmental remediation. However, it must be highlighted that despite their high performance, the industrial application of these smart separation systems is still at an embryonic stage. This is probably due to the scarce knowledge of their potential and the costs of membranes. Therefore, it is necessary to **promote their application worldwide by increasing their visibility, emphasizing** their advantages, and highlighting the possibility of integrating them with traditional membrane-based operations (such as membrane crystallization and membrane distillation) or as final purifying units in pressure-driven integrated membrane systems, to develop eco-friendly separation processes on a large scale. It is desired that this will be achieved with the aid of technological transfer action, finding an equilibrium between the benefits for humans and the environment and the investment costs.

References

1. Peña, O.I.G., Zavala, M.Á.L., Ruelas, H.C., Pharmaceuticals market, consumption trends and disease incidence are not driving the pharmaceutical research on water and wastewater. *Int. J. Environ. Res. Public Health*, 18, 1, 2021.
2. Nayak, A. and Bhushan, B., An overview of the recent trends on the waste valorisation techniques for food wastes. *J. Environ. Manage.*, 253, 352, 2019.
3. Koch, S., Epp, A., Lohmann, M., Böl, G.F., Pesticide residues in food: attitudes, beliefs, and misconceptions among conventional and organic consumers. *J. Food Prot.*, 80, 12, 2017.
4. Donato, L., Molecular recognition, in: *Encyclopedia of Membranes*, E. Drioli and L. Giorno (Eds.), pp. 1–2, Springer-Verlag, Berlin Heidelberg.
5. Holyoak, T., Molecular recognition: lock-and-key, induced fit, and conformational selection, in: *Encyclopedia of Biophysics*, C.K.R. Gordon (Ed.), pp. 1584–1588, Springer, Berlin, Heidelberg, 2013.
6. Donato, L. and Drioli, E., Imprinted membranes for sustainable separation processes. *Front. Chem. Sci. Eng.*, 15, 775, 2021.
7. Donato, L., Mazzei, R., *et al.*, Molecular recognition-driven membrane processes, in: *Smart membranes and sensors: synthesis, characterization, and applications*, A. Gugliuzza (Ed.), pp. 269–300269–300, Wiley-Scrivener, Beverly, 2014.
8. Donato, L., Nasser, I.I., Majdoub, M., *et al.*, Green chemistry and molecularly imprinted membranes. *Membranes*, 12, 472, 2022.
9. Yang, H., Liu, H.B., Tang, S.Z., *et al.*, Synthesis, performance, and application of molecularly imprinted membranes: A review. *J. Environ. Chem. Eng.*, 9, 106352, 2021.
10. Lu, J., Qin, Y., Wu, Y., *et al.*, Bidirectional molecularly imprinted membranes for selective recognition and separation of pyrimethamine: A double-faced loading strategy. *J. Membr. Sci.*, 101, 117917, 2020.
11. Zhu, T., Xu, D., Wu, Y., *et al.*, Surface molecularly imprinted electrospun affinity membranes with multimodal pore structures for efficient separation of proteins. *J. Mater. Chem. B*, 1, 6449, 2013.
12. Cui, J., Wu, Y., Meng, M., *et al.*, Bio-inspired synthesis of molecularly imprinted nanocomposite membrane for selective recognition and separation of artemisinin. *J. Appl. Polym. Sci.*, 133, 43405, 2016.
13. Ma, Y., Wang, H., Guo, M., Stainless steel wire mesh supported molecularly imprinted composite membranes for selective separation of Ebracteolata compound B from *Euphorbia fischeriana*. *Mol.*, 24, 465, 2019.
14. Son, S.H. and Jegal, J., Chiral separation of D,L-serine racemate using a molecularly imprinted polymer composite membrane. *J. Appl. Polym. Sci.*, 104, 1866, 2007.

15. Bukhari, A., Monier, M., Elsayed, N., Surface molecular imprinting of nylon fibers for chiral recognition of D-phenyl lactic acid. *Polym. Int.*, 68, 1460, 2019.
16. Michaels, A.S., Baddour, R.F., Bixler, H.J., *et al.*, Conditioned polyethylene as a permselective membrane: separation of isomeric xylenes. *Ind. Eng. Chem. Process Des. Dev.*, 1, 14, 1962.
17. Jantarat, C., Attakitmongkol, K., Nichsapa, S., *et al.*, Molecularly imprinted bacterial cellulose for sustained release delivery of quercetin. *J. Biomater. Sci., Polym. Ed.*, 31, 1961, 2020.
18. Suedee, R., Bodhibukkana, C., Tangthong, N., *et al.*, Development of a reservoir type transdermal enantioselective controlled delivery system for racemic propranolol using a molecularly imprinted polymer composite membrane. *J. Control. Release*, 129, 170, 2008.
19. Su, C., Li, Z., Zhang, D., *et al.*, A highly sensitive sensor based on a computer-designed magnetic molecularly imprinted membrane for the determination of acetaminophen. *Biosens. Bioelectron.*, 148, 111819, 2020.
20. Barahona, F., Diaz-Alvarez, M., Turiel, E., *et al.*, Molecularly imprinted polymer-coated hollow fiber membrane for the microextraction of triazines directly from environmental waters. *J. Chromatogr. A*, 1442, 12, 2016.
21. Donato, L., Greco, M.C., Drioli, E., Preparation of molecularly imprinted membranes and evaluation of their performance in the selective recognition of dimethoate. *Desalin. Water Treat.*, 30, 171, 2011.
22. Garcia Del Blanco, S., Donato, L., Drioli, E., Development of molecularly imprinted membranes for selective recognition of primary amines in organic medium. *Sep. Purif. Technol.*, 87, 40, 2012.
23. Mansour, M.S.M., Abdel-Shafy, H.I., Mehaya, F.M.S., Valorization of food solid waste by recovery of polyphenols using hybrid molecular imprinted membrane. *J. Environ. Chem. Eng.*, 6, 4160, 2018.
24. Wulff, G. and Sarhan, A., Über die anwendung von enzymanalog gebauten polymeren zur racemattrennung. *Angew. Chem.*, 84, 364, 1972.
25. Wulff, G., Sarhan, A., Zabrocki, K., Enzyme-analogue built polymers and their use for the resolution of racemates. *Tetrahedron Lett.*, 8, 4329, 1973.
26. Chen, L., Wang, X., Lu, W., *et al.*, Molecular imprinting: perspectives and applications. *Chem. Soc. Rev.*, 45, 2137, 2016.
27. Martin-Esteban, A., Molecularly-imprinted polymers as a versatile, highly selective tool in sample preparation. *TRAC.*, 45, 169, 2013.
28. Mosbach, K. and Ramström, O., The emerging technique of molecular imprinting and its future impact on biotechnology. *Nat. Biotechnol.*, 14, 163, 1996.
29. Sellergren, B., Lepisto, M., Mosbach, K., Highly enantioselective and substrate selective polymers obtained by molecular imprinting utilizing non-covalent interactions. NMR and chromatographic studies on the nature of recognition. *J. Am. Chem. Soc.*, 110, 5853, 1988.

30. Cormack, P.A.G. and Elorza, A.Z., Molecularly imprinted polymers: synthesis and characterization. *J. Chromatogr. B*, 804, 173, 2005.
31. Wang, C.C. and Lee, W.C., Chromatographic characteristics of cholesterol-imprinted polymers prepared by covalent and non-covalent imprinting methods. *J. Chromatogr. A*, 962, 69, 2002.
32. Trotta, F., Biasizzo, M., Caldera, F., Molecularly imprinted membranes. *Membranes*, 2, 440, 2012.
33. Díaz-Álvarez, M., Turiel, E., Martín-Esteban, A., Recent advances and future trends in molecularly imprinted polymers-based sample preparation. *J. Sep. Sci.*, 46, 12, 2023.
34. Boysen, R.I., Schwarz, L.J., Nicolau, D.V., *et al.*, Molecularly imprinted polymer membranes and thin films for the separation and sensing of biomacromolecules. *J. Sep. Sci.*, 40, 314, 2017.
35. Donato, L., *Imprinted polymeric membranes*, pp. 54–56, De Gruyter, Berlin, Boston, 2023.
36. Yoshikawa, M., Tharpa, K., Dima, S.O., Molecularly imprinted membranes: past, present, and future. *Chem. Rev.*, 116, 11500, 2016.
37. De Luca, G., Donato, L., Tasselli, F., Garcia del Blanco, S., Drioli, E., On the cause of controlling affinity to small molecules of imprinted polymeric membranes prepared by non-covalent approach: a computational and experimental investigation. *J. Phys. Chem. B.*, 46, 12, 2011.
38. Reddy, P.S., Kobayashi, T., Abe, M., *et al.*, Molecular imprinted nylon-6 as a recognition material of amino acids. *Eur. Polym. J.*, 38, 521, 2002.
39. Faizal, C.K.M. and Kobayashi, T., Tocopherol-targeted membrane adsorbends prepared by hybrid molecular imprinting. *Polym. Eng. Sci.*, 6, 1085, 2008.
40. Kochkodan, V., Hilal, N., Melnik, V., *et al.*, Selective recognition of organic pollutants in aqueous solutions with composite imprinted membranes. *Adv. Colloid Interface Sci.*, 159, 180, 2010.
41. Nasser, I.I., Algieri, C., Garofalo, A., *et al*, Hybrid imprinted membranes for selective recognition of quercetin. *Sep. Purif. Technol.*, 163, 331, 2016.
42. Liu, Y., Meng, M.J., Yao, J.T., *et al.*, Selective separation of phenol from salicylic acid effluent over molecularly imprinted polystyrene nanospheres composite alumina membranes. *Chem. Eng. J.*, 286, 622, 2016.
43. Kobayashi, T., Reddy., P.S., Ohta, M., *et al.*, Molecularly imprinted polysulfone membranes having acceptor sites for donor dibenzofuran as novel membrane adsorbents: charge transfer interaction as recognition origin. *Chem. Mat.*, 14, 2499, 2002.
44. Ma, X., Chen, R., Zheng, X., *et al.*, Preparation of molecularly imprinted CS membrane for recognizing naringin in aqueous media. *Polym. Bull.*, 66, 853, 2011.
45. Ramamoorthy, M. and Ulbricht, M., Molecular imprinting of cellulose acetate-sulfonated polysulfone blend membranes for rhodamine B by phase inversion technique. *J. Membr. Sci.*, 217, 207, 2003.

46. Zhou, Z., Cui, K., Mao, Y., *et al.*, Green preparation of D-tryptophan imprinted self-supported membrane for ultrahigh enantioseparation of racemic tryptophan. *R.C.S. Adv.*, 6, 1099992, 2016.
47. Algieri, C., Drioli, E., Guzzo, L., *et al.*, Bio-mimetic sensors based on molecularly imprinted membranes. *Sensors*, 14, 13863, 2014.
48. Zhou, Z., He, L., Mao, Y., *et al.*, Green preparation and selective permeation of D-tryptophan imprinted composite membrane for racemic tryptophan. *Chem. Eng. J.*, 310, 63, 2017.
49. Donato, L., Figoli, A., Drioli, E., Novel composite poly (4-vinylpyridine)/ polypropylene membranes with recognition properties for (S)-naproxen. *J. Pharm. Biomed. Anal.*, 37, 1003, 2005.
50. Sueyoshi, Y., Utsunomiya, A., Yoshikawa, M., *et al.*, Chiral separation with molecularly imprinted polysulfone-aldehyde derivatized nanofiber membranes. *J. Membr. Sci.*, 89, 401, 2012.
51. Kunitake, T. and Lee, S.W., Molecular imprinting in ultrathin titania gel films *via* surface gel process. *Anal. Chim. Acta*, 504, 1, 2004.
52. Moein, M.M., Javanbakht, M., Karimi, M., *et al.*, A new strategy for surface modification of polysulfone membrane by *in situ* imprinted sol-gel method for the selective separation and screening of L-tyrosine as a lung cancer biomarker. *Analyst*, 140, 1939, 2015.
53. Jingjing, Z., Faguang, M., Rongxin, L., *et al.*, Porous MOFs-based self-assembled membrane with specific rebinding nanocages for selective recognition and separation at molecular level. *Desalination*, 572, 117124, 2024.
54. Lu, J., Qin., Y., Zhang, Q., *et al.*, Multilayered ion-imprinted membranes with high selectivity towards Li based on synergistic effect of 12-crown-4 and polyether sulfone. *Appl. Surf. Sci.*, 427, 931, 2018.
55. Dima, S.O., Dobre, T., Stoica-Guzun, A., *et al.*, Molecularly imprinted bio-membranes based on cellulose nano-fibers for drug release and selective separations. *Macromol. Symp.*, 359, 124, 2016.
56. Xing, W., Ma, Z., Wang, C., *et al.*, Metal-organic framework based molecularly imprinted nanofiber membranes with enhanced selective recognition and separation performance: a multiple strengthening system. *Sep. Purif. Technol.*, 278, 119624, 2022.
57. Saad, E.M., Gohary, N.A.E., El-Shenawy, B.M., *et al.*, Fabrication of magnetic molecularly imprinted beaded fibers for rosmarinic acid. *Nanomaterials*, 10, 1478, 2020.
58. Pirdadeh-Beiranvand, M., Afkhami, A., Madrakian, T., Magnetic molecularly imprinted electrospun nanofibers for selective extraction of nilotinib from human serum. *Anal. Bioanal. Chem.*, 412, 1629, 2020.
59. Hejji, L., Ali, Y.A.E.H., Azzouz, A., *et al.*, Recent insights into molecularly imprinted membrane technology for removal of pollutants from environmental water: from organic molecules to metal ions. *J. Wat. Proc. Eng.*, 58, 104852, 2024.

60. Chen, J., Wei, M., Meng, M., Advanced development of molecularly imprinted membranes for selective separation. *Molecules*, 28, 5764, 2023.
61. Ratnaningsih, E., Kadja, G.T.M., *et al.*, Molecularly imprinted affinity membrane: a review. *ACS Omega*, 7, 23009, 2022.
62. Lah, N.F.C., Ahmad, A.L., Low, S.C., Molecular imprinted membrane biosensor for pesticide detection: perspectives and challenges. *Polym. Adv. Technol.*, 32, 17, 2021.
63. Torres-Cartas, S., Catalá-Icardo, M., Meseguer-Lloret, S., *et al.*, Recent advances in molecularly imprinted membranes for sample treatment and separation. *Separations*, 7, 69, 2020.
64. Wang, R., Liu, C., Chen, Z., Imprinted composite membranes. *Prog. Chem.*, 32, 989, 2020.
65. Lu, J., Qin, Y., Wu, Y., *et al.*, Recent advances in ion imprinted membranes: separation and detection *via* ion-selective recognition. *Environ. Sci. Wat. Res. Technol.*, 5, 1626, 2019.
66. Ghorani, B., Tucker, N., Yoshikawa, M., Approaches for the assembly of molecularly imprinted electrospun nanofibre membranes and consequent use in selected target recognition. *Food Res. Int.*, 78, 448, 2015.
67. Zaidi, S.A., Recent developments in molecularly imprinted polymer nanofibers and their applications. *Anal. Methods*, 7, 7406, 2015.
68. Algieri, C., Drioli, E., Ahmed, C., *et al.*, Emerging tools for recognition and/or removal of dyes from polluted sites: molecularly imprinted membranes. *J. Membr. Sep. Tech.*, 3, 243, 2014.
69. Yoshikawa, M., Separation with molecularly imprinted membranes. *Kob. Ronb. S.*, 71, 223, 2014.
70. Ulbricht, M., Membrane separations using molecularly imprinted polymers. *J. Chromatogr. B Analyt. Technol. Biomed. Life Sci.*, 804, 113, 2004.
71. Donato, L., *Imprinted polymeric membranes*, pp. 36–44, De Gruyter, Berlin, Boston, 2023.
72. Piletsky, S.A., Panasyuk, T.L., Piletskaya, E.V., *et al.*, Receptor and transport properties of imprinted polymer membranes. A review. *J. Membr. Sci.*, 157, 263, 1999.
73. Bai, M., Qiang, L.Q., Meng, M., *et al.*, Upper surface imprinted membrane prepared by magnetic guidance phase inversion method for highly efficient and selective separation of artemisinin. *Chem. Eng. J.*, 405, 126889, 2021.
74. Kashani, T. and Jahanshahi, M., Nanopore molecularly imprinted polymer membranes for environmental usage: selective separation of 2,4-dichlorophenoxyacetic acid as a toxic herbicide from water. *Polym. Technol. Eng.*, 55, 1700, 2016.
75. Zhao, X., Cheng, Y., Xu, H., *et al.*, Design and preparation of molecularly imprinted membranes for selective separation of acetoside. *Front. Chem.*, 8, 775, 2020.

76. Meng, M. and Liu, Y., Introduction of an ordered porous polymer network into a ceramic alumina membrane *via* non-hydrolytic sol–gel methodology for targeted dynamic separation. *RSC Adv.*, 4, 38630, 2014.
77. Donato, L., Tasselli, F., De Luca, G., *et al.*, Novel hybrid molecularly imprinted membranes for targeted 4,4′-methylendianiline. *Sep. Purif. Technol.*, 116, 184, 2013.
78. Yan, M., Wu, Y., Zhang, K., *et al.*, Multifunctional-imprinted nanocomposite membranes with thermo-responsive biocompatibility for selective/controllable recognition and separation application. *J. Colloid Interface Sci.*, 582, 991, 2021.
79. Cheng, H., Zhu, X., Yang, S., *et al.*, A pH-controllable imprinted composite membrane for selective separation of podophyllotoxin and its analog. *J. Appl. Polym. Sci.*, 128, 363, 2013.
80. Chou, C.Y., Chang, S., Huang, W.T., Effects of monodentate and bidentate ligands on adsorption characteristics of Cu-ion-imprinted hybrids. *Res. Chem. Intermed.*, 45, 6043, 2019.
81. Liu, Q.S., Zheng, T., Wang, P., *et al.*, Adsorption isotherm, kinetic and mechanism studies of some substituted phenols on activated carbon fibers. *Chem. Eng. J.*, 157, 348, 2010.
82. Foo, K.F. and Hamee, B.H., Insights into the modeling of adsorption isotherm systems. *Chem. Eng. J.*, 156, 2, 2010.
83. Demirtaş, E., Odabaşı, M. Preparation and application of naringenin imprinted cryogel membranes for selectively separation of naringenin from natural environment. *Polym. Bull.*, 81, 11795, 2024.
84. Zhang, Y., Shan, X., Gao, X., Development of a molecularly imprinted membrane for selective separation of flavonoids. *Sep. Purif. Technol.*, 76, 337, 2011.
85. Chen, H., Zhang, H., Tian, J., *et al.*, Recovery of high value-added nutrients from fruit and vegetable industrial wastewater. *Compr. Rev. Food Sci. Food Saf.*, 18, 1388, 2019.
86. Conidi, C., Donato, L., Algieri, C., *et al.*, Valorization of chestnut processing by-products: a membrane-assisted green strategy for purifying valuable compounds from shells. *J. Clean. Prod.*, 378, 134564, 2022.
87. Moraes Segundo, J.D.P., Moraes, M.O.S., Brito, W.R., *et al.*, Molecularly imprinted membrane produced by electrospinning for β-caryophyllene extraction. *Materials*, 18, 15, 2022.
88. Ouarda, E.L., Hani, J.J.G.G., Palacios-Santander, J.M., *et al.*, Molecularly imprinted alginate-PVP membrane for the selective fluorescence detection of flavonoids in foods: a practical case for rutin. *Sens. Actuators B. Chem.*, 403, 135232, 2024.
89. Wang, W.Y., Qin, B., Shao, G., *et al.*, A nanocellulose molecularly imprinted membrane: preparation, characterization and application in targeted separation of taxane 10-deacetylbaccatin III. *Int. J. Biol. Macromol.*, 253, 126794, 2023.

90. Wang, Y., Li, L., Cheng, G., *et al.*, Preparation and recognition properties of molecularly imprinted nanofiber membrane of chrysin. *Polymers.*, 14, 2398, 2022.
91. Lu, J. and Xu, B., Bio-inspired design on EGCG-selective membrane: an anchoring/imprinting strategy based on bi-interactions. *J. Membr. Sci.*, 658, 120750, 2022.
92. Nasir, A.M., Ishak, N.H., Said, M.S.M., *et al.*, One-pot synthesis of molecular-imprinted membrane for selective extraction of caffeic acid. *Polym. Bull.*, 77, 3953, 2020.
93. Fan, J.P., Cheng, Y.T., Zhang, X.H., *et al.*, Preparation of a novel mixed non-covalent and semi-covalent molecularly imprinted membrane with hierarchical pores for separation of genistein in *Radix Puerariae Lobatae*. *React. Funct. Polym.*, 146, 104439, 2020.
94. Kamarudin, S.F., Ahmad, M.N., Dzahir, I.H.M., *et al.*, Development of quercetin imprinted membrane-based PVDF substrate. *Polym. Bull.*, 76, 4313, 2019.
95. Ghasemi, S. and Nematollahzadeh, A., Molecularly imprinted ultrafiltration polysulfone membrane with specific nano-cavities for selective separation and enrichment of paclitaxel from plant extract. *React. Funct. Polym.*, 126, 9, 2018.
96. Wei, M.H., Chen, H.Y., Wang, S., *et al.*, Synthesis and characterization of hybrid molecularly imprinted membrane with blending SiO_2 nanoparticles for ferulic acid. *J. Inorg. Organomet. Polym.*, 27, 586, 2017.
97. Liu, Q., Zhao, Y., Pan, J., *et al.*, A novel chitosan base molecularly imprinted membrane for selective separation of chlorogenic acid. *Sep. Purif. Technol.*, 164, 70, 2016.
98. Huang, Z., Zhang, P., Yun, Y., Preparing molecularly imprinted membranes by phase inversion to separate kaempferol. *Polym. Adv. Technol.*, 28, 373, 2017.
99. Zhu, X.Y., Zheng, Z.J., Xie, J., *et al.*, Selective separation of magnolol using molecularly imprinted membranes. *J. Sep. Sci.*, 35, 315, 2012.
100. Donato, L., Tasselli, F., Drioli, E., Molecularly imprinted membranes with affinity properties for folic acid. *Sep. Sci. Tech.*, 45, 2273, 2010.
101. Wang, P., Chen, S., Zhu, X., *et al.*, Daidzein-imprinted membranes using co-functional monomers. *J. Chromatogr. A*, 1216, 7639, 2009.
102. Wang, P., Hu, W., Su, W., Molecularly imprinted poly (methacrylamide-co-methacrylic acid) composite membranes for recognition of curcumin. *Anal. Chim. Acta*, 615, 54, 2008.
103. Tasselli, F., Donato, L., Drioli, E., Evaluation of molecularly imprinted membranes based on different acrylic copolymers. *J. Membr. Sci.*, 320, 167, 2008.
104. Faizal, C.K.M., Hoshina, Y., Kobayashi, T., Scaffold membranes for selective adsorption of α-tocopherol by phase inversion covalently imprinting technique. *J. Membr. Sci.*, 322, 503, 2008.

105. Wei, M., Wang, S., Jiang, W., *et al.*, Preparation and characterization of dual-template molecularly imprinted membrane with high flux based on blending the inorganic nanoparticles. *J. Inorg. Organomet. Polym.*, 28, 295, 2018.
106. Miean, K.H. and Mohamed, S., Flavonoid (myricetin, quercetin, kaempferol, luteolin, and apigenin content of edible tropical plants. *J. Agri. Food. Chem.*, 496, 3106, 2001.
107. Zhang, W., Zhang, Q., Zhang, X., *et al.*, Preparation and evaluation of molecularly imprinted composite membranes for inducing crystallization of oleanolic acid in supercritical CO_2. *Anal. Methods*, 8, 5651, 2016.
108. Lu, J., Qin, Y., Zhang, Q., *et al.*, Antibacterial, high-flux and 3D porous molecularly imprinted nanocomposite sponge membranes for cross-flow filtration of emodin from analogues. *Chem. Eng. J.*, 360, 483, 2019.
109. Kupchan, S.M. and Karim, A., Tumor inhibitors. Aloe emodin: antileukemic principle isolated from Rhamnus Frangula. *Lloydia.*, 39, 223, 1975.
110. Singh, S., Shukla, S., Tandia, N., *et al.*, Antibiotic residues: a global challenge. *Pharm. Sci. Mon.*, 5, 184, 2014.
111. Du, W., Sun, M., Guo, P., *et al.*, Molecularly imprinted membrane extraction combined with high-performance liquid chromatography for selective analysis of cloxacillin from shrimp samples. *Food Chem.*, 259, 73, 2018.
112. Cañada-Cañada, F., Muñoz de la Peña, A., Espinosa-Mansilla, A., Analysis of antibiotics in fish samples. *Anal. Bioanal. Chem.*, 395, 987, 2009.
113. Canadas, R., Garcinuno, R.M., Gonzalez, G.P., *et al.*, Development of a molecularly imprinted polymeric membrane for determination of macrolide antibiotics from cow milk. *Polymer*, 249, 124843, 2023.
114. Chen, W.J., Shang, P.P., Fang, S.B., *et al.*, Origin of macromolecular crowding: analysis of recognition mechanism of dual-template molecularly imprinted polymers by in silico prediction. *J. Chromatogr. A*, 1662, 462695, 2022.
115. Dong, Z., Lu, J., Wu, Y., *et al.*, Antifouling molecularly imprinted membranes for pretreatment of milk samples: selective separation and detection of lincomycin. *Food Chem., 333*, 127477, 2020.
116. Lee, T.P., Saad, B., Nakajima, L., *et al.*, Preparation and characterization of hybrid molecularly imprinted polymer membranes for the determination of citrinin in rice. *Sain. Malays*, 48, 1661, 2019.
117. Díaz-Álvarez, M. and Martín-Esteban, A., Hollow fiber membrane-protected molecularly imprinted microspheres for micro solid-phase extraction and clean-up of thiabendazole in citrus samples. *J. Chromatogr. A*, 39, 1531, 2018.
118. Aihebaier, S., Muhammad, T., Wei, A., *et al.*, Membrane-protected molecularly imprinted polymer for the microextraction of indole-3-butyric acid in mung bean sprouts. *ACS Omega*, 4, 16789, 2019.
119. Wu, Y.T., Zhang, Y.H., Zhang, M., *et al.*, Selective and simultaneous determination of trace bisphenol A and tebuconazole in vegetable and juice samples by membrane-based molecularly imprinted solid phase extraction and HPLC. *Food Chem.*, 164, 527, 2014.

120. Borrelli, C., Barsanti, S., Silvestri, D., *et al.*, Selective depletion of riboflavine from beer using membranes incorporating imprinted polymer particles. *J. Food Process. Preserv.*, 35, 112, 2011.
121. Li, L., Liu, H., Lei, X., *et al.*, Electrospun nanofiber membranes containing molecularly imprinted polymer (MIP) for rhodamine B (RhB). *Adv. Chem. Eng. Sci.*, 2, 266, 2012.
122. Nascimento, T.A., Fdz-Polanco, F., Peña, M., Membrane-based technologies for the up-concentration of municipal wastewater: a review of pre-treatment intensification. *Sep. Purif. Rev.*, 49, 1, 2020.
123. Mansour, I., Salah J., Antonis, A., Zorpas, P.D., Membrane technology for sustainable water resources management: challenges and future projections. *Sustain. Chem. Pharm.*, 25, 100590, 2022.
124. Khalidi-Idrissi, A., Souabi, S., Madinzi, A., *et al.*, Recent advances in the treatment of wastewater contaminated with pharmaceutical pollutants: a critical review. *E. M. J. E.*, 9, 23, 2024.
125. Masinga, P., Simbanegavi, T.T., Makuvara, Z., *et al.*, Emerging organic contaminants in the soil–plant-receptor continuum: transport, fate, health risks, and removal mechanisms. *Environ. Monit. Assess.*, 196, 367, 2024.
126. Biswas, P., Vellanki, B.P., Kazmi, A.A., Investigating a broad range of emerging contaminants in a set of anthropogenically impacted environmental compartments. *Sci. Total Environ.*, 824, 153757, 2022.
127. Wu, Y., Lu, J., Meng, M., *et al.*, Bioinspired synthesis of pDA/SiO_2-based porous ciprofloxacin imprinted nanocomposite membrane by a polydopamine-assisted organic inorganic method. *Chem. Eng. J.*, 309, 263, 2017.
128. Wang, X., Liu, Z., Lu, J., *et al.*, Highly selective membrane for efficient separation of environmental determinands: enhanced molecular imprinting in polydopamine embedded porous sleeve. *Chem. Eng. J.*, 449, 137825, 2022.
129. Xing, W., Yan, Y., Wang, C., *et al.*, MOFs self-assembled molecularly imprinted membranes with photoinduced regeneration ability for long-lasting selective separation. *Chem. Eng. J.*, 437, 135128, 2022.
130. Lu, J., Qin, Y.Y., Wu, Y.L., *et al.*, Mimetic-core-shell design on molecularly imprinted membranes providing an antifouling and high-selective surface. *Chem. Eng. J.*, 417, 128085, 2021.
131. Sun, C., Lu, J., Wu, Y., *et al.*, Imitated core-shell molecularly imprinted membranes for selective separation applications: a synergetic strategy by polydopamine and SiO_2. *J. Taiwan Inst. Chem. Eng.*, 115, 304, 2020.
132. Trotta, F., Baggiani, C., Luda, M.P., *et al.*, A molecular imprinted membrane for molecular discrimination of tetracycline hydrochloride. *J. Membr. Sci.*, 254, 13, 2005.
133. Xing, W., Wu, Y., Lu, J., *et al.*, Biomass-based synthesis of green and biodegradable molecularly imprinted membranes for selective recognition and separation of tetracycline. *Nano*, 15, 2050004, 2020.

134. Liu, X., He, H., Zhao, B., *et al.*, Preparation and selective adsorption performance of the carboxymethyl salix psammophila wood powder-imprinted membrane for tetracycline. *Langmuir*, 40, 31, 2024.
135. Wu, Y., Lu, J., Xing, W., *et al.*, Double-layer-based molecularly imprinted membranes for template dependent recognition and separation: an imitated core-shell-based synergistic integration design. *Chem. Eng. J.*, 397, 125371, 2020.
136. Wu, Y., Xing, W., Ma, F., *et al.*, Three-dimensional basswood-based membrane with well-designed multilevel/hierarchical imprinting surface: a high-efficiency selective separation system. *Chem. Eng. J.*, 398, 125636, 2020.
137. Wu, Y., Zhang, K., Lin., R., *et al.*, Dual-imprinted organic/inorganic nanocomposite membranes with highly selective polydopamine-intimated nanostructures for pharmaceutically active compound separation. *J. Colloid Interface Sci.*, 604, 704, 2021.
138. Ma, F., Wu, Y., Guo, K., *et al.*, Double-imprinted mixed matrix membranes with polydopamine induced nanocomposite for selective recognition and separation applications. *J. Environ. Chem. Eng.*, 10, 107069, 2022.
139. Wu, Y., Lin, R., Ma, F., *et al.*, Dual-imprinted mixed matrix membranes for selective recognition and separation: a synergetic imprinting strategy based on complex initiation system. *J. Colloid Interface Sci.*, 606, 87, 2022.
140. Wu, Y., Xia, H., Guo, Q., *et al.*, Mixed matrix nanocomposite imprinted membranes with multilevel hierarchical structure for high-efficiency selective recognition and separation: a bioinspired dual-imprinted strategy. *Micropor. Mesopor. Mat.*, 326, 111393, 2021.
141. Söylemez, M.A., Synthesis and characterization of tetracycline-imprinted membranes: a detailed positron annichilation lifetime spectroscopy investigation. *J. Mol. Recognit.*, 34, 2895, 2021.
142. Zhao, J., Wu, Y., Wang, C., *et al.*, Insights into high-efficiency molecularly imprinted nanocomposite membranes by channel modification for selective enrichment and separation of norfloxacin. *J. Taiwan Inst. Chem. Eng.*, 89, 198, 2018.
143. Wu, J., Gao, J., Hou, Z., *et al.*, Accelerating the design of β-CD-PVDF-based molecularly imprinted nanocomposite membrane for selective separation: a surface functional monomer-directing strategy. *Nano*, 15, 2050138, 2020.
144. Lu, J., Qin, Y., Wu, Y., *et al.*, Bidirectional molecularly imprinted membranes for selective recognition and separation of pyrimethamine: a double-faced loading strategy. *J. Membr. Sci.*, 601, 117917, 2020.
145. Hani, O.L., García-Guzman, J.J., Palacios-Santander, J.M., *et al.*, Development of a molecularly imprinted membrane for selective, high-sensitive, and on-site detection of antibiotics in waters and drugs: application for sulfamethoxazole. *Chemosphere*, 350, 141039, 2024.
146. Meng, M., He, Z., Yan, L., *et al.*, Fabrication of a novel cellulose acetate imprinted membrane assisted with chitosan-wrapped multi-walled carbon

nanotubes for selective separation of salicylic acid from industrial wastewater. *J. Appl. Polym. Sci.*, 132, 42, 2015.

147. He, Z., Meng, M., Yan, L., *et al.*, Fabrication of new cellulose acetate blend imprinted membrane assisted with ionic liquid ([BMIM]Cl) for selective adsorption of salicylic acid from industrial wastewater. *Sep. Purif. Technol.*, 145, 63, 2015.
148. Yu, X., Mi, X., He, Z., *et al.*, Fouling resistant CA/PVA/TiO_2 imprinted membranes for selective recognition and separation salicylic acid from waste water. *Front. Chem.*, 5, 2, 2017.
149. Yan, M. and Wu, Y., Fabrication and evaluation of bioinspired pDA@ TiO_2-based ibuprofen-imprinted nanocomposite membranes for highly selective adsorption and separation applications. *New J. Chem.*, 44, 10703, 2020.
150. Ma, F., Yan, M., Lin, X., *et al.*, Biomimetic dual-layer imprinted UiO-66-based basswood membranes for ibuprofen recognition and separation. *ACS Sustain. Chem. Eng.*, 11, 6373, 2023.
151. Yan, M., Fei, H., Zhen, J., *et al.*, New insights into high-performance nanocomposite membranes with threefold-imprinted layers for selective recognition and separation. *Langmuir*, 38, 9321, 2022.
152. Yan, J., Sui, Y., Yan, M., *et al.*, Wood-based multi-layered porous imprinted membranes with dual metal composite nanostructures: synergistic promotion of rebinding capacity, selectivity and permeability. *J. Environ. Chem. Eng.*, 10, 107774, 2022.
153. Jiang, W., Conkle, J.L., Luo, Y., *et al.*, Occurrence, distribution, and accumulation of pesticides in exterior residential areas. *Environ. Sci. Technol.*, 50, 12592, 2016.
154. Niu, B., Cai, J., Song, W., *et al.*, Intermediate accumulation and toxicity reduction during the selective photoelectrochemical process of atrazine in complex water bodies. *Wat. Res.*, 205, 117663, 2021.
155. Sergeyeva, T.A., Brovko, O.O., Piletska, E.V., *et al.*, Porous molecularly imprinted polymer membranes and polymeric particles. *Anal. Chim. Acta*, 582, 311, 2007.
156. Sergeyeva, T.A., Piletsky, S.A., Piletska, E.V., *et al.*, *In situ* formation of porous molecularly imprinted polymer membranes. *Macromol.*, 36, 7352, 2003.
157. Gkementzoglou, C., Kotrotsiou, O., Kiparissides, C., Synthesis of novel composite membranes based on molecularly imprinted polymers for removal of triazine herbicides from water. *Ind. Eng. Chem. Res.*, 52, 14001, 2013.
158. Gkementzoglou, C., Kotrotsiou, O., Koronaiou, M., *et al.*, Development of a sandwich-type filtration unit packed with MIP nanoparticles for removal of atrazine from water sources. *Chem. Eng. J.*, 287, 233, 2016.
159. Turiel, E., Diaz-Alvarez, M., Martin-Esteban, A., Supported liquid membrane protected molecularly imprinted beads for the solid-phase micro-extraction of triazines from environmental waters. *J. Chromatogr. A*, 1432, 1, 2016.

160. Shah, N., Shah, M., Rehan, T., *et al.*, Molecularly imprinted polymer composite membranes: from synthesis to diverse applications. *Heliyon*, 10, 16, e36189, 2024.
161. Lah, N.F.C., Ahmad, A.L., Zaulkiflee, N.D., Polyethersulfone ultrafiltration membranes modified with MAA-EGDMA imprinted polymers for selective atrazine separation. *J. Water Proc. Eng.*, 47, 1027612, 2022.
162. Xing, W., Ma, Z., Wang, C., *et al.*, Hierarchically porous molecularly imprinted membranes with multiple transfer channels for micropollutants selective separation. *Desalination*, 547, 116226, 2023.
163. Lu, J., Qin, Y., Li, C., *et al.*, Irregular dot array nanocomposite molecularly imprinted membranes with enhanced antibacterial property: synergistic promotion of selectivity, rebinding capacity and flux. *Chem. Eng. J.*, 405, 126716, 2021.
164. Soylemez, M.A. and Güven, O., Radiation induced *in-situ* synthesis of membranes for removal of 2,4-dichlorophenoxy acetic acid from real water samples. *Radiat. Phys. Chem.*, 171, 108708, 2020.
165. Wang, Y., Ruan, H., Zhang, J., Recyclable and selective PVDF-based molecularly imprinted membrane combining mussel-inspired biomimetic strategy for dimethomorph elimination. *Chem. Eng. J.*, 478, 147322, 2023.
166. Mkhize, D.S., Nyoni, H., Quinn, L.P., Molecularly imprinted membranes (MIMs) for selective removal of polychlorinated biphenyls (PCBs) in environmental waters: fabrication and characterization. *Environ. Sci. Pollut. Res.*, 24, 11694, 2017.
167. Ghasemi, S. and Nematollahzadeh, A., Molecularly imprinted polymer membrane for the removal of naphthalene from petrochemical wastewater streams. *Adv. Polym. Technol.*, 37, 2288, 2018.
168. Garg, A. and Chopra, L., Dye waste: A significant environmental hazard. *Mater. Today Proceed.*, 48, 1310, 2022.
169. Al-Thoamy, R., Ali, S.S., Li, F., *et al.*, A critical review on the treatment of dye-containing wastewater: ecotoxicological and health concerns of textile dyes and possible remediation approaches for environmental safety. *Ecotoxicol. Environ. Saf.*, 231, 113160, 2022.
170. Yang, Z.P., Yan, J.L., Zhang, C.J., *et al.*, Enhanced removal of bilirubin on molecularly imprinted titania film. *Colloids Surf., B*, 8, 187, 2011.
171. Melvin, Ng, H.K., Leo, C.P., Abdullah, A.Z., Selective removal of dyes by molecular imprinted TiO_2 nanoparticles in polysulfone ultrafiltration membrane. *J. Environ. Chem. Eng.*, 5, 3991, 2017.
172. Zhao, R., Li, X., Sun, B., *et al.*, Preparation of molecularly imprinted sericin/poly(vinyl alcohol) electrospun fibers for selective removal of methylene blue. *Chem. Res. Chin. Univ.*, 33, 986, 2017.
173. Asman, S., Yusof, N.A., Abdullah, A.H., *et al.*, Synthesis and characterization of hybrid molecularly imprinted polymer (MIP) membranes for removal of methylene blue (MB). *Molecules*, 17, 1916, 2012.

174. Lu, L., Yue, X., Lin, F., *et al.*, Template-synthesized ultra-thin molecularly imprinted polymers membrane for the selective pre-concentration of dyes. *J. Mater. Chem. A*, 3, 10959, 2015.
175. Bakhshizadeh, S., Sabzehmeidani, M.M., Ghaedi, M., *et al.*, Preparation and characterization of mixed matrix membranes based on PVDF blend and hydrophilic molecularly imprinted MIL-101 (Cr) as filler for efficient selective removal of dye. *J. Environ. Chem. Eng.*, 10, 108864, 2022.
176. Elnabi, M.K.A., Elkaliny, N.E., Elyazied, M.M., *et al.*, Toxicity of heavy metals and recent advances in their removal: a review. *Toxics*, 11, 580, 2023.
177. Sall, M.L., Diaw, K.D., Sall, D.G., *et al.*, Toxic heavy metals: impact on the environment and human health, and treatment with conducting organic polymers, a review. *Environ. Sci. Pollut. Res.*, 27, 29927, 2020.
178. Charkiewicz, A.E. and Backstrand, J.R., Lead toxicity and pollution in Poland. *Int. J. Environ. Res. Pub. Health*, 17, 4385, 2020.
179. Fatoki, J.O. and Badmus, J.A., Arsenic as an environmental and human health antagonist: A review of its toxicity and disease initiation. *J. Haz. Mat. Adv.*, 5, 100052, 2022.
180. Shah, N., Shah, M., Rehan, T., et al., Molecularly imprinted polymer composite membranes: from synthesis to diverse applications. *Heliyon*, 10, 16, e36189, 2024.
181. Tanveer, M., Hasanuzzaman, M., Wang, L., Lithium in environment and potential targets to reduce lithium toxicity in plants. *J. Plant Growth Regul.*, 38, 1574, 2019.
182. Ezugbe, E.O. and Rathilal, S., Membrane technologies in wastewater treatment: a review. *Membranes*, 10, 89, 2020.
183. Oller, I., Cuevas, S.M., Agüera, A., *et al.*, Monitoring and removal of organic micro-contaminants by combining membrane technologies with avanced oxidation processes. *Curr. Org. Chem.*, 22, 1103, 2018.
184. Chen, L., Dai, J., Hu, B., *et al.*, Recent progresses on the adsorption and separation of ions by imprinting routes. *Sep. Purif. Rev.*, 49, 265, 2020.
185. Zheng, X., Zhang, Y., Bian, T., *et al.*, Selective extraction of gadolinium using free-standing imprinted mesoporous carboxymethyl chitosan films with high capacity. *Cellulose*, 26, 1209–1219, 2019.
186. Donato, L., *Imprinted polymeric membranes*, pp. 174–176. De Guyter, Berlin, Boston, 2023.
187. Roy, E., Patra, S., Madhuri, R., *et al.*, A single solution for arsenite and arsenate removal from drinking water using cysteine@ ZnS: TiO_2 nanoparticle modified molecularly imprinted biofouling-resistant filtration membrane. *Chem. Eng. J.*, 304, 259, 2016.
188. Xi, Y., Shi, H., Liu, R., Insights into ion imprinted membrane with a delayed permeation mechanism for enhancing Cd^{2+} selective separation. *J. Hazard. Mater.*, 416, 125772, 2021.

189. Hong, Y., Wang, X., Fu, D., Synthesis of cadmium(II) ion-imprinted composite membrane with a pyridine functional monomer and characterization of its adsorption performance. *e-Polymers*, 23, 2022810, 2023.
190. Haider, F.U., Cai, L.-Q., Coulter, J.A., *et al.*, Cadmium toxicity in plants: impact and remediation strategies. *Ecotoxicol. Environ. Saf.*, 211, 22, 2021.
191. Li, P., Wang, X., Wang, G., *et al.*, Synthesis and evaluation of ion-imprinted composite membranes of Cr(VI) based on β-diketone functional monomers. *RSC Adv.*, 11, 38915, 2021.
192. Liu, Y., Hu, D., Hu, X., *et al.*, Preparation and characterization of chromium(VI) ion-imprinted composite membranes with a specifically designed functional monomer. *Anal. Lett.*, 53, 1113, 2020.
193. Chen, J.H., Li, G.P., Liu, Q.L., *et al.*, Cr (III) ionic imprinted polyvinyl alcohol/sodium alginate (PVA/SA) porous composite membranes for selective adsorption of Cr (III) ions. *Chem. Eng. J.*, 165, 465, 2010.
194. Zhao, J., Zhao, J., He, X., *et al.*, Emission behavior, enol–keto tautomerism and bioactivity of hydroxy-substituted cyclic chalcone derivatives. *J. Mater. Chem. C*, 9, 1000, 2021.
195. Tivendale, N.D., Davies, N.W., Horne, J., *et al.*, Analysis of the enol–keto tautomers of indole-3-pyruvic. *Acid. Aust. J. Chem.*, 68, 345, 2015.
196. Zavahir, S., Riyaz, N.S., Elmakki, T., *et al.*, Ion-imprinted membranes for lithium recovery: A review. *Chemosphere*, 354, 141674, 2024.
197. Das, D., Abarajitha, R., Kay, P., Selective recovery of lithium from spent coin cell cathode leachates using ion imprinted blended chitosan microfibers: pilot scale studies provide insights on scalability. *J. Hazard Mat.*, 431, 128535, 2022.
198. Qu, G., Yang, J., Ran, Y., *et al.*, Adsorption performance and mechanism of TiO_2/PVDF-based lithium-ion imprinted membrane in leaching solution of spent lithium-ion batteries. *J. Clean. Prod.*, 442, 140982, 2024.
199. Wang, Y., Zhu, S., Liao, Y., *et al.*, A calcium ion-imprinted porous film prepared from a cellulose-alginate composite. *J. Polym. Res.*, 21, 612, 2014.
200. Alizadeh, T., Akhoundian, M., Hoseinnejad, B., Synthesis of dodecanol-anchored nano-sized magnesium imprinted polymer and its application as an efficient ionophore in Mg^{2+}-selective PVC membrane electrode. *Anal. Bioanal. Electrochem.*, 13, 436, 2021.
201. Wang, X., Zhang, L., Ma, C., *et al.*, Enrichment and separation of silver from waste solutions by metal ion imprinted membrane. *Hydrometallurgy*, 100, 82, 2009.
202. Garcia, L.V., Ho, Y.C., Thant, M.M., *et al.*, Lithium in a sustainable circular economy: a comprehensive review. *Processes*, 11, 418, 2023.
203. Swain, B., Recovery and recycling of lithium: a review. *Sep. Purif. Technol.*, 72, 388, 2017.
204. He, G., Li, Z., Liu, Y., *et al.*, A novel lithium ion imprinted membrane with robust adsorption capacity and anti-fouling property based on the uniform multilayered interlayer. *Desalination*, 539, 115973, 2022.

205. Sun, D., Zhou, T., Lu, Y., *et al.*, Ion-imprinted antifouling nanocomposite membrane for separation of lithium ion. *Kor. J. Chem. Eng.*, 39, 2482, 2022.
206. Yang, J., Qu, G., Liu, C., *et al.*, An effective lithium ion-imprinted membrane containing 12-crown ether-4 for selective recovery of lithium. *Chem. Eng. Res. Des.*, 184, 639, 2022.
207. Li, Z., He, G., Zhao, G., *et al.*, Y., Preparation of a novel ion-imprinted membrane using sodium periodate-oxidized polydopamine as the interface adhesion layer for the direction separation of Li^+ from spent lithium-ion battery leaching solution. *Sep. Purif. Sci.*, 277, 119519, 2021.
208. Yu, C., Lu, J., Dai, J., *et al.*, Bio-inspired fabrication of ester-functionalized imprinted composite membrane for rapid and high-efficient recovery of lithium ion from seawater. *J. Colloid Interface Sci.*, 572, 340, 2020.
209. Liu, W., Yan, G., Zhang, E., Extraction of lithium ions from acidic solution using electrochemically imprinted membrane. *Desalination*, 496, 114751, 2020.
210. Cui, J., Zhang, Y., Wang, Y., *et al.*, Fabrication of lithium ion imprinted hybrid membranes with antifouling performance for selective recovery of lithium. *New J. Chem.*, 42, 118, 2018.
211. Sun, D., Meng, M., Qiao, Y., *et al.*, Synthesis of ion imprinted nanocomposite membranes for selective adsorption of lithium. *Sep. Purif. Technol.*, 194, 64, 2018.
212. Sun, D., Zhu, Y., Meng, M., *et al.*, Fabrication of highly selective ion imprinted macroporous membranes with crown ether for targeted separation of lithium ion. *Sep. Purif. Technol.*, 175, 19, 2017.
213. Lu, J., Qin, Y., Zhang, Q., *et al.*, Multilayered ion- imprinted membranes with high selectivity towards Li^+ based on the synergistic effect of 12-crown-4 and polyether sulfone. *Appl. Surf. Sci.*, 427, 931, 2018.
214. Wuang, W., Yuan, J., Jang, Y., *et al.*, Study on selective adsorption and separation of dysprosium by a multi-template oriented imprinted composite membrane. *Desal. Wat. Treat.*, 230, 310, 2021.
215. Liu, E., Xu, X., Zheng, X., *et al.*, An ion imprinted macroporous chitosan membrane for efficiently selective adsorption of dysprosium. *Sep. Purif. Technol.*, 189, 288, 2017.
216. Lu, J., Wu, Y., Lin, X., Gao, J. *et al.*, Anti-fouling and thermosensitive ion-imprinted nanocomposite membranes based on grapheme oxide and silicon dioxide for selectively separating europium ions. *J. Hazard Mat.*, 353, 244, 2018.
217. Chen, J., Bai, H., Xia, J., *et al.*, Electrochemical sensor for detection of europium based on poly-catechol and ion-imprinted sol-gel film modified screen-printed electrode. *J. Electroanal. Chem.*, 824, 32, 2018.
218. Zheng, X., Liu, E., Zhang, F., *et al.*, Selective adsorption and separation of gadolinium with three-dimensionally interconnected macroporous imprinted chitosan films. *Cellulose*, 24, 977, 2017.

219. Liu, E., Lin, X., Zhang, D., *et al.*, Preparation of an ion imprinted chitosan-based porous film with an interpenetrating network structure for efficient selective adsorption of Gd (III) new. *J. Chem.*, 45, 725, 2012.
220. Wu, Y., Lin, R., Ma, F., *et al.*, Three-dimensional macroporous wood-based selective separation membranes decorated with well-designed Nd (III)-imprinted domains: A high-efficiency recovery system for rare earth element. *J. Colloid Int. Sci.*, 587, 703, 2021.
221. Binnemans, K., McGuiness, P., Jones, P.T., Rare-earth recycling needs market intervention. *Nat. Rev. Mater.*, 6, 459, 2021.

9

Few-Layer Materials in Porous Membranes for Advanced Water Desalination

M. Frappa, G. Di Luca, E. Drioli and A. Gugliuzza*

Research Institute on Membrane Technology-National Research Council (CNR-ITM), Rende, Italy

Abstract

The constant growth of the human population, together with drastic climate change, is compromising the availability of an essential natural resource, water, which is needed for drinking and domestic use, agriculture, industry, and the safety of biodiversity. Saline water represents more than 97% of the water on the planet, but only 2.5% is used as freshwater. The main obstacle is the costs, including energy, required for process implementation enabling a sustainable large access to water. Seawater desalination is a reliable method for water supply and since the post-war period, it has been practiced successfully with classic distillation techniques such as multistage flashing (MSF) and multi-effects distillation (MED), and then membrane operations such as nanofiltration (NF) and reverse osmosis (RO) and derived processes. In recent decades, increasingly efficient desalination techniques have been proposed through the development of new membrane-based technologies such as Membrane Distillation (MD). However, the scalability of this kind of operation requires more fruitful productivity-efficiency trade-offs and lower energy consumption to be within the framework of green and circular economies. In the case of membrane operations, desalination procedures have been optimized through the design of progressively high-performing membranes, which are regarded to work as the engine for each separation process. This has solicited the exploration of different classes of materials, taking the chance to drive their specific properties towards the design of next-generation membranes needed to make desalination competitive on a large scale.

Among others, the portfolio of 2D materials has grown exponentially for different uses, but more recently has been oriented to water desalination. Few-layer materials have been proposed to fill polymeric porous membranes for assisted

**Corresponding author*: a.gugliuzza@itm.cnr.it

Annarosa Gugliuzza and Wanqin Jin (eds.) Porous Membranes: Breakthroughs in Manufacturing and Applications, (275–312)

productivity and durability or to generate additional nanochannels through which ions bringing different charges can be filtered. This chapter gives an overview of the attractive use of few-layer materials confined in membranes for water desalination.

After an introduction on the most used and promising membrane operation dedicated to water desalination and a brief discussion about environmental issues, a wide section is dedicated to the production of few-layer materials and their related use in water desalination. Specifically, exfoliation strategies are discussed together with the use of 2D materials as nanofillers in composite membranes and intercalated channels in layered membranes. Effects on productivity, resistance to wetting and fouling, thermal efficiency, and capability to discriminate from mono- and divalent ions, as well as the recovery of reusable crystals from brine, are analyzed and discussed. An assessment of the role of few-layer materials within the framework of circular economy is provided.

Keywords: Water, desalination, 2D materials, ion filtering, membrane distillation, graphene, chalcogenides

9.1 Introduction

Water is a life-giving element for health, well-being, societal, and economic prosperity [1]. It is a fact that this natural element is essential for human, ecosystem, and biodiversity survival [2]. However, stress due to environmental pollution, overuse, and mismanagement of natural resources, together with changing land-sea exploitation and sudden climate changes, is reducing the availability of this vital resource.

An accessible and scalable water treatment infrastructure is therefore of utmost importance to provide suitable tools for yielding constant food and energy production and driving the global economy of each single Country towards safety, health, and prosperity [3]. Membrane technology has already provided interesting solutions to critical issues regarding water scarcity [4]. It relies on a wide range of applications, which make the recovery of freshwater and drinkable water successful from waste and hypersaline streams through practices, that fall in the logic of recycling strategy [4, 5]. Water desalination is a practical route for producing reusable water from streams with different contents of Total Dissolved Solids (TDS), including streams with TDS over 100,000 ppm, seawater (TDS over 45,000 ppm) and brackish water (TDS over 10,000 ppm). Normally, drinkable water contains TDS in the range of 500 to 1,000 ppm [6]. Depending on degree of salinity and chemical composition of the streams, different membrane operations can be therefore selected and, if required, integrated with other technologies to produce water at a desired degree of salinity. Nanofiltration (NF) and Reverse Osmosis (RO) are two well-established pressure-driven membrane technologies dedicated to water treatment [4]. However, RO is predominant because it has the highest capability to remove salts from water. It uses semipermeable membranes with excellent separation performance and good chemical stability [7]. The water is pushed into the membrane module by a pump, which exerts a pressure higher than the osmotic pressure of the feed water so as to allow pure water to pass through the membrane, while the remaining part comes out with a high salt concentration. This is the retention of all the components that do not cross the membrane. However, the performance of RO membrane water permeability is limited by imposed transport processes and thermodynamic constraints. The recovery ratio is limited, and increased electrical energy consumption is required at higher feed concentrations [7]. Membrane distillation (MD) is regarded as a more sustainable alternative process for water desalination [8, 9]. This kind of operation exploits the partial pressure difference between the two sides of a microporous hydrophobic membrane to generate a mass transfer of volatile components like water vapor through the membrane pores. It is enough to apply a little difference in temperature or concentration across the membrane or vacuum at the permeate side to drive the separation of pure water from other non-volatile components such as salts. However, some limitations to scale up also persist for this technology due to low productivity-selectivity trade-off, membrane flooding, and thermal energy consumption [10]. Despite commercial membranes exhibiting good selectivity, MD processes lack fruitful production of freshwater, long-term durability, and resistance to thermal polarization as well as wetting phenomena. Membrane design is expected to give a major propulsion to this eco-friendly technology due to the chance to adjust some intrinsic properties of materials for the amplification of mass transport, resistance to wetting, and contrast to undesired polarization events. The use of new families of advanced materials enables new chemical functions and morphological features in the membranes, making them potentially more suitable to equip MD devices than commercial ones [11]. Among others, 2D materials have been successfully explored for overcoming some critical issues that affect water desalination, such as improved contrast to flux decline due to flooding and fouling, amplified productivity through assisted transport mechanisms, and enhanced thermal efficiency as a consequence of reduced thermal polarization events [10, 12–15]. However, few-layer materials have also been targeted to be powerful for a quicker recovery of better-quality crystals from brine due to the establishment of water sequestration mechanisms from ion-water clusters and the restoration of ion-ion forces necessary to form and grow crystal nuclei [16–18]. Another interesting success related to the use of few-layer materials is the development of a new concept of nanofiltration based on ion filtering membranes, which enable a discrimination between mono- and divalent ions [19, 20]. In this case, 2D material flakes are intercalated in layered membranes, providing additional subchannels wherein sieving filtration or charge exclusion mechanisms take place, depending on the chemical functionalities and/or ionic charge combined in the membranes.

Hereafter, a brief assessment about environmental issues and useful membrane strategies for water purification is provided. Water desalination, with a focus on membrane distillation, is discussed as a practical way for recovering reusable water, while attempts to use 2D materials for desalination purposes are discussed, taking into account their sustainable use in relation to criteria of the circular economy.

9.2 Environmental Issues: Pollutant Source and Useful Membrane Strategies

Water is a basic element for all our activities and is essential in many industrial processes. In agriculture, freshwater is required to irrigate the fields of crops, but pesticides and fertilizers often compromise the waste quality and do not allow its reuse [21]. Today, based on the latest advanced technologies, smart solutions make optimum use of intelligent spraying. However, water scarcity continues to be a critical matter for extensive fruit and vegetable production. The use of desalted and decontaminated water could be a practical route for providing reusable water for agricultural irrigation. Integrated membrane desalination processes can solve the problems related to the production of freshwater from saline streams, also yielding brine as a fruitful source of salts to reuse in agriculture as natural fertilizers [12, 16].

Industrial production also generates massive wastewater streams and is one of the principal causes of pollution for the environment and water. For example, the electronics and textile industries lead to contamination due to electronic waste (e-waste) and heavy metals [22, 23], including nickel, cadmium, zinc, cobalt, and copper. These metals are found in water at ppb levels in the form of cations such as Zn^{2+}, Ni^{2+}, Cd^{2+}, Pb^{2+}, Co^{2+}, Cu^{2+}, and arsenic (Ar^{3-}, Ar^{3+}, and/or Ar^{5+}). Membrane processes can provide environmental remediation by removing heavy metals from groundwater. Selective ion exchange resins, named Lewatit TP 207 or Lewatit TP 208, have been proposed to remove heavy metals [22]. Kim *et al.* proposed a NO_3^--selective carbon electrode coated with nitrate-selective A520E resin particles for desalination of municipal wastewater and NO_3^- recovery [24]. Al Rashdi *et al.* [25] proposed nanofiltration (NF) based on a commercial membrane (NF270), for the extraction of copper with a rate of 100 % at low metal concentration and 58% at high concentration. Different RO and NF processes for removing heavy metals from water have been analyzed [26]. It was found that composite membranes equipping reverse osmosis plants allow a high extraction rate with a recovery of more than 95% of metal ions and a low Chemical Oxygen Demand (COD) value (10 mg/L^{-1} at permeate side), while by nanofiltration (NF), the recovery decreases down to 78.9% with a COD value in the permeate of 35 mg/L^{-1}. Despite the lower performance, nanofiltration would be more suitable for large-scale industrial practice due to the higher permeate flow at lower operating pressure.

Arsenic is one of the most hazardous chemical elements that contaminate water. It is a toxic compound that can be found naturally on Earth in small concentrations, as well as in air or water [27]. It appears in three different allotropic forms (yellow, black, and grey). The stable form is a silver-grey, brittle crystalline solid that quickly tarnishes in the air and burns at high temperatures to form a white cloud of arsenic trioxide. The arsenic in the atmosphere comes from various sources: volcanoes release about 3000 tons per year, and human activity is responsible for 80,000 tons of arsenic per year released by the combustion of fossil fuels. The problem of arsenic in water has been repeatedly debated in various bodies because it is classified by the International Agency for Research on Cancer as a first-class carcinogenic element. For this reason, the European Union has imposed a limit of 10 μg per liter for mineral waters. Inorganic arsenic is well absorbed by the gastrointestinal system and in the lungs generally over 50% of the dose taken. Organic arsenical compounds are generally considered to be poorly absorbable and their absorption is relative to their water solubility. Furthermore, once absorbed, these compounds are easily eliminated with stools and urine; in fact, they are subject to detoxifying liver biomethylation, therefore organic arsenicals are less toxic and more easily excreted. Inorganic arsenic is considered carcinogenic for the lungs, skin, kidneys, and liver, especially in chronic intoxication. It is relevant to observe that arsenic is also of natural origin and is released from sediments in groundwater due to the anoxic conditions of the subsoil. So, water contamination due to the presence of arsenic is a current problem in different areas of the world where potable water is extracted from groundwater basins. Different technologies have been proposed to remove arsenic such as adsorption, chemical coagulation–precipitation, ion-exchange, and membrane processes. In the last case, MD allows 100% rejection of non-volatile compounds such as arsenic and other heavy metals, making the recovery of usable water safer and more effective [28].

Another huge issue for drinking water is the presence of micro and nanoplastics [29]. The improper use of plastics is one of the major sources of pollution for oceans, seas, lakes, and rivers. Plastics in the environment tend to be gradually decomposed into not non-biodegradable micro and nanoplastics, which are dangerous for humans and ecosystems. In this case, microplastic removal from water is found in membrane separation processes, including MD, a practical and useful tool for purification [30].

9.3 Water Desalination: From Traditional to Advanced Membrane Operations

The removal of organic/inorganic pollutants and salts from ocean and seawater remains the main challenge in producing freshwater [2]. Desalination is the most attractive solution to overcome some shortcomings related to water scarcity and low quality [1]. Membrane technology proposes different options to remove salt and other contaminants from aqueous streams, including membrane bioreactor (MBR), nanofiltration (NF), forward osmosis (FO), reverse osmosis (RO), ion exchange resin (IXR), membrane distillation (MD), and membrane crystallization (MCr) [4–6]. RO processes, together with conventional multistage flashing (MSF) and multi-effects distillation (MED) systems, continue to dominate the desalination market [31].

In RO processes, the separation takes place thanks to diffusion and dissolution mechanisms, which occur at different degrees and allow action up to the ionic level. The performance of the process depends on the membrane type, as well as working conditions, including temperature and concentration. From an energetics point of view, RO requires a consumption of electrical energy between 3 and 4 kWh/m^3, with an emission of 1.4 to 3.6 kg CO_2 per cubic meter of produced water. Traditional thermal desalination technologies, on the other hand, produce emissions of 8 to 20 kg CO_2/m^3, with the exception of stand-alone MED at 3.4 kg CO_2/m^3. This broad range depends on the fuel used to produce the electricity. For this reason, Western and developed countries prefer RO systems due to their efficient power consumption, while the Middle East and Gulf countries prefer MSF and MED systems due to the abundant source of available oil.

Despite some critical issues to be solved, membrane distillation technology continues to be regarded as a competitive desalination process with great potential for the production of freshwater from hypersaline streams. This technology takes numerous advantages over traditional and RO desalination operations, as listed in Table 9.1 [31, 32].

The main benefits derive from the chance to work at a temperature lower than the water boiling point and the possibility to use renewable energy, e.g., solar, geothermal, or waste heat, as greener power sources. Low hydrostatic pressure is also required and, only if necessary, hypersaline streams could be pretreated. Theoretical rejection factors of 100% are expected due to the fact that only volatile species can be removed while water can be treated at nearly any salinity up to saturation. Compact and remote-control systems can be easily realized by using no harmful and recyclable plastics against corrosion inconvenient.

However, limitations for MD scale up are lower productivity and thermal energy efficiency. While fluxes up to 10–15 $Lm^{-2}h^{-1}$ are usually estimated for more conventional commercial membranes, higher productivity can be obtained with membranes functionalized with emerging materials for desalination such as 2D compounds [10, 33, 34]. Concerning energetics aspects, MD uses thermal energy to heat the brine, cool water at the permeate side, and run the pumps. So, it is classified as an energy-intensive technology even if the electrical consumption is significantly reduced.

Table 9.1 Merit/issues addressed by MD vs. other conventional technologies (RO, MED, MSF). Reprinted from [32] (2024), with permission from ELS Publishing.

Merit parameters	MD	RO	MED	MSF
Operated with low grade energy	++++	+	+++	++
No-volatile compounds rejection	++++	++	++++	++++
High salinity degree	++++	+	+++	++++
Compact devices	++++	++++	++	++
Minimum pretreatment	++	+	++	++++
Low suffering from polarization concentration	++	+	++++	++++
Energy efficiency	+	++++	++	+
Low lifespan costs	+	++++	+++	++

A recent study proposed an energetic analysis for MD, RO, and other conventional techniques (MSF and MED) during the production of freshwater in Qatar (Table 9.2) [9]. MSF was found to be an energy-intensive technology working with fossil fuels and rejected waste heat; MED was found to be more energetically efficient even if it uses boiling water; RO, as a pressure-driven process, required major electrical energy consumption rather than thermal energy. According to energetic efficiency accomplishments, MD needs to minimize and recover the heat loss due to conduction and convection across the membrane.

Nevertheless, membrane distillation remains particularly attractive because it exploits a difference in partial pressure across the membrane, allowing water vapor to pass through the membrane pores and condense at the permeate side or externally. As mentioned, this driving force can be generated by a difference in temperature or concentration, or by applying a vacuum at the permeate side. The hydrophobicity of the membrane, which puts in contact two liquid-liquid or liquid-gas phases, is expected to stop the passage of liquid water through the pores, thus preventing flooding. A future scale up of the MD process requires careful assessment of the productivity of high-quality water at low energy consumption. These criteria depend on both the membrane properties and process parameters. The main operating conditions that influence the performance are represented by the concentration, temperature, and flow rate of the feed. The feed temperature is a parameter that strongly influences the transmembrane flux because the vapor pressure increases with temperature, consequently increasing the driving force of the process. An increase in

Table 9.2 Desalination technologies used in Qatar and membrane distillation systems. Reprinted from [32] (2024), with permission from ELS Publishing.

Desalination methods	**MD**	**RO**	**MED**	**MSF**
Production of freshwater (m^3/day)	Na	5790	28,384	974,222
Thermal energy requirement ((kWh/m^3)	100	0	40–65	53–70
Electrical energy requirement (kWh/m^3)	1.5–3.65	4.0–6.0	2.0–2.5	2.5–5.0
Temperature (°C)	60–90	<60	65–70	90–100
Recovery factor (%)	60–80	30–50	15–50	15–50

the feed concentration causes a reduction in the permeate flow due to the decreased vapor pressure, as expressed by Raoult's law in eq. 9.1:

$$P_i = p_i^* \chi_i^* \tag{9.1}$$

Where P_i is the partial pressure of the component i, p_i^* the equilibrium vapor pressure of the pure component i, {\displaystyle i} and χ_i^* {\displaystyle x_{i}} the mole fraction of the component {\displaystyle i} i in solution.

Materials forming the membrane are envisaged to contrast wetting and fouling, which are the major causes of flux decline, energy consumption, and short-term processes [35]. The common materials used for making MD membranes hydrophobic are polytetrafluoroethylene (PTFE), polypropylene (PP), or polyvinylidene fluoride (PVDF). The chemistry of these polymers is expected to make the membrane surface waterproof, but often is not enough to contrast the liquid intrusion inside the membrane pores. So, the morphological factor becomes fundamental to amplify the resistance to liquid spreading through the design of topography and morphology with the ability to amplify water, and more generally, liquid repellence [35]. In this context, critical factors are pore size (preferably less than 0.5 mm), unimodal pore distribution, and surface asperity that contribute to increased surface roughness and reduced liquid-surface adhesion during the MD process. Other important morphological factors are thickness and overall porosity.

The flux is inversely proportional to the thickness, so thinner membranes exhibit major productivity [35]. However, its excessive reduction favors heat conduction, thus reducing the driving force for direct contact membrane distillation (DCMD) processes essentially. It is necessary to calibrate a suitable thickness to increase mass transport and keep a constant driving force. Concerning with porosity, it is important to consider that high pore size and uniform pore distribution provide a high interfacial area, reducing the resistance to mass transfer. A pore diameter between 0.1 and 1 μm is usually recommended, together with a porosity higher than 80%. However, larger pores can facilitate the intrusion of liquid inside the pores, causing membrane flooding. The overall porosity can also affect the mechanical resistance and thermal conductivity of the membrane. Excessive porosity can make the membrane brittle and not useful for MD purposes. Also, porosity is crucial for controlling the polarization temperature. The latter is caused by heat loss through the membrane due to heat advection by vapor and heat conduction through the pores and polymer network of the membrane. Thermal polarization seriously affects the thermal efficiency of MD processes, considering that the temperature at the membrane surface on the feed side tends to be reduced with respect to the bulk feed temperature. So, additional thermal energy is needed to restore the driving force and keep a constant flux [9]. As a result, major thermal energy consumption is required.

9.4 2D Materials for Next Generation Water Desalination

Based on previous discussions, factors to improve MD technology include reducing thermal energy consumption, enhancing the durability of the operation, and restoring the membrane's initial functions. Different approaches have been proposed to improve membrane performance, including blending, coatings, layering, and other tools to manipulate the materials on different length scales [36]. The challenge has mainly been to improve pore size and wetting and fouling resistance, while also shifting the productivity-efficiency trade-off beyond the state of the art. With this intent, various hydrophobic/hydrophilic membranes have been fabricated by using macromolecules (SMM) and polyethermide to modify the membrane surface [37]. Nanoparticles such as SiO_2, Al_2O_3, ZnO, and carbon nanotubes have been proposed to make the surfaces superomniphobic [38, 39]. Among the various families of compounds, 2D materials have recently been regarded as attractive and emerging for the design of membranes dedicated to water desalination [10, 12, 39]. These kinds of materials show singular mechanical strength, different electrical and thermal conductivities, high surface area-to-volume ratio, quantum mechanical effects, etc. [40]. Until now, their scale for wafer-scale applications has been limited by time-consuming and expensive fabrication procedures, such as chemical vapor deposition (CVD) and physical vapor deposition (PVD) which allow the formation of few-layer and low-defect density 2D films on substrates [41]. However, the attractive properties of few-layer 2D materials have been recently addressed without necessarily requiring monolayer precision, in the reinforcement of composites, production of conductive inks, fabrication of coatings, and construction of purification and delivery systems.

9.5 Techniques of Exfoliation

Over the years, various strategies have been proposed for the production of few-layer materials [41], including electrochemical exfoliation, micro-mechanical cleavage, ball milling, shear exfoliation, and recently wet jet milling technique (Figure 9.1) [40].

In each case, the target is to get individual sheets from bulk materials providing energy enough to disable weak van der Waals forces, which hold the layers together along the *z* direction while covalent bonds fix the 2D material structure on the *x-y* plane. The major concern remains with the breakage of these covalent bonds which causes defects, holes, buckles, wrinkles, and oxidation in the layers. However, these untoward events could provide a powerful chance to optimize water desalination through dedicated membrane processes.

9.5.1 Electrochemical Exfoliation

Electrochemical exfoliation takes place in solution in the presence of electrical current by cathodic reduction or anodic oxidation. Ions are introduced within an electrolyte and directed into the bulk 2D material so that

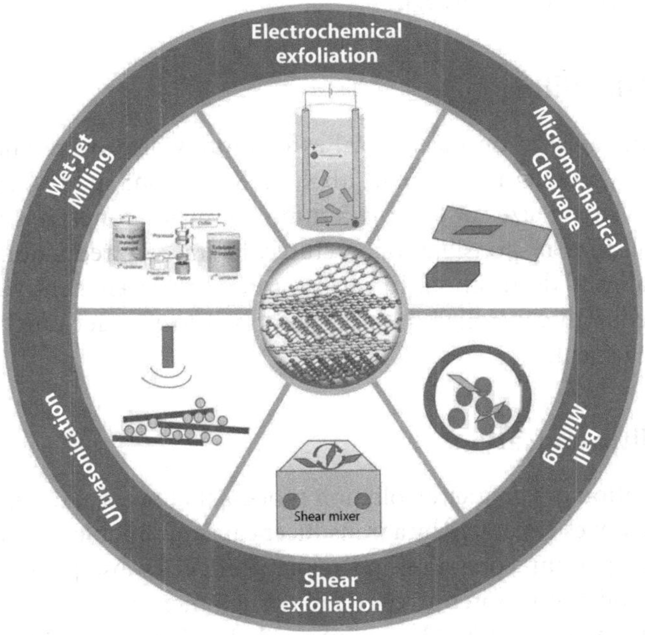

Figure 9.1 Representative scheme of exfoliation techniques for layered materials.

intercalation between layers occurs, overcoming van der Waals bonds, and generating few layers. This approach can be applied to various types of 2D materials, from graphene to MoS_2, WS_2, ZrS_2, Sb_2Te_3, phosphorene, black phosphorous, and so on. Generally, this approach allows for the massive production of few layers with suitable quality for graphene, black phosphorous, and metal chalcogenides. However, additional steps are often required such as sonication and use of organic radical assistance causing dispersion, broad distribution of thickness of the produced layers, chemical residues, and structural damages.

9.5.2 Micromechanical Cleavage

Micromechanical cleavage is the first procedure proposed to exfoliate graphene. It consists of using Scotch tape to peel flakes from bulk crystals repeatedly. Then, thinner flakes are transferred on a substrate by pressing. This is a simple and cheap exfoliation technique to obtain few-layer to centimeter-thick graphene, MoS_2, hBN, WS_2, and others without affecting the intrinsic electron transport properties. However, this technique is niche and difficult to scale up for the continuous production of monolayers. No precision and repeatability are given for thickness and lateral size due to residues of adhesive and interfacial adhesion occurring between the few layers and the substrate.

9.5.3 Ball Milling

Ball milling is based on mechanical exfoliation, which overcomes van de Waals forces producing graphene, hBN, MoS_2, and SnS_2 nanosheets of 1–5 layers with lateral sizes of 0.5 to 1.5 mm. The major concern is related to strong collisions responsible for fractures, lateral size reduction, defects, and crystallinity destruction. In this regard, a higher efficient yield could be obtained with controlled milling speed, ball-to-powder ratio, ball size, and milling agent.

9.5.4 Ultrasonication

Ultrasonication is a gentler exfoliation process that takes place in solution. Ultrasonic waves generated by a transducer cause air bubbles, which locally increase temperature, pressure, and velocity jet. As a subsequence, the generated hydrodynamic force gives rise to deagglomeration of bulk materials producing graphene, *h*BN, MoS_2, $MoSe_2$, $MoTe_2$, TaS_2, $TaSe_2$, $NiTe_2$, $NbSe_2$, Bi_2Te_3, and WS_2 flakes. The few-layer nanosheets consist of 5 to 10 layers and lateral sizes up to hundreds of millimeters. The technique is somewhat cheap even if the material-solvent interaction can require different degrees of energy for exfoliation. Also, residues of solvents and/or surfactants can affect the electric properties of the resulting flakes.

9.5.5 Shear Exfoliation

Shear exfoliation exploits the viscosity of solutions to apply shear forces across the 2D bulk material. In this way, van der Waals bonds are broken, low-defect flakes with thicknesses of 0.3 to 15 nm, and lateral sizes from 500 nm to 1.5 μm are produced. The yield is generally low, while residues of surfactants and polar solvents that stabilize 2D materials are hard to remove. This affects the final quality of the flakes.

9.5.6 Wet Jet Milling

Wet Jet Milling (WJM) requires the dispersion of the 2D bulk material in a solvent. After dispersion, the mixture is exposed to hydrodynamic forces generated by high-pressure compression of the dispersant fluid phase. The process allows the production of more than 2 L/h^{-1} of 10 g/L^{-1} of single- and few-layer 2D crystal flakes in dispersion, thus making the scaling-up more affordable. The WJM process improves the production of defect-free and high-quality 2D-crystal dispersions on a large scale, opening the way for future exploitation in different commercial applications. The lateral size of the flakes is larger than that obtained with ultrasonication or other mechanical exfoliation procedures.

9.6 Few-Layers 2D Materials-Based Membranes and Water Treatment

Recently, graphene (G), reduced graphene (rG), graphene oxide (GO) [38], as well as molybdenum disulfide (MoS_2), hexagonal boron nitride (h-BN), transition metal chalcogenides (TMDs) [10, 12, 42–45], metal organic frameworks (MOFs) [35], MXenes [46], layered double hydroxides [45–48], layers silicates [49], and other classes of materials have been successfully tested in membrane processes, including ion filtration and, more generally, water desalination. Exfoliated materials have been demonstrated to provide a higher surface area between flakes/nanosheets and diffusing molecules, leading to assisted transport through reversible adsorption/desorption mechanisms. The result has been an important intensification of the production of freshwater, recovery of added-value products such as salts by time-consuming operations, and prolonged operational time as a subsequence of contrasted wetting and fouling events.

In this context, the literature refers to some interesting studies, including a modified interfacial polymerization (IP) reaction proposed to fabricate fouling-resistant membranes with few-layered amine functionalized-boron nitride $BN(NH_2)$ nanosheets [50]. Membranes with $BN(NH_2)$ nanosheets have been tested in nanofiltration to separate NOM from water, providing fluxes increased by 59% with an optimization of 50% of the fouling resistance, long-term operational stability, and enhanced recoverability with no leaching during filtration [50]. Hydrophobic GO membranes have been also prepared by intercalation with silica nanoparticles and proposed for longer water desalination due to optimized resistance to wetting and fouling (Figure 9.2) [51].

Molybdenum disulfide (MoS_2) has been demonstrated to have a smooth surface without any oxygenated functional groups, reducing the resistance

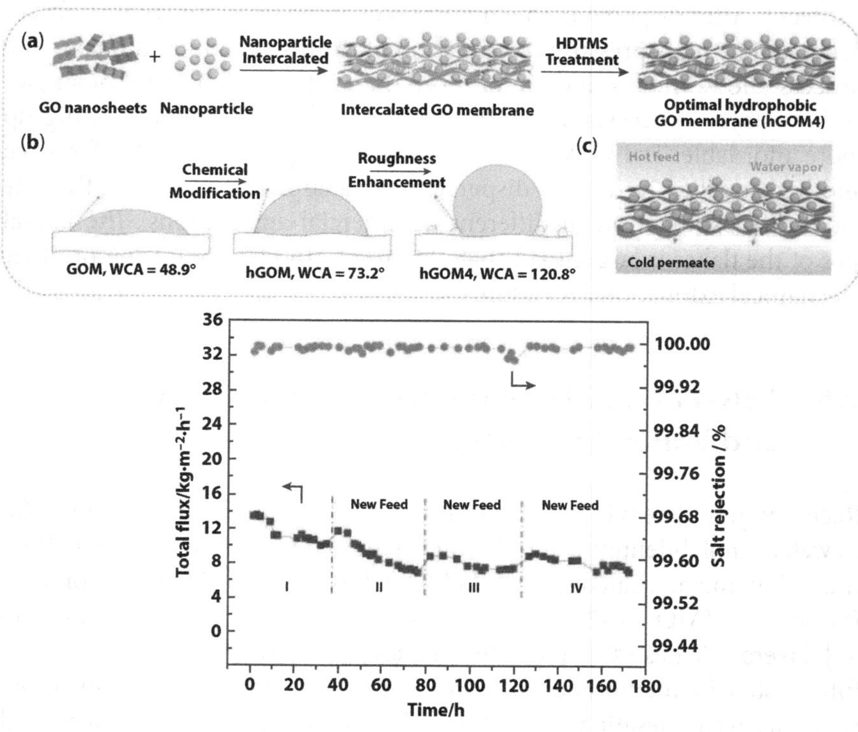

Figure 9.2 GO nanosheets intercalated with silica nanoparticles to improve resistance to flooding and prolong the operational time during water desalination by membrane distillation. Reprinted from [51] (2020), with permission from Elsevier.

of water molecules to transport [43]. The pore structure of MoS_2 allows for achieving superior separation performance with a flux of 3–5 times higher than that estimated for GO membrane prepared under similar conditions. Oxidation and exfoliation of molybdenum disulfide (O-MoS_2) have improved NF process performance, resulting in an excellent anti-fouling performance [45]. Metal–organic framework-based sub-nanochannel (MOFSNC) membranes have been proposed to separate K^+, Na^+, and Li^+ from Ca^{2+} and Mg^{2+} with a mono/divalent ion selectivity of 10^3. By increasing pH from 3 to 8, the ion selectivity moved from 10^2 to 10^4, due to a reduction of energy barrier for monovalent cations as an effect of ion–carboxyl interactions [52]. Wu *et al.* [53] have filled polyamide membranes with mesoporous silica nanoparticles *via* interfacial polymerization of trimesoyl chloride (TMC) and piperazine (PIP) in the presence of mMSN. To prevent the reaction of amino groups with the acid chloride group of TMC during the polymerization, silica nanoparticles have been chemically incorporated into the nanocomposite membrane *in situ*, leading to better compatibility of materials. mMSN nanoparticles have improved the anti-fouling property, yielding long-term stability and enhanced water flux. MXene layered membranes have also been proposed for water treatment [54]. These membranes show narrow nanochannels between the neighboring MXene nanosheets, which cause exceptional water permeance. Water-transport pathways are due to intercalated nanoparticles, which separate the slit pores between the MXene nanosheets, providing greater interlayer distance and numerous free nanochannels. The low toxicity of degradation products, antibacterial properties, and chemical functions have made MXene an interesting material for membrane processes dedicated to ion filtration from water [55]. Several studies have confirmed their effectiveness in removing metal ions, salts, dyes, and radionuclides, killing *E. coli*, *B. subtilis*, and antibiotics [56, 57]. Furthermore, surface chemistry enables them to be variously functionalized to impart further capacities. Coated polytetrafluoroethylene (PTFE) membranes with Ti_3C_2Tx MXene have been proposed for water desalination. A water vapor flux of 0.77 kg/m^2 h under intermittent lighting and a photothermal efficiency of 65.3% have been obtained for a feed salinity of 0.36 g/L [58]. MIL-100(Fe)/1%GO-400 magnetic nanocomposite membranes have been proposed for environmental remediation [59]. The combination of GO and Fe-MOFs has been proven to enhance the adsorption capacity of As(V), water stability, and magnetic recovery ability. Other few-layer materials, such as transition metals chalcogenides (TMC), which consist of binary compounds of a transition metal (M) and a chalcogenide element (X) of general formula MnXm, have been proposed for purification purposes [60]. MX_2 stoichiometries commonly have a sheet configuration stacked one above the other in sheets having thickness of three atoms with the metal in the center and above and below the chalcogenides. The MX bond is covalent while the sheets are held together by weak interactions, which makes exfoliation feasible. TMDs exhibit four crystalline polytypes: a trigonal prismatic 2H phase, an octahedral 1T phase, and two variations of the T1 structure, such as monoclitic 1T' and orthorhombic Td phases. Individually exfoliated TMCs have a flat honeycomb structure in the case of light chalcogenides or slightly wavy in the case of heavy ones. Similarly, there is a family of semimetal chalcogenides (SMC), differing in that the transition metal is replaced by a semimetal. Their most common stoichiometry is M_2X_2, arranged in layers of four atoms thick in X-M-M-X configuration. Exfoliated Bi_2Se_3 has been proposed for the first time to make membrane distillation processes more productive and thermally efficient [10]. Also, the ability of this compound to accelerate the crystallization of NaCl is regarded as a practical and efficient route for treating the brine from desalination, in full agreement with the rules of circular chemistry and biodiversity safe [61].

9.7 A Focus on Graphene-Based Membranes for Water Desalination

In recent years, graphene has been the most studied and used material in different research fields. This is due to its unique proprieties such as excellent electronic conductivity, hydrophobic and anti-wetting nature, anti-fouling properties, great mechanical strength, low density, and selective sorption of water vapors [62, 63]. Graphene consists of single-atom-thick sheets of sp^2-bonded carbon. The combination of high-performance graphene with economical and relatively durable polymer membranes is, hence, a promising approach to support sustainable growth in industrial and research development. Graphene is produced from exfoliation of graphite by exfoliation [64, 65]. The quality of graphene depends strictly on the number of layers and lateral size of the flakes.

9.7.1 Defective Graphene Confined in Polymeric Porous Membranes

Liquid phase exfoliation is one of the most commonly used techniques to break van der Waals bond among the stacked layers of graphite. However, an innovative wet jet milling (WJM) technique for high-quality graphene exfoliation has been recently developed to prepare in easy and scalable way few-layer graphene with a better control of thickness and lateral size [63]. A recent comparative study demonstrated that defective few-layer graphene from WJM exhibits a larger lateral size of 490 nm against 100 nm, with a thickness of ~1.5 nm. This results in a larger surface area and amplified assisted diffusion of water vapor during membrane distillation process with fluxes up to 16 $Lm^{-2}h^{-1}$ [12]. Few-layer graphene has also been deposited on PVDF porous substrates to induce multi-scale rough structural effects on transport properties. As a result, enhanced surface properties have contrasted flooding events, leading to a significant improvement in the efficiency of the MD process compared to pristine membranes (Figure 9.3) [66].

It is relevant to observe that nanofillers like exfoliated graphene can provide assisted transport due to presence of defects, holes, and oxidation sites generated during exfoliation [22]. While these defects are undesirable for

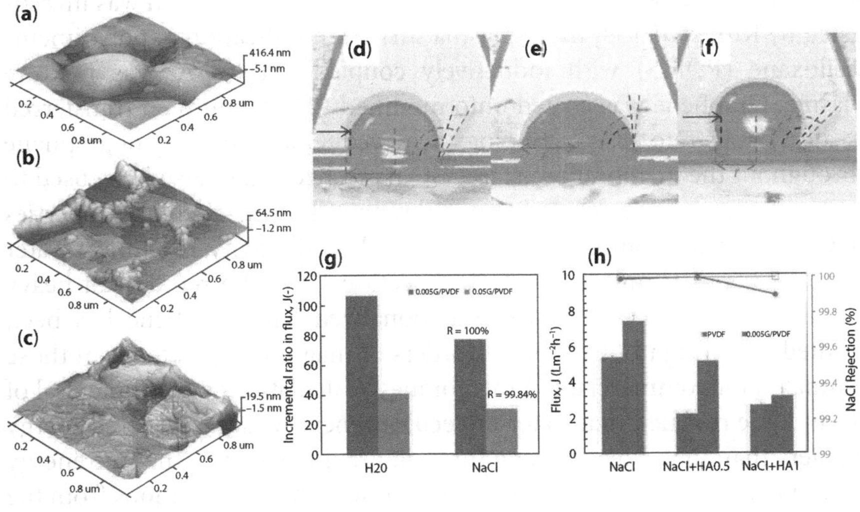

Figure 9.3 AFM images of membrane with different amount of few-layer graphene deposited onto the surface: (a) pristine PVDF, (b) 0.005G/PVDF, (c) 0.05G/PVDF. The time-dependent changes in contact radius and contact angle of evaporating droplet on (d) pristine PVDF, (e) 0.05G/PVDF, and (f) 0.005G/PVDF surface. Transport through the functional membranes facing pure water, NaCl 0.6 M and mixtures of NaCl and HA (0.6 M/0.5 mg mL^{-1} and 0.6 M/1.0 mg mL^{-1}): (g) Incremental ratio in the flux for 0.005G/PVDF and 0.05G/PVDF membranes with respect to pristine PVDF membrane at $T_f = 33$ °C; (h) Flux (J) and salt rejection (R) estimated at $T_{feed} = 40$ °C for pristine and 0.005G/PVDF membranes. Reprinted from [66] (2022), with permission from MDPI (CC BY 4.0).

more traditional applications where the precision of monolayer is required, for some water desalination operations, they could be suitable tools for directing and shortening the recovery of better-quality crystals from brine. In this case, graphene has ability to establish reversible sorption-desorption interaction with water, promoting water sequestration from ion–water clusters and subsequent ion–ion aggregation [16, 17]. The result is a reduced nucleation time with an accelerated growth rate of crystals within narrow distributions and coefficient of variation of 27 to 32%. Graphene oxide (GO) also finds large applications in water desalination. GO paper has been prepared by filtration of a GO solution through a microfiltration membrane [67]. The obtained graphene has been studied by X-ray diffraction (XRD) pattern showing the layer-to-layer distance (*d*-spacing) of the GO paper was approximately 0.83 nm. The *d*-spacing of a GO paper depends on the degree of oxidation. Moreover, the study showed how increasing the degree of oxidation, the *d*-spacing could be increased to up to about 1 nm. Li *et al.* [68] have prepared graphene membranes to investigate the corrugation of GO nanosheets and have found that GO nanosheets could be used with great results in the nanofiltration process, with a rejection rate of 67% for direct yellow. A direct transfer of graphene pattern was investigated by Kim *et al.* [69] using plasma surface modification of polydimethylsiloxane (PDMS) with inductively coupled $C_4F_8/O_2/Ar$ gas mixture plasma. Graphene was coated onto modified PDMS and then transferred to a PET substrate. This technology achieved 85% uniformity of graphene throughout the membrane surface. GO nanosheets have also been used to coat commercial polyamide RO membranes. The anti-swelling properties of GO have been combined with its fruitful productivity, reaching a water flux of 33.5 LMH and 98.5% NaCl rejection [70]. The removal of heavy metal ions by RO membranes functionalized with graphene has been studied, resulting in fluxes of 2–5 orders of magnitude greater than those estimated for commercial RO membranes. Calculations of the potential of mean force revealed that water molecules generally face lower free energy barriers than ions when passing through graphene pores. The free energy barriers for ions can be explained as the combining contributions from the ion dehydration effect and the surface electrostatic interaction. Simulation studies have suggested that OH-graphene membranes achieve 100% of ion rejection but with the lowest water permeability. NH-graphene shows relatively low ion rejection with intermediate permeability, while boron-doped graphene shows high water permeation and acceptable ion rejection. Overall, the functionalized nanoporous graphene membranes exhibit potential application in the removal of heavy metal ions with different productivity-rejection trade-offs.

9.7.2 Graphene Membranes to Membrane Distillation Processes

Graphene-based membranes have been proposed to amplify the production of freshwater by membrane distillation [71–81]. Table 9.3 resumes a comparative analysis of MD processes implemented by using membranes without and with few-layer graphene and derivates, worked under different conditions.

Recently, functionalized graphene oxide has been proposed for MD by Zahirifar *et al.* [79]. Octadecylamine (ODA) has been linked to GO flakes and then coated onto PVDF membrane surfaces. In this case, the extent hydrocarbon chain of ODA reduced the surface energy and amplified the roughness factor, leading to enhanced hydrophobicity and resistance to wetting with benefits for flux stability. Bhadra *et al.* [82] have investigated graphene-based membranes in DCMD by immobilizing graphene oxide on PTFE membranes. The improvement in the productivity-efficiency trade-off was ascribed to selective sorption, non-capillary effects, and reduced temperature polarization. Graphene quantum dots (GQDs)/polyvinylidene fluoride (PVDF) nanofibrous membranes have been proposed for air gap membrane distillation (AGMD) by Jafari *et al.* [33]. AGMD experiments have been conducted with 3.5wt% NaCl solution for 8 h, showing how the graphene may prevent wetting and fouling after long operational times. Water fluxes of 17.6 kg/m^2 h and salt rejection of about 99.7% membrane have been measured for membranes containing 0.25 wt% GQDs compared to neat PVDF nanofibrous membranes. A further study on reduced graphene oxide (rGO) in PVDF membranes has been successfully performed by Abdel-Karim *et al.* [73] by changing the degree of reduction of graphene oxide (rGO) nanoplatelets. The fillers incorporated into PVDF matrices have produced effects on MD performance due to different oxygen content. Specifically, 0.5 wt% rGO enhanced the performance of PVDF membranes by approximately 169%. Furthermore, continuous testing for up to 96 h have confirmed the stable performance of the functional membranes over time, indicating optimized resistance to wetting and fouling phenomena. A super-amphiphobic membrane based on poly(vinylidene fluoride-*co*-hexafluoropropylene) (PVDF-HFP) filled with reduced graphene oxide (rGO) has been designed for MD operations by electrospinning and the grafted with low-surface-tension fluoroalkylsilane (POTS) [83]. Enhanced stability and durability with flux of 27.94 kg m^{-2} h^{-1} and salt rejection near 100% have been obtained. Li *et al.* [74] have proposed hydrophobic nanofibrous composites membrane for vacuum membrane distillation (VMD). These composite membranes have

Table 9.3 Flux and rejection estimated for pristine and graphene-based PVDF membranes.

Membrane	MD conf.	Feed NaCl	Gr	$T_{Feed/Perm}$ T°C	Flow rate	Flux LMH	R %	Note	Ref.
PVDF	DCMD	0.6 M	n	50/20	0.28/0.28 m/s	1.85	<90 using 10–15 polymer conc.	20 wt.% PVDF	[71]
PVDF+No woven	DCMD	0.6 M	n	50/20	0.28/0.28 m/s	12.4	< 90 using	Supported membranes	[71]
PVDF + GQDs1 0.05 wt%	AGMD	0.6 M	y	60/20	30 L/h	6	PVDF + GQD1P 99.9 to 99.8 after 8 h	Graphene quantum dots (GQDs) in PVDF ES membranes	[33]
PVDF	AGMD	0.6 M	n	85/20	n.a.	3.4	99.9	PVDF filled with GO and GO/3-(aminopropyl) triethoxysilane (APTS)	[72]
PVDF + GO 0.1%	AGMD	0.6 M	GO	85/21	n.a.	3.8	99.7		
PVDF + GO 0.3%	AGMD	0.6 M	GO	85/22	n.a.	4.7	99.9		
PVDF + GO 0.5%	AGMD	0.6 M	GO	85/23	n.a.	5.4	99.6		
PVDF + GO-APTS 0.1%	AGMD	0.6 M	GO	85/24	n.a.	4.5	99.8		
PVDF + GO-APTS 0.3%	AGMD	0.6 M	GO	85/25	n.a.	6.25	99.9		
M_15.5-rGO/0.5 PVDF	AGMD	3.5 wt%	GO	80/20	385 ± 5 mL/min	≈7	99.9	Mixed matrix membranes (MMMs) with rGO nanosheets in PVDF	[73]

(*Continued*)

Table 9.3 Flux and rejection estimated for pristine and graphene-based PVDF membranes. (*Continued*)

Membrane	MD conf.	Feed NaCl	Gr	$T_{Feed/Perm}$ T°C	Flow rate	Flux LMH	R %	Note	Ref.
G3-PVDF	VMD	0.6 M	y	50 (31.3 kPa vacuum permeate side)	0.85 L/min	≈35	99.9	ES PVDF on PP substrate and functionalized with GO nanosheets	[74]
2% GO-PSF	DCMD	2.5 wt%	GO	90/20	5.7 L/min	22.8	99.85	Graphene oxide-polysulfone (GO-PSF) mixed-matrix membranes	[75]
GO-PVP/PVDF	DCMD	3.5 wt%	GO	65/20	100 rpm	≈22	≈99.9	Graphene oxide (GO) located on PVDF pore surface	[34]
GP2 0.16%	DCMD	6 L RO brine 57,000 ppm	y	85/n.a.	75/48 L/h	16.7	99.5	Membranes with graphene nanoplatelets GNPs	[76]
rGO/PVDF-HFP	DCMD	60 g/L	GO	75/25	1/0.75 L/min	20.37	99.97	(rGO/PVDF-HFP) flatsheet membranes by electrospinning technique.	[77]

(*Continued*)

Table 9.3 Flux and rejection estimated for pristine and graphene-based PVDF membranes. (*Continued*)

Membrane	MD conf.	Feed NaCl	Gr	$T_{Feed/Perm}$ T°C	Flow rate	Flux LMH	R %	Note	Ref.
PVDF/Gr 0.87 wt%	DCMD	35 g/L	y	75/2	120 mL/min	59.5	99.7	Monolayered graphene oxide sheets in PVDF	[78]
M2	AGMD	3.5 wt%	y	60/15	15/15 L/h	7.5	>96	Dual-layer membrane coated by octadecylamine and graphene GO-ODA	[79]
M2	AGMD	3.5 wt%	y	80/15	15/15 L/h	16.5	>96		
G/PVDF 0.5 wt%	AGMD	RO brine from CSG produced water	y	60/20	n.a.	20.5	99.99	PVDF filled with graphene flakes	[80]

been made of a hydrophobic polyvinylidene fluoride (PVDF) nanofibrous layer and a hydrophobic polypropylene (PP) nonwoven fabric (NWF) substrate. The PVDF layer has been directly fabricated on the surface of the PP substrate by the electrospinning technique and functionalized with GO. Also, in this case, enhanced anti-wetting properties have been combined with increased flux. To address severe swelling occurring in membranes containing GO, polymer-intercalated GO membranes have been proposed [84]. This intercalation offers superior resistance to swelling and allows water to pass through narrowed channels within the polymer network situated between the 2D nanochannels of GO interlaminations. A flux of 52 $Lm^{-2}h^{-1}$ and total salt rejection rates have been successfully obtained when treating complex hypersaline wastewater.

Gugliuzza *et al.* [85] envisaged chemisorption mechanisms in graphene-assisted water diffusion and ion-ion aggregation, yielding insightful indications about the role of defective graphene in membranes designed for scalable MD and MCr.

9.8 Few-Layered Graphene Nanochannels Like Ion Filtering

As is well known, membrane processes designed for ion filtering require membranes with well-shaped, sized, and functionalized pores [32]. In this context, 2D materials have been explored to realize porous membranes with ion-filtering properties [86–88]. This is because the separation of mono- and divalent ions is somewhat hard when using traditional nanocomposite membranes due to the fact that both types of ions exhibit a sub-nanometer hydrated radii (K^+: 3.31 Å; Na^+: 3.58 Å; Li^+: 3.82 Å; Mg^{2+}: 4.28 Å), even though divalent ions are characterized by higher charges [90]. Traditionally, two mechanisms—size sieving and electrostatic effects—are proposed to separate mono and divalent ions [52, 89]. In the first approach, membranes are engineered to reduce the pathways for ion transport. In the electrostatic mechanism, the repulsive charge of the membrane is utilized to facilitate ion removal through electrostatic interactions.

The productivity-selectivity trade-off for the separation process has not yet been scored. Single-layered graphene and graphene oxide have been proposed to produce porous membranes for selective molecular adsorption and diffusion [91]. However, precisely tuning GO interlayer spacing within the sub-nanometer range is somewhat hard, which limits high salt rejection in pressure-driven filtration processes, reducing water permeance. Some authors [90] have inspired stacked charged GO graphene membranes to biological ion channels, producing surface charges on GO membranes for controllable ion transport without affecting water permeance (Figure 9.4).

By Layer-by-layer (LbL) techniques, the charge on pre-stacked GO sheets has been switched, thus producing repulsive forces against co-ions with the same charge of the membrane GO surface, while weak electrostatic attraction has been suppressed. In this way, water is free to permeate through the membrane whereas AB_2 or A_2B-types salts are retained. The involvement of size, structural defects, and functional edges has also been tuned in separation processes involving layered GO membranes. However, much more practical ways have been proposed to produce handler membranes where synergistic effects of size-sieving and electrostatic repulsion have been combined in charged interlayer channels of 2D materials membranes for selective transport of monovalent ions over divalent ions. The transport through GO has been envisaged to take place by nanocapillary

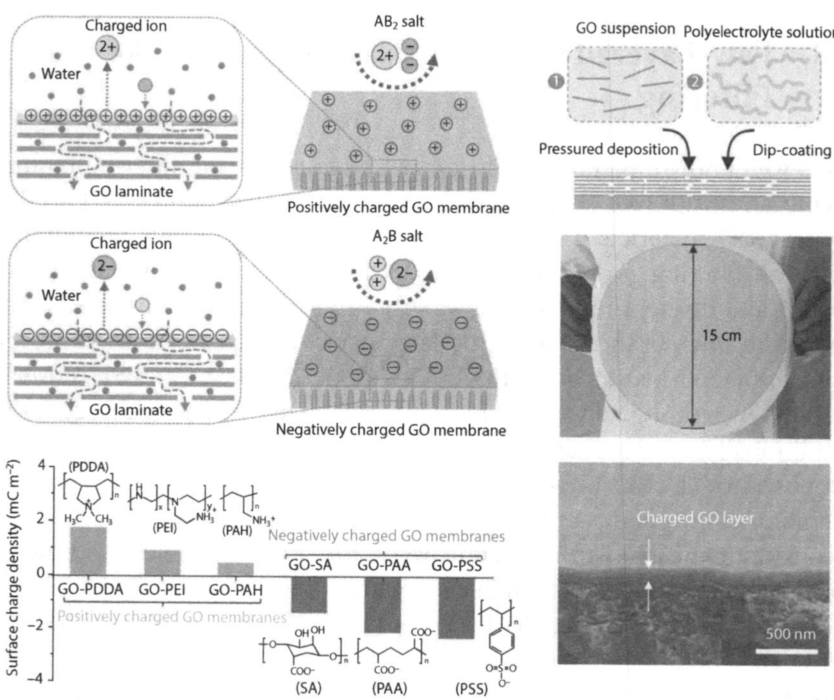

Figure 9.4 Design of surface-charged graphene oxide (GO) membrane. A schematic of the design of surface-charged GO membranes by coating polyelectrolytes (PA, PAA, PSS) on the surface of GO laminates to realize controllable ion transport. Reprinted from [90] (2019) with permission from Nature (CC BY).

interlayer spacing between GO sheets by assisted non-covalent interactions that support molecular diffusion. It has been demonstrated that functional groups improve interlayer stacking so that small solutes can be separated according to size sieving mechanisms. However, hydroxylation of chemical moieties can locally do hybridization of carbon atoms from sp^2 to sp^3 providing interaction between diffusing molecules and membrane [91]. Huang *et al.* [92] have proposed to intercalate GO nanosheets with charged polyethyleneimine (PEI) to separate mono- and divalent ions. A layer-by-layer procedure based on spin coating has been used to generate electrostatically stable GO-PEI-bilayered membranes at different contents of PEI and GO using PAN as a substrate. A sieving effect can be ascribed to positively charged PEI into the interlayer of negatively charged GO laminates. An increase in PEI amount produces a further increase in charge, leading to an enhanced repulsive effect with a higher permeation energy barrier for divalent ions in the GO interlayer channels and subsequently increased perm-selectivity of mono-/di-valent ions. In this way, the separation mechanism moves from sieving to electrostatic repulsion effect. Qiu *et al.* [93] have focused on the effect of edge functional groups, i.e., carboxyl (COOH), hydroxyl (OH), and hydrogen (H), on the filtration of LiCl, NaCl, and KCl solutions through layered membranes. It has been observed that water permeance is affected by the steric size of the edge functional groups and ions are stopped at the entrance of the channels. According to the dehydration mechanism, it has been observed a lower KCl and NaCl permeance through G-OH membranes due to the drastic dehydration of K^+ and Na^+, while higher LiCl permeance has been promoted by the attraction of hydrated Li^+ on hydrated Cl^-. Anyway, rGO and GO exhibit water permeance and NaCl rejection higher than commercial membranes traditionally used in seawater reverse osmosis (SWRO) and brackish water reverse osmosis (BWRO) membranes (Figure 9.5) [94].

9.9 Chalcogenides in Porous Confined Membranes for Water Desalination

Chalcogenide compounds are regarded as frontier materials for water desalination [95]. For the first time, bismuth telluride crystals have been used to functionalize PVDF membranes designed for membrane distillation [10] and membrane crystallization [18, 61]. In the first case, water permeation has been significantly amplified reaching values up to 28 $Lm^{-2}h^{-1}$ at 50 °C and by a flow rate of 100 $Lmin^{-1}$ for $PVDF/Bi_2Te_3$ containing

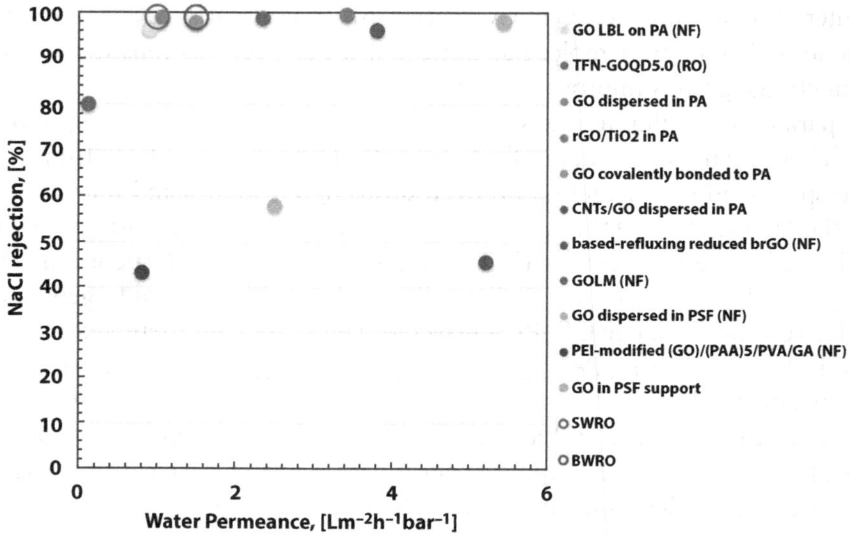

Figure 9.5 Water permeance and NaCl rejection of traditional polymeric RO membranes and GO-containing RO membranes (GOQD stands for GO quantum dots, while GOLM stands for GO laminate membranes). Reprinted from [94] (2017), with permission from Elsevier.

0.5 % of chalcogenide (Figure 9.6). The effects of fluid dynamics conditions at different chalcogenide loadings have been studied, confirming higher productivity for MD processes. This kind of material has also allowed making MD processes more thermally efficient, i.e., thermal efficiency up to 25.6%, resulting in a somewhat competitive productivity-efficiency trade-off.

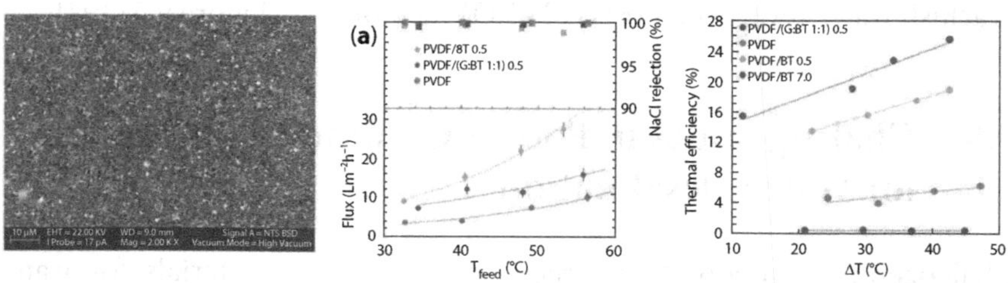

Figure 9.6 SEM picture collected on PVDF/ Bi_2Te_3 membrane surface (a); Flux vs NaCL rejection estimated for PVDF membranes filled with Bi_2Te_3 and a mixture of Bi_2Te_3/Graphene (b); Thermal efficiency calculated at different driving forces (DT) for membranes functionalized with Bi_2Te_3 and a mixture of Bi_2Te_3/Graphene during DCMD operations. Reprinted from [10] (2021), with permission from Elsevier.

This can be ascribed to the synergistic effects of the water sorption-desorption mechanism along with the ability of chalcogenide—with maximized Seebeck effect—to contrast thermal polarization and subsequently reduce the decline in flux. Fruitful outcomes have also been obtained with membranes functionalized with bismuth selenide and worked in membrane crystallization devices.

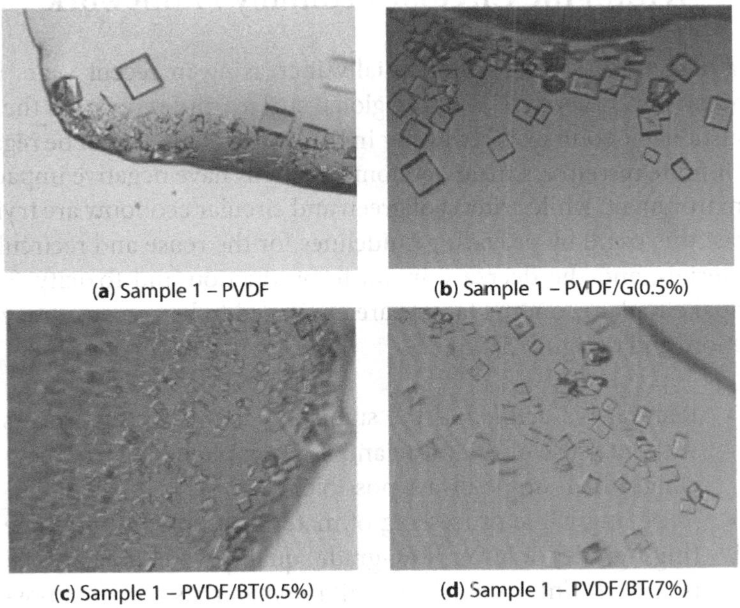

Figure 9.7 Optical pictures collected on the first detectable NaCl crystals obtained with PVDF membranes functionalized with few-layer Bi_2Te_3 (BT) and graphene (G). Reprinted from [61] (2021), with permission from MDPI (CC BY 4.0).

A remarkable reduction in nucleation time is associated with controlled nucleation and a higher growth rate of crystals exhibiting a cubic shape uniformly. This is due to a fast-reversible sequestration of water from ion-water clusters, thus producing strengthened attractive ion–ion Coulomb as a first step towards the formation of nuclei [16, 18]. Again, assisted water exclusion can be associated with Se vacancies, which are regarded as responsible for the highest ability of the membranes to promote assisted water diffusion. Generally speaking, this behavior has also been observed for defective graphene confined in PVDF membranes, which causes a quick turnover in sorption-desorption mechanisms, leading to massive nucleation of NaCl cubic crystals within a coefficient distribution around

26 %. Comparative studies on the crystallization of NaCl from brine using membranes functionalized with chalcogenides and graphene flakes have been carried out, yielding insights into the different kinetics controlling the formation and growth of crystals (Figure 9.7).

9.10 Water Desalination and Few-Layer Materials Within the Circular Economy Framework

Water demand has been exponentially increasing in recent years [1–2]. The ocean covers about 97% of the global surface and represents the main eco-sustainable source of freshwater in the world, but it cannot be regarded as an infinite resource. Linear economic systems have negative impacts on the environment, while criteria of green and circular economy are trying to contrast this trend by providing guidelines for the reuse and recirculation of materials, and the decrease in waste production and toxicity. So, the principal circular economy targets are envisaged to be recovery, recycling, reduction, and design [95, 96].

- Recovery is defined as the supplying source in the process without reuse of the waste and it is complementary to recycling, which is not always possible.
- Recycling is the processing of materials to obtain the same (high-grade) or lower (low-grade) quality when the material is reused. Unfortunately, not all the materials are often possible to be reused. Different considerations are necessary for recycling materials. The first case is when recycling occurs within the same materials (closed-loop) reused in the same system, while the second case is when a part of the product (open-loop) is used in a different system. Recycling depends on many factors, including uses, prices, type/properties of materials, displacements, and losses/impurities connected with recycling.
- Reduction of materials allows fewer inputs of energy, raw materials, material costs, and waste production. This definition can be extended to the definition of Reuse or Refuse.
- Design is the research of smart eco-sustainable production of a product.

Membrane processes enhance the management and recovery of natural resources, reducing negative environmental impacts. The most common desalination processes provide a minimal liquid discharge (MLD) (80–90% of total water recovery) or zero liquid discharge (ZLD) (100% water recovery) [97, 98]. As previously explained, Reverse Osmosis (RO) dominates the market for freshwater recovery from seawater. In 2018, a global cumulative contracted capacity reached 104.7 million m^3/day.

Large energy consumption with a unit water cost of ~\$0.5/m^3 and an overall recovery factor around 50 % make RO less competitive with respect to the criteria of circular economy [9]. So, thermal driven Membrane Distillation (MD) appears to be a more sustainable route for the production of desalted water with a recovery factor superior to 70% [4, 32].

However, the circular economy has some limitations, including physical flows of materials and energy [99]. Approximately 75% of global energy production is based on non-renewable and emissions-intensive fossil fuels, whose combustion is not adapted to the recycling phase. On the other hand, today there is a necessity to identify new frontier materials to make the process more fruitful and energetically efficient. Also, reuse, remanufacturing, and recycling tasks are needed for the physical flows of materials [100].

In this regard, few-layer materials have grown enormously in their tuned characteristics and different uses, including the propulsion to fruitful water desalination by membrane processes. In the last years, there has been a rapid transition of these materials towards commercialization and real-world implementation with expected sales up to \$1.64 billion by 2034 [101]. However, the question is whether the portfolio of 2D materials adheres to the criteria of sustainability demanded by the circular economy. The most used techniques of exfoliation require large energy consumption, the use of toxic solvents, and the production of by-products, resulting in the emission of 620g CO_2 equivalent per gram of graphene yield, as an example. Nevertheless, advanced and scalable techniques of exfoliation such as WJM are now proposed to produce few-layer materials by using greener solvents and bio-based compounds in place of traditional N-Methylpyrrolidone (NMP) and Dimethylformamide (DMF) without using specific catalytic substrates.

Another concern is the recovery of these materials from mineral mining, with the subsequent depletion of high-quality basins or related artificial synthesis, which results in increased carbon emission and loss of quality in the properties for these material types.

The recyclability of 2D materials, especially when used as fillers in polymer matrices, represents another critical issue due to the complexity and costs needed to recover them after the end of their lifecycle or in the case of incineration of the composite. Today, there is an urgency to identify new sustainable practices to purify and detach the materials from other compounds when mixed. Toxicity aspects for human health, biodiversity, and the environment must be examined due to the fact that these compounds are extremely resistant, durable, and difficult to biodegrade. Unquestionably, the excellent performance of few-layer materials in different applicative fields, including water desalination, must not disregard the assessment of sustainability and potential environmental concerns in the future. Research closely looks at the issues discussed, trying to identify new paths for decarbonizing synthesis procedures or recovering defect-free 2D materials from biomass waste, reducing water footprints by over 90% without compromising economic viability [102, 103].

Acknowledgments

We acknowledge the financial grant from 'the Italian Ministry of Foreign Affairs and International Cooperation' within the framework of the Great Relevance International Project Italy (MAECI)-China (NSFC) 2018-2020 - New Materials, with particular reference to Two-dimensional systems and Graphene (2DMEMPUR), MAE00691702020-06-26.

References

1. World Water Development Report, 2023. Available: https://www.unesco.org/reports/wwdr/2023/en (Accessed on 28 February 2024).
2. Pielke, R.A., Adegoke, J., Hossain, F., Niyogi, D., Environmental and Social Risks to Biodiversity and Ecosystem Health-A Bottom-Up, Resource-Focused Assessment Framework. *Earth*, 2, 440–456, 2021.
3. Tang, C.Y., Yang, Z., Guo, H., Wen, J.J., Nghiem, L.D., Cornelissen, E., Potable Water Reuse through Advanced Membrane Technology. *Environ. Sci. Technol.*, 52, 18, 10215–10223, 2018.
4. Gugliuzza, A. and Basile, A., Membrane contactors: Fundamentals, membrane materials and key operations, in: *Handbook of Membrane Reactors*, 1st, vol. 2, pp. 54–106, Woodhead Publishing Limited, Cambridge, 2013.
5. Talaeipour, M., Nouri, J., Hassani, A., Mahavi, A.H., An investigation of desalination by nanofiltration, reverse osmosis and integrated (hybrid NF/RO) membranes employed in brackish water treatment. *J. Environ. Health Sci. Eng.*, 15, 18, 1–9, 2017.

6. Alkaisi, A., Mossad, R., Sharifian-Barforoush, A., A Review of the Water Desalination Systems Integrated with Renewable Energy, in: *Energy Procedia*, 2017, doi:10.1016/j.egypro.2017.03.138.
7. Okamoto, Y. and Lienhard, J.H., How RO membrane permeability and other performance factors affect process cost and energy use: A review. *Desalination*, 470, 114064, 2019, doi:10.1016/j.desal.2019.07.004.
8. Ullah, R., Khraisheh, M., Esteves, R.J., McLeskey, J.T., AlGhouti, M., Gad-el-Hak, M., Tafreshi, H.V., Energy efficiency of direct contact membrane distillation. *Desalination*, 433, 56–67, 2018.
9. Castel, C., Favre, E., Membrane separations and energy efficiency, *J. Membr. Sci.*, 548, 345–357, 2018.
10. Frappa, M., Del Rio Castillo, A.E., Macedonio, F., Di Luca, G., Drioli, E., Gugliuzza, A., Exfoliated Bi_2Te_3-enabled membranes for new concept water desalination: Freshwater production meets new routes. *Water Res.*, 203, 117503, 2021.
11. Sinha, R.S., Singh, B.H., Dangayach, R., Singh, R., Deb, C.K., Ganesapillai, M., Chen, S.-S., Purkait, M.K., Recent Developments in Nanomaterials-Modified Membranes for Improved Membrane Distillation Performance. *Membranes*, 10, 7, 40, 2020.
12. Frappa, M., Del Rio Castillo, A.E., Macedonio, F., Politano, A., Drioli, E., Bonaccorso, F., Pellegrini, V., Gugliuzza, A., A few-layer graphene for advanced composite PVDF membranes dedicated to water desalination: a comparative study. *Nanoscale Adv.*, 2, 4728–4739, 2020.
13. Sun, L., Ying, Y., Huang, H., Song, Z., Mao, Y., Xu, Z., Peng, X., Ultrafast molecule separation through layered WS_2 nanosheet membranes. *ACS Nano*, 8, 6304–6311, 2014, doi:10.1021/nn501786m.
14. Zuo, J. and Chung, T.-S., Metal–Organic Framework-Functionalized Alumina Membranes for Vacuum Membrane Distillation. *Water*, 8, 12, 586, 2016, https://doi.org/10.3390/w8120586.
15. Qiu, S., Xue, M., Zhu, G., Metal-organic framework membranes: From synthesis to separation application. *Chem. Soc. Rev.*, 43, 6116–6140, 2014, doi:10.1039/c4cs00159a.
16. Perrotta, M.L., Macedonio, F., Tocci, E., Giorno, L., Drioli, E., Gugliuzza, A., Graphene stimulates the nucleation and growth rate of NaCl crystals from hypersaline solution via membrane crystallization. *Environ. Sci. Water Res. Technol.*, 6, 6, 1723–1736, 2020.
17. Perrotta, M.L., Macedonio, F., Giorno, L., Jin, W., Drioli, E., Gugliuzza, A., Tocci, E., Molecular insights on NaCl crystal formation approaching PVDF membranes functionalized with grapheme. *Phys. Chem. Chem. Phys.*, 22, 15, 7817–7827, 2020.
18. Macedonio, F., Politano, A., Drioli, E., Gugliuzza, A., Bi_2Se_3-assisted membrane crystallization. *Mater. Horiz.*, 5, 5, 912–919, 2018.

19. Ren, C.E., Hatzell, K.B., Alhabeb, M., Ling, Z., Mahmoud, K.A., Gogotsi, Y., Charge- and Size-Selective Ion Sieving Through Ti_3C_2Tx MXene Membranes. *J. Phys. Chem. Lett.*, 6, 4026–4031, 2015, doi:10.1021/acs.jpclett.5b01895.
20. Liu, Y., Wang, N., Cao, Z., Caro, J., Molecular sieving through interlayer galleries. *J. Mater. Chem. A*, 2, 1235–1238, 2014, doi:10.1039/c3ta13792a.
21. Jaramillo, M.F. and Restrepo, I., Wastewater Reuse in Agriculture: A Review about Its Limitations and Benefits. *Sustainability*, 9, 10, 1734, 2017, doi.org/10.3390/su9101734.
22. Rudnicki, P., Hubicki, Z., Kołodyńska, D., Evaluation of heavy metal ions removal from acidic waste water streams. *Chem. Eng. J.*, 252, 362–373, 2014, doi:10.1016/j.cej.2014.04.035.
23. Kaur, J., Bhatti, S.S., Bhat, S.A., Nagpal, A.K., Kaur, V., Katnoria, J.K., Evaluating Potential Ecological Risks of Heavy Metals of Textile Effluents and Soil Samples in Vicinity of Textile Industries. *Soil Syst.*, 5, 63, 2021, https://doi.org/10.3390/soilsystems5040063.
24. Kim, D.I., Gonzales, R.R., Dorji, P., Gwak, G., Phuntsho, S., Hong, S., Shon, H., Efficient recovery of nitrate from municipal wastewater via MCDI using anion-exchange polymer coated electrode embedded with nitrate selective resin. *Desalination.*, 484, 114425, 2020, doi:10.1016/j.desal.2020.114425.
25. Al-Rashdi, B.A.M., Johnson, D.J., Hilal, N., Removal of heavy metal ions by nanofiltration. *Desalination*, 315, 2–17, 2013, doi:10.1016/j.desal.2012.05.022.
26. Liu, M., Lü, Z., Chen, Z., Yu, S., Gao, C., Comparison of reverse osmosis and nanofiltration membranes in the treatment of biologically treated textile effluent for water reuse. *Desalination*, 281, 372–378, 2011, doi:10.1016/j.desal.2011.08.023.
27. Kumar, A., Rahman, M.S., Iqubal, M.A., Ali, M., Niraj, P.K., Anand, G., Kumar, P., Abhinav, A.K., Ghosh, Ground water arsenic contamination: A local survey in India. *Int. J. Prev. Med.*, 7, 100, 2016, doi:10.4103/2008-7802.188085.
28. Manna, A.K., Sen, M., Martin, A.R., Pal, P., Removal of arsenic from contaminated groundwater by solar-driven membrane distillation. *Environ. Pollut.*, 158, 805–811, 2010, doi:10.1016/j.envpol.2009.10.002.
29. Shen, M., Song, B., Zhu, Y., Zeng, G., Zhang, Y., Yang, Y., Wen, X., Chen, M., Yi, H., Removal of microplastics via drinking water treatment: Current knowledge and future directions. *Chemosphere*, 251, 126612, 2020, doi:10.1016/j.chemosphere.2020.126612.
30. Garfí, M., Cadena, E., Sanchez-Ramos, D., Ferrer, I., Life cycle assessment of drinking water: Comparing conventional water treatment, reverse osmosis and mineral water in glass and plastic bottles. *J. Clean. Prod.*, 137, 997–1003, 2016, doi:10.1016/j.jclepro.2016.07.218.
31. Christie, K.S.S., Horseman, T., Lin, S., Energy efficiency of membrane distillation: Simplified analysis, heat recovery, and the use of waste-heat. *Environ Int.*, 138, 105588, 2020. doi.org/10.1016/j.envint.2020.105588.

32. Gugliuzza, A., Review of techniques for future manufacturing of ordered and functional porous films: conceivable membranes for water desalination by membrane distillation processes. *Adv. Manuf.*, 1, 2, 29–30, 2024.
33. Jafari, A., Kebria, M.R.S., Rahimpour, A., Bakeri, G., Graphene quantum dots modified polyvinylidenefluride (PVDF) nanofibrous membranes with enhanced performance for air Gap membrane distillation. *Chem. Eng. Process. Process Intensif.*, 126, 222–231, 2018, doi:10.1016/j.cep.2018.03.010.
34. Qiu, H., Peng Y., L., Villacorta Hernandez, B., Zhu, Z., Pore channel surface modification for enhancing anti-fouling membrane distillation. *Appl. Surf. Sci.*, 443, 217–226, 2018, doi:10.1016/j.apsusc.2018.03.004.
35. Di Luca, G., Chen, G., Jin, W., Gugliuzza, A., Aliquots of MIL-140 and Graphene in Smart PNIPAM Mixed Hydrogels: A Nanoenvironment for a More Eco-Friendly Treatment of NaCl and Humic Acid Mixtures by Membrane Distillation. *Membranes*, 13, 4, 437, 2023.
36. Tai, Z.S., Dzarfan Othman, M.H., Koo, K.N., Jaafar, J., Critical review on membrane designs for enhanced flux performance in membrane distillation. *Desalination*, 553, 116484, 2023, doi.org/10.1016/j.desal.2023.116484.
37. Qtaishat, M., Matsuura, T., Khayet, M., Khulbe, K.C., Comparing the desalination performance of SMM blended polyethersulfone to SMM blended polyetherimide membranes by direct contact membrane distillation. *Desalin. Water Treat.*, 5, 1–3, 91–98, 2009, https://doi.org/10.5004/dwt.2009.584.
38. Zhang, X., Zhou, J., Zou, X., Wang, Z., Chu, Y., Wang, S., Preparation of Nano-SiO_2/Al_2O_3/ZnO-Blended PVDF Cation-Exchange Membranes with Improved Membrane Permselectivity and Oxidation Stability. *Materials*, 11, 12, 2465, 2018, https://doi.org/10.3390/ma11122465.
39. El Haj Assad, M., Bani-Hani, E., Al-Sawafta, I., Issa, S., Hmida, A., Gupta, M., Atiqure, R.S.M., Hidouri, K., Applications of nanotechnology in membrane distillation: a review study. *Desalin. Water Treat.*, 192, 61–77, 2020, doi.org/10.5004/dwt.2020.25821.
40. Del Rio Castillo, A.E., Pellegrini, V., Ansaldo, A., Ricciardella, F., Sun, H., Marasco, L., Buha, J., Dang, Z., Gagliani, L., Lago, E., Curreli, N., Gentiluomo, S., Palazon, F., Prato, M., Oropesa-Nuñez, R., Toth, P.S., Mantero, E., Crugliano, M., Gamucci, A., Tomadin, A., Polini, M., Bonaccorso, F., High-yield production of 2D crystals by wet-jet milling. *Mater. Horiz.*, 5, 890–904, 2018, doi:10.1039/c8mh00487k.
41. Islam, M.A., Serles, P., Kumral, B., Guerra Demingos, P., Qureshi, T., Meiyazhagan, A.K., Puthirath, A.B., Bin Abdullah, M.S., Rafat Faysal, S., Ajayan, P.M., Panesar, D., Veer Singh, C., Filleter, T., Exfoliation mechanisms of 2D materials and their applications featured. *Appl. Phys. Rev.*, 9, 041301, 2022, doi.org/10.1063/5.0090717.
42. Lei, W., Mochalin, V.N., Liu, D., Qin, S., Gogotsi, Y., Chen, Y., Boron nitride colloidal solutions, ultralight aerogels and freestanding membranes through one-step exfoliation and functionalization. *Nat. Commun.*, 6, 8849, 2015.

43. Yang, S. and Zhang, K., Few-layers MoS_2 nanosheets modified thin film composite nanofiltration membranes with improved separation performance. *J. Memb. Sci.*, 595, 117526, 2020, doi:10.1016/j.memsci.2019.117526.
44. Haldar, D., Duarah, P., Purkait, M.K., MOFs for the treatment of arsenic, fluoride and iron contaminated drinking water: A review. *Chemosphere*, 251, 126388, 2020, doi:10.1016/j.chemosphere.2020.126388.
45. Yang, S., Jiang, Q., Zhang, K., Few-layers 2D O–MoS_2 TFN nanofiltration membranes for future desalination. *J. Memb. Sci.*, 604, 118052, 2020, doi:10.1016/j.memsci.2020.118052.
46. Al-Hamadani, Y.A.J., Jun, B.-M., Yoon, M., Taheri-Qazvini, N., Snyder, S.A., Jang, M., Heo, J., Yoon, Y., Applications of MXene-based membranes in water purification: A review. *Chemosphere*, 254, 126821, 2020, doi:10.1016/j.chemosphere.2020.126821.
47. Chaudhary, P. and Ingole, P.P., In-Situ solid-state synthesis of 2D/2D interface between Ni/NiO hexagonal nanosheets supported on g-C_3N_4 for enhanced photo-electrochemical water splitting. *Int. J. Hydrogen Energy*, 45, 32, 160170, 2020, doi:10.1016/j.ijhydene.2020.04.011.
48. Liu, T., Zhou, H., Graham, N., Yu, W., Sun, K., 2D kaolin ultrafiltration membrane with ultrahigh flux for water purification. *Water Res.*, 156, 425–433, 2019, doi:10.1016/j.watres.2019.03.050.
49. Baniasadi, H. and Seppal, J., Novel long-chain aliphatic polyamide/surface-modified silicon dioxide nanocomposites: *in-situ* polymerization and properties, *Mater. Today Chem.*, 20, 100450, 2021, doi.org/10.1016/j.mtchem.2021.100450.
50. Abdikheibari, S., Lei, W., Dumée, L.F., Barlow, A.J., Baskaran, K., Novel thin film nanocomposite membranes decorated with few-layered boron nitride nanosheets for simultaneously enhanced water flux and organic fouling resistance. *Appl. Surf. Sci.*, 488, 565–577, 2019, doi.org/10.1016/j.apsusc.2019.05.217.
51. Mao, Y., Huang, Q., Meng, B., Zhou, K., Liu, G., Gugliuzza, A., Drioli, E., Jin, W., Roughness-enhanced hydrophobic graphene oxide membrane for water desalination via membrane distillation. *J. Membr. Sci.*, 611, 118364, 2020.
52. Lu, J., Zhang, H., Hou, J., *et al.*, Efficient metal ion sieving in rectifying subnanochannels enabled by metal–organic frameworks. *Nat. Mater.*, 19, 767–774, 2020, https://doi.org/10.1038/s41563-020-0634-7.
53. Wu, H., Tang, B., Wu, P., Optimizing polyamide thin film composite membrane covalently bonded with modified mesoporous silica nanoparticles. *J. Membr. Sci.*, 428, 341–348, 2013, doi.org/10.1016/j.memsci.2012.10.053.
54. Sun, Y., Lu, J., Li, S., Dai, C., Zou, Z., Jing, W., MXene-based membranes in water treatment: Current status and future prospects. *Sep. Purif. Tech.*, 331, 125640, 2024, doi.org/10.1016/j.seppur.2023.125640.
55. Liu, G., Guo, Y., Meng, B., Wang, Z., Liu, G., Jin, W., Two-dimensional MXene hollow fiber membrane for divalent ions exclusion from water. *Chin. J. Chem. Eng.*, 41, 7, 2022, 10.1016/j.cjche.2021.09.022.

56. Sun, Y., Yi, F., Li, R.H., Min, X., Qin, H., Cheng, S.Q., Liu, Y., Inorganic-organic hybrid membrane based on pillararene-intercalated MXene nanosheets for efficient water purification. *Angew. Chem. Int. Ed.*, 61, e202200482, 2022, 10.1002/anie.202200482.
57. Huang, Z., Zeng, Q., Liu, Y., Xu, Y., Li, R., Hong, H., Shen, L., Lin, H., Facile synthesis of 2D TiO2@MXene composite membrane with enhanced separation and antifouling performance. *J. Membr. Sci.*, 640, 119854, 2021, 10.1016/j.memsci.2021.119854.
58. Mustakeem, M., El-Demellawi, J.K., Obaid, M., Ming, F., Alshareef, H.N., Ghaffour, N., MXene-Coated Membranes for Autonomous Solar-Driven Desalination. *ACS Appl. Mater. Interf.*, 14, 4, 5265–5274, 2022, doi.org/10.1021/acsami.1c20653.
59. Ji, W., Li, W., Wang, Y., Zhang, T.C., Yuan, S., Fe-MOFs/graphene oxide-derived magnetic nanocomposite for enhanced adsorption of As(V) in aqueous solution. *Sep. Purif. Tech.*, 334, 126003, 2024, doi.org/10.1016/j.seppur.2023.126003.
60. Pinto, F.M., de Conti, M.C.M.D., Pereira, W.S., Sczancoski, J.C., Medina, M., Corradini, P.G., de Brito, J.F., Nogueira, A.E., Góes, M.S., Ferreira, O.P., *et al.*, Recent Advances in Layered MX_2-Based Materials (M=Mo, W and X=S, Se, Te) for Emerging Optoelectronic and Photo(electro)catalytic Applications. *Catalysts*, 14, 6, 388, 2024, https://doi.org/10.3390/catal14060388.
61. Frappa, M., Macedonio, F., Gugliuzza, A., Jin, W., Drioli, E., Performance of PVDF Based Membranes with 2D Materials for Membrane Assisted-Crystallization Process. *Membranes*, 302, 11, 1–16, 2021.
62. Li, X., Cai, W., An, J., Kim, S., Nah, J., Yang, D., Piner, R., Velamakanni, A., Jung, I., Tutuc, E., Banerjee, S.K., Colombo, L., Ruoff, R.S., Large-area synthesis of high-quality and uniform graphene films on copper foils. *Science*, 324, 1312–1314, 2009, DOI: 10.1126/science.1171245.
63. Del Rio-Castillo, A.E., Merino, C., Díez-Barra, E., Vázquez, E., Selective suspension of single layer graphene mechanochemically exfoliated from carbon nanofibres. *Nano Res.*, 7, 7, 963–972, 2014, doi:10.1007/s12274-014-0457-4.
64. Chai, L., Cui, X.-J., Qi, Y.-Q., Teng, N., Hou, X.-L., Deng, T.-S., A new strategy for the efficient exfoliation of graphite into grapheme. *New Carbon Mater.*, 36, 6, 1179–1186, 2021, doi.org/10.1016/S1872-5805(21)60100-2.
65. Achee, T.C., Sun, W., Hope, J.T., *et al.*, High-yield scalable graphene nanosheet production from compressed graphite using electrochemical exfoliation. *Sci. Rep.*, 8, 14525, 2018, https://doi.org/10.1038/s41598-018-32741-3.
66. Gontarek-Castro, E., Di Luca, G., Lieder, M., Gugliuzza, A., Graphene-Coated PVDF Membranes: Effects of Multi-Scale Rough Structure on Membrane Distillation Performance. *Membranes*, 12, 5, 511, 2022, https://doi.org/10.3390/membranes12050511.
67. Dikin, D.A., Stankovich, S., Zimney, E.J., Piner, R.D., Dommett, G.H.B., Evmenenko, G., *et al.*, Preparation and characterization of graphene oxide paper. *Nature*, 448, 457–460, 2007.

68. Li, D., Muller, M.B., Gilje, S., Kaner, R.B., Wallace, G.G., Processable aqueous dispersions of graphene nanosheets. *Nat. Nanotechnol.*, 3, 101–105, 2008.
69. Kim, Y.G., Lim, N., Kim, J., Kim, C., Lee, J., Kwon, K.-H., Study on the surface energy characteristics of polydimethylsiloxane (PDMS) films modified by $C_4F_8/O_2/Ar$ plasma treatment. *Appl. Surf. Sci.*, 477, 198–203, 2019, doi.org/10.1016/j.apsusc.2017.11.009.
70. Kim, S., Ou, R., Hu, Y., Li, X., Zhang, H., Simon, G.P., Wang, H., Non-swelling graphene oxide-polymer nanocomposite membrane for reverse osmosis desalination. *J. Membr. Sci.*, 562, 47–55, 2018, doi.org/10.1016/j.memsci.2018.05.029.
71. Hou, D., Fan, H., Jiang, Q., Wang, J., Zhang, X., Preparation and characterization of PVDF flat-sheet membranes for direct contact membrane distillation. *Sep. Purif. Technol.*, 135, 211–222, 2014, doi:10.1016/j.seppur.2014.08.023.
72. Leaper, S., Abdel-Karim, A., Faki, B., Luque-Alled, J.M., Alberto, M., Vijayaraghavan, A., Holmes, S.M., Szekely, G., Badawy, M.I., Shokri, N., Gorgojo, P., Flux-enhanced PVDF mixed matrix membranes incorporating APTS-functionalized graphene oxide for membrane distillation. *J. Memb. Sci.*, 554, 309–323, 2018, doi:10.1016/j.memsci.2018.03.013.
73. Abdel-Karim, A., Luque-Alled, J.M., Leaper, S., Alberto, M., Fan, X., Vijayaraghavan, A., Gad-Allah, T.A., El-Kalliny, A.S., Szekely, G., Ahmed, S.I.A., Holmes, S.M., Gorgojo, P., PVDF membranes containing reduced graphene oxide: Effect of degree of reduction on membrane distillation performance. *Desalination*, 452, 196–207, 2019.
74. Li, H., Shi, W., Zeng, X., Huang, S., Zhang, H., Qin, X., Improved desalination properties of hydrophobic GO-incorporated PVDF electrospun nanofibrous composites for vacuum membrane distillation. *Sep. Purif. Technol.*, 230, 115889, 2020, doi:10.1016/j.seppur.2019.115889.
75. Camacho, L.M., Pinion, T.A., Olatunji, S.O., Behavior of mixed-matrix graphene oxide – Polysulfone membranes in the process of direct contact membrane distillation. *Sep. Purif. Technol.*, 240, 116645, 2020, doi:10.1016/j.seppur.2020.116645.
76. Mansour, S., Giwa, A., Hasan, S.W., Novel graphene nanoplatelets-coated polyethylene membrane for the treatment of reject brine by pilot-scale direct contact membrane distillation: An optimization study. *Desalination*, 441, 9–20, 2018, doi:10.1016/j.desal.2018.04.026.
77. Chen, T., Soroush, A., Rahaman, M.S., Highly Hydrophobic Electrospun Reduced Graphene Oxide/Poly(vinylidene fluoride-co-hexafluoropropylene) Membranes for Use in Membrane Distillation. *Ind. Eng. Chem. Res.*, 57, 43, 14535–14543, 2018, doi:10.1021/acs.iecr.8b03584.
78. Athanasekou, C., Sapalidis, A., Katris, I., Savopoulou, E., Beltsios, K., Tsoufis, T., Kaltzoglou, A., Falaras, P., Bounos, G., Antoniou, M., Boutikos, P., Romanos, G.E., Mixed Matrix PVDF/Graphene and Composite-Skin PVDF/Graphene Oxide Membranes Applied in Membrane Distillation. *Polym. Eng. Sci.*, 59, E262–278, 2019, doi:10.1002/pen.24930.

79. Zahirifar, J., Moosavian, S.M.A., Hadi, A., Khadiv-Parsi, P., Karimi-Sabet, J., Fabrication of a novel octadecylamine functionalized graphene oxide/PVDF dual-layer flat sheet membrane for desalination via air gap membrane distillation. *Desalination.*, 428, 227–239, 2018, doi:10.1016/j.desal.2017.11.028.
80. Woo, Y.C., Kim, Y., Shim, W.-G., Tijing, L.D., Yao, M., Nghiem, L.D., Choi, J.-S., Kim, S.-H., Shon, H.K., Graphene/PVDF flat-sheet membrane for the treatment of RO brine from coal seam gas produced water by air gap membrane distillation. *J. Memb. Sci.*, 513, 74–84, 2016, doi:10.1016/j.memsci.2016.04.014.
81. Fouladivanda, M.R., Karimi-Sabet, J., Abbasi, F., Moosavian, M.A., Step-by-step improvement of mixed-matrix nanofiber membrane with functionalized graphene oxide for desalination via air-gap membrane distillation, *Sep. Technol.*, 256, 117809, 2020, DOI:10.1016/j.seppur.2020.117809
82. Bhadra, M., Roy, S., Mitra, S., Desalination across a graphene oxide membrane via direct contact membrane distillation. *Desalination*, 378, 37–43, 2016, doi.org/10.1016/j.desal.2015.09.026.
83. Chen, T., Ma, W., Lee, J., Jassby, D., Saifur Rahaman, M., Development of robust and superamphiphobic membranes using reduced graphene oxide (rGO)/PVDF-HFP nanocomposite mats for membrane distillation. *Environ. Sci.: Nano*, 8, 2883–2893, 2021, doi.org/10.1039/D1EN00288K.
84. Zhu, Z., Song, M., Qu, F., Zhou, Y., Yang, Y., Qi, J., Li, J., Engineering Multinanochannel Polymer-Intercalated Graphene Oxide Membrane for Strict Volatile Sieving in Membrane Distillation. *Environ. Sci. Technol.*, 16, 58, 2, 1399–1409, 2024. doi: 10.1021/acs.est.3c08452.
85. Gugliuzza, A., Macedonio, F., Politano, A., Drioli, E., Prospects of 2D Materials-Based Membranes in Water Desalination. *Chem. Eng. Trans.*, 73, 265–270, 2019. doi:10.3303/ CET1973045.
86. Martinez-Jimenez, F.D., Musteata, V.E., Cespedes-Zuluaga, S., Blankert, B., Picioreanu, C., Two-dimensional model of ion transport in composite membranes active layers with TEM-scanned morphology. *Desalination*, 565, 116876, 2023, doi.org/10.1016/j.desal.2023.116876.
87. Kim, S., Choi, H., Kim, B., Lim, G., Kim, T., Lee, M., Ra, H., Yeom, J., Kim, M., Kim, E., Hwang, J., Lee, J.S., Shim, W., Extreme Ion-Transport Inorganic 2D Membranes for Nanofluidic Applications. *Adv. Mater.*, 35, 2206354, 2023, doi.org/10.1002/adma.202206354.
88. Tansel, B., Significance of thermodynamic and physical characteristics on permeation of ions during membrane separation: hydrated radius, hydration free energy and viscous effects. *Separ. Purif. Technol.*, 86, 119–126, 2012.
89. Werber, J., Osuji, C., Elimelech, M., Materials for next-generation desalination and water purification membranes. *Nat. Rev. Mater.*, 1, 16018, 2016. https://doi.org/10.1038/natrevmats.2016.18
90. Zhang, M., Guan, K., Ji, Y., *et al.*, Controllable ion transport by surface-charged graphene oxide membrane. *Nat. Commun.*, 10, 1253, 2019, https://doi.org/10.1038/s41467-019-09286-8.

91. Pérez, M., Elías, J., Sosa, M., Vallejo, M., Hybridization bond states and band structure of graphene: a simple approach. *Eur. J. Phys.*, 43, 045401, 2022.
92. Huang, Q., Liu, S., Guo, Y., Liu, G., Jin, W., Separation of mono-/di-valent ions via charged interlayer channels of graphene oxide membranes. *J. Membr. Sci.*, 645, 120212, 2022, doi.org/10.1016/j.memsci.2021.120212.
93. Qiu, R., Xiao, J., Chen, X.D., Selomulya, C., Zhang, X., Woo, M.W., Comparison of the effects of edge functionalized graphene oxide membranes on monovalent cation selectivity. *J. Membr. Sci.*, 620, 118892, 2021.
94. Gugliuzza, A., Politano, A., Drioli, E., The advent of graphene and other two-dimensional materials in membrane science and technology. *Curr. Opin. Chem. Eng.*, 16, 78–85, 2017, doi.org/10.1016/j.coche.2017.03.003.
95. Pomoni, D.I., Koukou, M.K., Vrachopoulos, M.G., Vasiliadis, L., Circular economy: A multilevel approach for natural resources and wastes under an agri-food perspective. *Water-Energy Nexu*, 7, 103–123, 2024, doi.org/10.1016/j.wen.2023.12.003.
96. Bongaarts, J., Summary for policymakers of the global assessment report on biodiversity and ecosystem services of the Intergovernmental Science-Policy Platform on Biodiversity and Ecosystem Services. *Popul. Dev. Rev.*, 45, 680–681, 2019, doi:10.1111/padr.12283.
97. Zapelloni, G., García Rellán, A., Bello Bugallo, P.M., Sustainable production of marine equipment in a circular economy: deepening in material and energy flows, best available techniques and toxicological impacts. *Sci. Total Environ.*, 687, 991–1010, 2019, doi:10.1016/j.scitotenv.2019.06.058.
98. Najafi, A., Jafarian, A., Darand, J., Thermo-economic evaluation of a hybrid solar-conventional energy supply in a zero liquid discharge wastewater treatment plant. *Energy Convers. Manag.*, 188, 276–295, 2019, doi:10.1016/j.enconman.2019.03.059.
99. Korhonen, J., Honkasalo, A., Seppälä, J., Circular Economy: The Concept and its Limitations. *Ecol. Econ.*, 143, 37–46, 2018. dx.doi.org/10.1016/j.ecolecon.2017.06.041.
100. Dumée, L.F., Circular Materials and Circular Design—Review on Challenges Towards Sustainable Manufacturing and Recycling. *Circ. Econ. Sust.*, 2, 9–23, 2022, https://doi.org/10.1007/s43615-021-00085-2.
101. Editorial: Moving towards the market. *Nat. Mater.*, 18, 519, 2019, doi.org/10.1038/s41563-019-0394-4.
102. Dong, W.-X., Qu, Y.-F., Liu, X., Chen, L.-F., Biomass-derived two-dimensional carbon materials: Synthetic strategies and electrochemical energy storage applications. *Flat Chem.*, 37, 100467, 2023, doi.org/10.1016/j.flatc.2022.100467.
103. Saha, J.K. and Dutta, A., A Review of Graphene: Material Synthesis from Biomass Sources. *Waste Biomass Valor*, 13, 1385–1429, 2022, https://doi.org/10.1007/s12649-021-01577-w.

10

Sub-Nanometer Channels in Two-Dimensional-Material Membranes for Gas Separation

Song Liu, Long Cheng, Gongping Liu and Wanqin Jin*

State Key Laboratory of Materials-Oriented Chemical Engineering, College of Chemical Engineering, Nanjing Tech University, Nanjing, China

Abstract

Compared with traditional separation methods, membrane technology with lower energy consumption has broad application prospects in the field of gas separation. However, traditional polymer membrane materials suffer from a trade-off between selectivity and permeability. Two-dimensional (2D) materials can maximize gas flux due to their atom-level thickness. Meanwhile, the in-plane nanopores or sub-nanometer interlayer channels of laminates can separate gas mixtures based on molecular sieving, which is expected to surpass the upper bound of traditional polymer membranes. The most representative 2D material is graphene and its derivative, graphene oxide (GO). Therefore, this review focuses on graphene-based materials as an example to introduce the characteristics of three main types of membrane structures and their applications in different gas pairs. In addition, other representative 2D-material membranes will be briefly introduced as well. Finally, the challenges and opportunities faced by the development of 2D-material membranes are discussed.

***Keywords*:** Two-dimensional materials, gas separation, membrane, sub-nanometer channels

**Corresponding author*: wqjin@njtech.edu.cn

Annarosa Gugliuzza and Wanqin Jin (eds.) Porous Membranes: Breakthroughs in Manufacturing and Applications, (313–336) © 2025 Scrivener Publishing LLC

10.1 Introduction

Gas separation membrane technology has been developed since the 1940s. So far, membrane separation has been used in separations such as carbon dioxide from natural gas [1], air separation [2], C_2-C_4 light hydrocarbon/nitrogen separation [3], and hydrogen purification [4]. Polymer membranes are called first-generation gas separation membranes. However, they suffer from the trade-off between selectivity and permeability [5]. The discovery of graphene triggered a great deal of attention toward 2D materials in numerous scientific fields [6–10]. 2D materials have also been pursued as promising membrane materials owing to their unique atomic thickness and micrometer lateral dimensions [10, 11]. In recent years, various 2D materials, including graphene and graphene oxide (GO) [12], metal-organic frameworks (MOFs) [13], layered double hydroxides (LDHs) [14], transition metal dichalcogenides (TMDCs) [15], and MXene [16, 17], have served as superb building blocks for new-generation high-performance membranes [11, 12]. By assembling nanosheets into membranes with sub-nanometer channels, 2D-material membranes have exhibited superior performance in precise molecular separation, such as gas separation. According to the difference in membrane structures, as shown in Figure 10.1, Two-dimensional-material membranes (2DMMs) can be categorized into three types: i) porous nanosheet membranes, ii) laminar membranes, and iii) nanosheet-based mixed matrix membranes. The three types of membranes have obvious advantages and disadvantages in preparation and performance. For example, porous nanosheet membranes have high

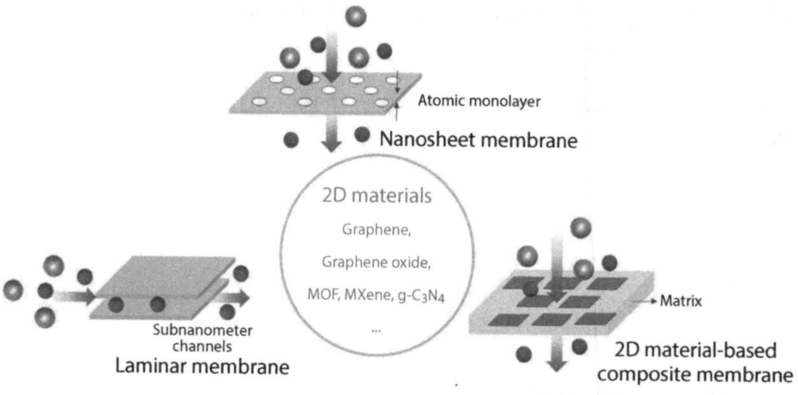

Figure 10.1 Main types of 2D materials membranes: (i) porous nanosheet membranes; (ii) assembled nanosheet laminar membranes; and (iii) 2D material-based composite membranes.

flux, but are difficult to prepare, while laminar membranes can be easily obtained by solution processing [11, 12]. Graphene has been selected as the representative nanosheet for introducing these three membrane structures. Moreover, other 2D material-based membranes for gas separation are discussed as well. This review focuses on the advancements in constructing and manipulating the sub-nanometer channels of 2D-material membranes for gas separation. Finally, we conclude the review with the remaining challenges and the new opportunities that will be opened up for 2D-material membranes in gas separation.

10.2 Three Main Types of Membrane Structures

10.2.1 Porous Monolayer Graphene Membrane

The atomic thickness of graphene helps to minimize the resistance to gas transport. However, single-layer graphene is completely impermeable. This is due to the fact that graphene's π-orbitals form a dense, delocalized cloud that blocks the gap within its aromatic rings [18]. Although there were many theoretical studies about the impermeable nature of perfect graphene through simulations, for the first time, Bunch *et al.* [19] (Figure 10.2a) experimentally demonstrated that a defect-free graphene sheet is impermeable to even small gas molecules such as helium. This work

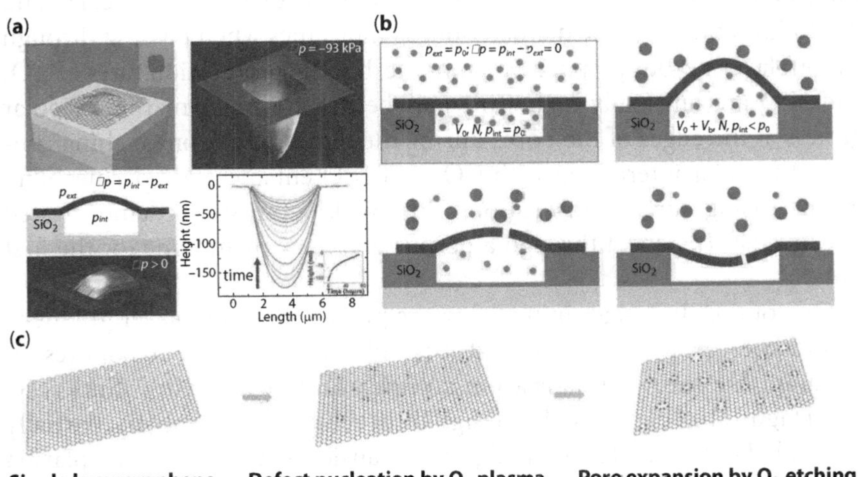

Figure 10.2 (a) Impermeable defect-free graphene sheet; (b) Schematic of measuring gas leak rates in a macroscopic porous graphene membrane on a SiO_2 substrate; and (c) Evolution of graphene lattice after subsequent exposures to O_2 plasma and O_3.

triggered many other studies to create selective pores in graphene to allow the separation of gas mixtures, and many simulation studies predicting the gas separation performance of various types of drilled pores in graphene. Dai *et al.* [20] used first-principles density functional theory to calculate the separation of H_2/CH_4 in single-layer graphene with precise pore formation achieved by removing two adjacent six-membered C rings. It is found that when N atoms are used to modify the edge of the pores, the pore size is 3.0 Å. At this size, the monolayer graphene has a selectivity of H_2/CH_4 as high as 10^8, and H_2 permeability reaches 1 mol·m^{-2}·s^{-1} Pa^{-1}. A single graphene layer for developing a porous graphene membrane is usually produced *via* the Scotch tape method or chemical vapor deposition (CVD) approach [21]. Thus, it is difficult to avoid defects in the graphene during the actual synthesis process, and it is technically challenging to produce pores with narrow pore sizes and high pore density without major defects. These irregular, large-sized defects become gas transport channels, causing the actual separation selectivity to be far lower than the theoretical value.

Koenig *et al.* [22] (Figure 10.2b) used ultraviolet/ozone to make pores with gas selectivity in the monolayer or double-layer graphene. They also measured the gas permeability of the graphene sheet that was not etched. They found that the permeation rate of H_2 (or CO_2) through a graphene membrane with a pore size of 0.34 nm is much higher than that of N_2, Ar, and CH_4. In comparison, another membrane with a pore size of 0.49 nm is able to reject larger gas molecules (such as SF_6), and the gas selectivity is extremely high. Interestingly, the permeation rate of gases (He, Ne, H_2, and Ar) follows the dynamic diameter of gas molecules when passing through a monolayer porous graphene membrane, but the permeation rates of CO_2 and N_2O are unusually high. This may be due to the formation of polar groups at the edges of the pores during the ultraviolet/ozone etching process. The polar interaction with CO_2 and N_2O enhances gas permeability by increasing surface adsorption. This work successfully demonstrated selective gas transport through a porous graphene membrane for the first time.

One of the challenges in realizing the potential of monolayer graphene membranes for gas sieving is that it is difficult to produce nanopores on the graphene sheet with angstrom accuracy and high density. Zhao *et al.* [23] (Figure 10.2c) realize this design by developing a synergistic, partially decoupled defect nucleation and pore expansion strategy using O_2 plasma and O_3 treatment. A high density (2.1×10^{12} cm^{-2}) of H_2 sieving pores was achieved while limiting the percentage of CH_4 permeating pores to 13 to 22 parts per million. As a result, a record-high gas mixture separation performance was achieved (H_2 permeance, 1340 to 6045 GPU; H_2/CH_4

separation factor, 15.6 to 25.1; H_2/C_3H_8 separation factor, 38.0 to 57.8). This highly scalable pore etching strategy will accelerate the development of single-layer graphene-based membranes.

Nonetheless, there remain many challenges in extending this approach to create m^2-scale membranes needed for practical applications in the gas separation industry. Although the pore-size distribution is impressively narrow, it needs to be further narrowed to improve the gas separation factor. Also, the synthesis technique faces substantial challenges in increasing the membrane area to m^2-scale because larger graphene sheets almost inevitably have more defects.

10.2.2 Laminar Membranes

A feasible strategy to solve the above-mentioned scaling-up problems is to fabricate GO laminates. The laminar membranes assembled from nanosheets provide a more practical approach for preparing 2D-material membranes. The 2D sub-nanometer interlayer channels between nanosheets could provide pathways for rapid and selective gas transport.

Kim *et al.* [24] prepared several layers of polycrystalline graphene on a highly permeable glassy polymer substrate (poly(1-trimethylsilyl-1-propyne)), which shows selective gas transport. Since defect-free graphene is inherently impermeable to any gas molecule, they initially planned to prepare a gas barrier polymer film by depositing several layers of graphene. However, the graphene sheet prepared by the CVD method still has many defects or large pores, and many wrinkles and ripples will be formed during the transfer process. Gas molecules could be separated by structural defects and randomly stacked 2D sub-nanometer channels, but the gas permeance would drop sharply as the number of graphene layers increases.

As one of the derivatives of graphene, GO is an important member of the 2D material family and an indispensable unit for the construction of 2D material membranes [11]. Like graphene, GO nanosheets possess atomic thickness, and their lateral dimensions can reach tens of microns. In addition, GO nanosheets contain rich oxygen-containing functional groups, such as hydroxyl, carboxyl, and epoxy groups [25]. The 2D sub-nanometer channels formed between the stacked GO nanosheets could allow gas transport, and the oxygen-containing functional groups on the nanosheets were able to interact with targeted components to promote their permeation. Various coating methods are investigated for creating ultrathin GO layers on porous substrates, such as vacuum filtration, drop-casting, layer-by-layer deposition, spray coating, and spin casting [12].

Yu *et al.* [26] vacuum-filtered GO nanosheets onto the surface of anodic aluminum oxide (AAO) (Figure 10.3a). They prepared GO membranes with thicknesses ranging from 1.8 to 18 nm and with a H_2 permeability of $1{\times}10^{-7}$ $mol{\cdot}m^{-2}{\cdot}s^{-1}{\cdot}Pa^{-1}$, and record-high H_2/CO_2 selectivity of 3400 and H_2/N_2 selectivity of 1000. They found that ultrathin GO membranes have much higher gas permeability than relatively thick GO membranes (18 nm thick), indicating that the gas permeance of such laminated GO membranes strongly depends on membrane thickness. Kim *et al.* [24] used two methods to prepare a few-layer GO membrane and compared the two structures. The first method is to contact the surface of the substrate with the gas-liquid interface of the GO solution followed by spin-coating, and the second method is spin-casting of a GO solution onto the substrate surface (Figure 10.3b). The results showed that the GO nanosheets are not only subjected to mutual repulsion at the edges during the stacking process but also face-to-face capillary attraction generated by spin coating. In method 1, the electrostatic repulsion between the edges of GO makes the stacking of GO nanosheets relatively isolated and disordered. In contrast, in method 2, the GO solution-membrane contact occurs only during spin-casting which makes the interlocked GO laminates. Since the capillary interaction

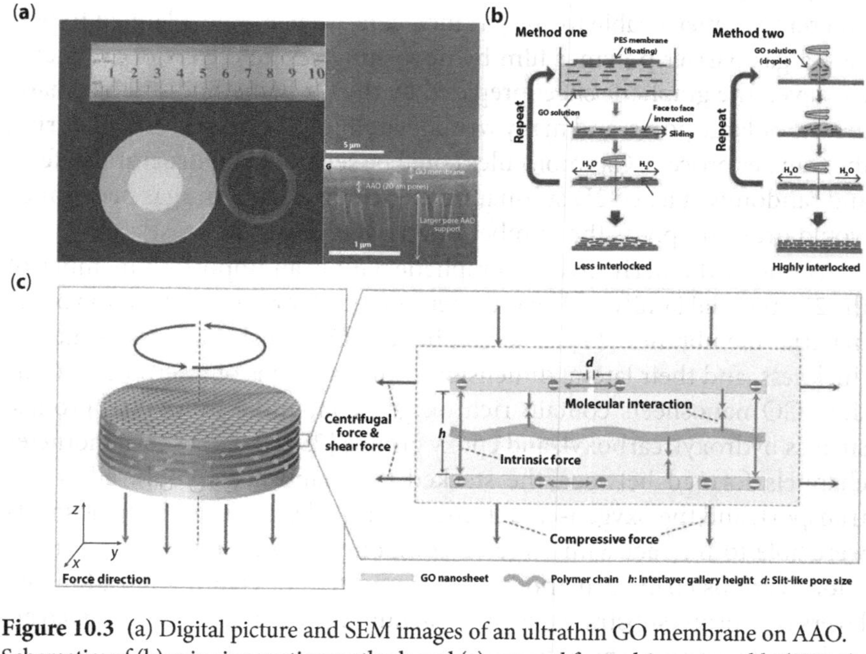

Figure 10.3 (a) Digital picture and SEM images of an ultrathin GO membrane on AAO. Schematics of (b) spinning coating methods and (c) external force-driven assembly (EFDA) approach for fabricating 2D channels.

between the surfaces of the GO sheets overcomes the electrostatic force between the edges of GO, the GO sheets are stacked more densely.

It is challenging to precisely manipulate the GO channels within sub-nanometer size that is necessary for gas separation. Our group proposed to assemble GO nanosheets into laminates with 2D sub-nanometer channels via the reasonable design of external forces [27]. As shown in Figure 10.3c, the designed external force is applied to both the outside and inside of the GO laminate. Together, they overcome the inherent repulsive electrostatic interaction between GO nanosheets, aiming to eliminate non-selective stacking defects and stimulate the highly ordered assembly of GO nanosheets. This method contributed to 2D sub-nanometer channels with interlayer height of ~0.4 nm, realizing the fast and selective transport of gases. Compared with commercial membranes, the H_2 permeability of external force-driven assembly (EFDA) GO membranes has increased by 2 to 3 orders of magnitude, and H_2/CO_2 selectivity has increased by 3 times. This work has taken an important step to realize the precise molecular separation of 2D-material membranes.

In addition to the above methods, cross-linking is also a common method to control the sub-nanometer interlayer channels of GO membranes. Jiang *et al.* [28] used borate as a cross-linking agent to achieve fast CO_2 transport and excellent CO_2/CH_4 selectivity. Yu *et al.* [29] prepared a hollow fiber GO membrane cross-linked by piperazine. The molecular brush of piperazine, which has an affinity to CO_2 and promotes CO_2 transport, endows GO membranes with high CO_2 permeability and CO_2/N_2 selectivity. Cheng *et al.* [30] developed ultrathin GO membranes with properly reduced interlayer height for H_2/CO_2 separation. Cysteamine, which contains both amino and thiol and can react with oxygen-containing groups distributed between the interlayer gallery, was employed to reduce GO nanosheets and narrow the interlayer height for sieving gases. Compared with pristine GO membrane, the CGO membrane exhibited a two-fold enhancement in H_2/CO_2 selectivity.

It can be seen that, compared with graphene, GO nanosheets have a large number of oxygen-containing functional groups, which makes it possible to stably disperse in many solvents. There are more ways to assemble it into a membrane and regulate the membrane structure. Although it is easier to prepare the laminated membrane structure, the tortuous pathway tends to increase the transport resistance of gases compared with porous nanosheet membranes, leading to relatively low gas flux.

10.2.3 Nanosheet-Based Mixed-Matrix Membranes

GO is a nanosheet decorated with carboxyl groups on the edges and hydroxyl and epoxy groups on the surface. Compared with traditional filler materials such as zeolite, it can form covalent bonds or non-covalent interactions (e.g., hydrogen bonding) with the polymer matrix, leading to better dispersity in the polymer matrix. In addition, the atomic thickness of GO can effectively reduce the thickness of the mixed matrix membrane to obtain higher gas flux. Due to the excellent processability of the polymer, this kind of membrane shows high potential for practical implementation.

We incorporated GO nanosheets into polyether block amide (PEBA) to fabricate mixed matrix membranes with excellent CO_2/N_2 separation performance [31]. Commercial PEBA chains are rich in -N-H-, H-N-C=O, and O-C=O groups that can form hydrogen bonds with GO nanosheets. As shown in Figure 10.4a, driven by the hydrogen bonds between GO and PEBA, the GO nanosheets are assembled into a multilayer GO laminate with molecular sieving interlayer spacing and linear diffusion paths.

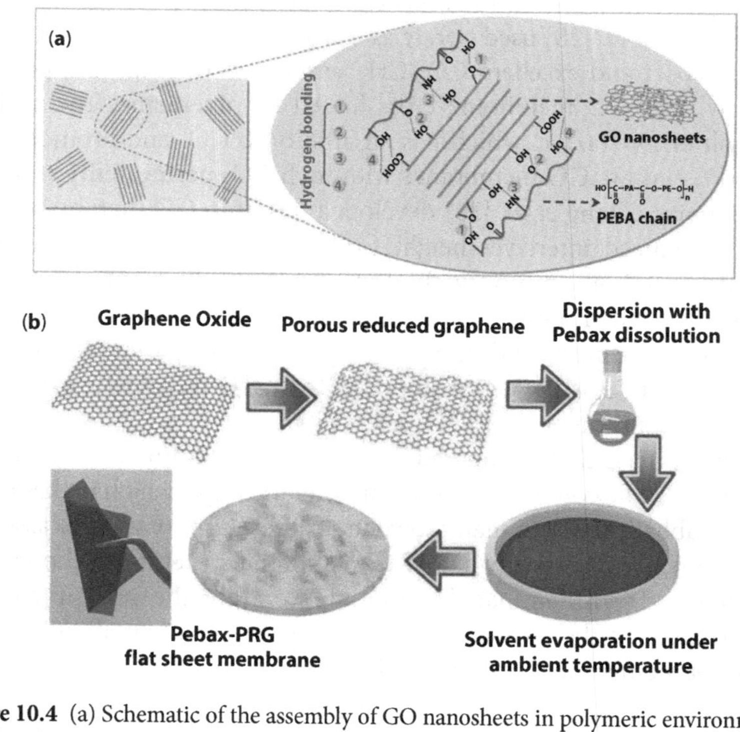

Figure 10.4 (a) Schematic of the assembly of GO nanosheets in polymeric environment based on hydrogen bonds between oxygen-containing groups on GO and the PEBA chain, (b) Schematic of the fabrication process of Pebax-PRG membranes.

The introduction of GO laminates enhances the selective diffusion of gas molecules, while the strong affinity of GO to CO_2 enhances the adsorption selectivity of the membrane. The prepared GO-based mixed-matrix membrane (MMM) shows excellent CO_2 preferential permeation (100 Barrer) with a very high and stable CO_2/N_2 separation factor (91), exceeding the upper bound of state-of-art membranes. Subsequently, we [32] found that the lateral size of GO plays a crucial role in the mobility of polymer chains and the length of gas transport channels. By optimizing the lateral dimensions, the obtained MMM showed an enhanced CO_2 permeability of 110 Barrer. At the same time, the CO_2/N_2 mixed gas selectivity is maintained at ~80. This new type of membrane preparation strategy has unique advantages such as a simple preparation process and high processability, exhibiting great potential for practical CO_2 capture.

Inspired by the above work, Wu *et al.* [33] embedded GO functionalized with polyethylenimine-polyethylene glycol (PEI-PEG) into PEBA to prepare the mixed matrix membrane. Due to the introduction of PEG and PEI, the affinity of MMM to CO_2 is significantly enhanced. Due to the simultaneous enhancement of diffusivity selectivity and solubility selectivity, the mixed matrix membrane exhibits efficient CO_2 separation performance. In addition to modifying the surface of GO nanosheets, Zhang *et al.* (Figure 10.4b) [34] reduced GO by wet-chemical methods to prepare partially porous reduced GO (PRG), which was incorporated into PEBA polymer to fabricate MMM. Due to the existence of nanopores, the as-fabricated MMM exhibits excellent CO_2/N_2 separation performance, with CO_2 permeability of 119 Barrer and CO_2/N_2 selectivity of ~104.

Ha *et al.* [35] prepared a GO/PDMS composite membrane by cross-linking the oxygen-containing groups (epoxy, hydroxyl, and carboxyl) on GO with amino-terminated poly(dimethylsiloxane) (PDMS). The content of GO in the composite membrane is relatively high (from 1 to 8 wt%). The gas permeability decreases while the CO_2/N_2 selectivity increases sharply with the increase of GO content. This is attributed to the enhanced selectivity of diffusion, rather than the enhanced selectivity of solubility.

Although the research on GO-based mixed matrix membranes has made great progress, there are still insufficient membranes with extremely high gas flux and gas selectivity to meet commercial needs. The atomic GO nanosheets make it possible to prepare ultra-thin mixed matrix membranes. It is also an effective way to modify the GO surface or perforate in-plane to improve the performance of mixed matrix membranes.

10.3 Other Two-Dimensional Material Membranes

In recent years, apart from graphene and GO, other 2D materials, including 2D MOFs [13, 36, 37], LDH [14, 38, 39], MXene [16, 17], and g-C_3N_4 [40, 41], have also been employed to fabricate gas separation membranes. In 2018, Ding *et al.* [16] realized the effective separation of H_2/CO_2 mixed gas by MXene membranes for the first time. In this work, as shown in Figure 10.5a, MXene nanosheets are highly oriented and compactly stacked on the porous AAO substrate via the vacuum filtration method, and then peeled off to obtain a free-standing laminar membrane. The distance between adjacent nanosheets is only 0.35 nm, which can separate gas mixtures based on molecular sieving. Therefore, small-sized He and H_2 molecules can quickly permeate through the membrane, while the permeability of large-sized N_2 molecules is two orders of magnitude lower. In addition, CO_2 has a large quadrupole moment and it interacts strongly with MXene nanosheets. Therefore, although the molecular dynamic diameter of CO_2

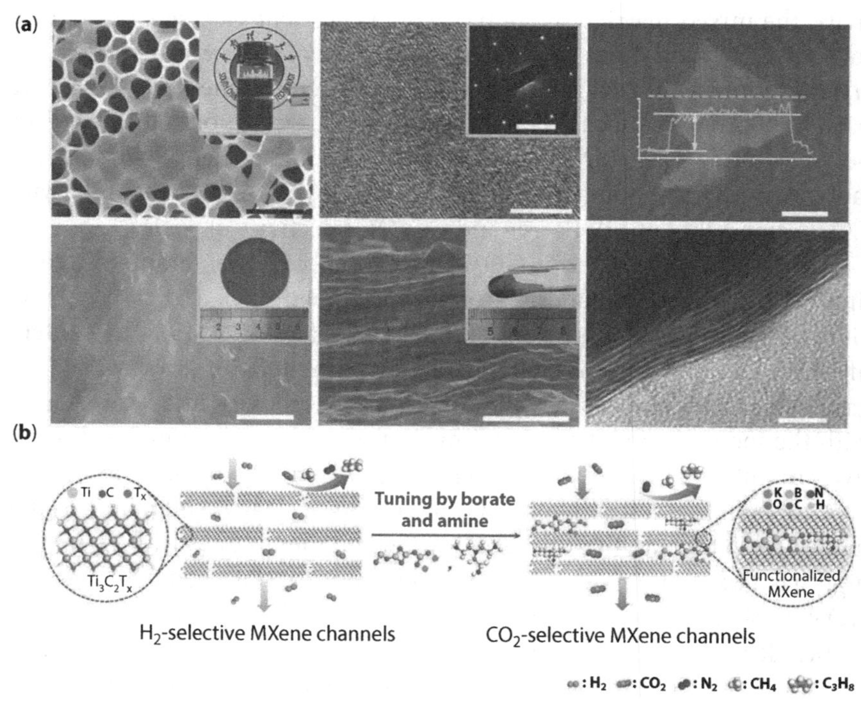

Figure 10.5 (a) Morphology and structure of exfoliated MXene ($Ti_3C_2T_X$) nanosheets and stacked MXene membrane. (b) Schematic of the microstructure and gas transport of H_2-selective and CO_2-selective MXene nanofilms.

is smaller than that of N_2 molecules, its permeability is only half of the latter. Based on molecular sieving, MXene membrane exhibits excellent gas separation selectivity (H_2/CO_2 selectivity: 167, H_2/N_2 selectivity: 78). As shown in Figure 10.5b, our group introduced borate and polyethyleneimine molecules to precisely manipulate the nanochannels between the MXene nanosheets [17]. Two types of CO_2-philic molecules were introduced to cross-link MXene nanosheets, which effectively controls the gas transport behavior in the membrane. The mechanism within these nano-confined MXene films are investigated, revealing the transformation from "diffusion-controlled" to "solution-controlled" channels after chemical tuning. As a result, the as-fabricated MXene membranes were converted from H_2-selective permeation to CO_2-selective permeation with CO_2 permeance of 350 GPU and CO_2/CH_4 selectivity of 15.3.

Compared with GO, it is relatively difficult to synthesize other 2D nanosheets with few defects and assemble them into membranes for gas separation. For example, the MOF structure contains a large number of medium-strength coordination bonds. It is difficult to prepare 2D MOF nanosheets with a well-defined structure and a large aspect ratio without destroying the coordination bonds. On the other hand, it is very difficult to assemble soft MOF nanosheets into a dense and defect-free membrane. In 2014, Yang *et al.* [13] made a breakthrough in assembling 2D ZIF nanosheets into nanosheet membranes for H_2/CO_2 separation. As shown in Figure 10.6, an exfoliation strategy combining wet ball milling and ultrasonication was employed to successfully prepare $Zn_2(bim)_4$ nanosheets with good dispersion, intact structure, and large aspect ratio. A hot drop coating technology was used to stack the nanosheets on the porous alumina substrate in a compact and parallel orientation. The as-fabricated membrane achieved excellent H_2/CO_2 separation performance with a selectivity of 290 and H_2 permeance of 2790 GPU, as well as excellent thermal and hydrothermal stability.

Graphitic carbon nitride (g-C_3N_4) nanosheets possess suitable pore size, high pore density in plane, and CO_2-philic property due to the rich amine groups, which have attracted significant attention. To exfoliate bulky g-C_3N_4 into nanosheets, the interlayer interaction needs to be broken by physical or chemical methods. However, in this exfoliation process, the in-plane atomic structure is simultaneously destroyed with large nonselective defects owing to the weak bond of N-$(C)_3$ between polymerized tri-s-triazine units, which greatly hinders the application of g-C_3N_4 in gas separation. For the first time, we synthesized g-C_3N_4 nanosheets with tunable CO_2 adsorption properties and nanostructures [40], and incorporated into PEBA membrane for CO_2 separation. The results demonstrated that

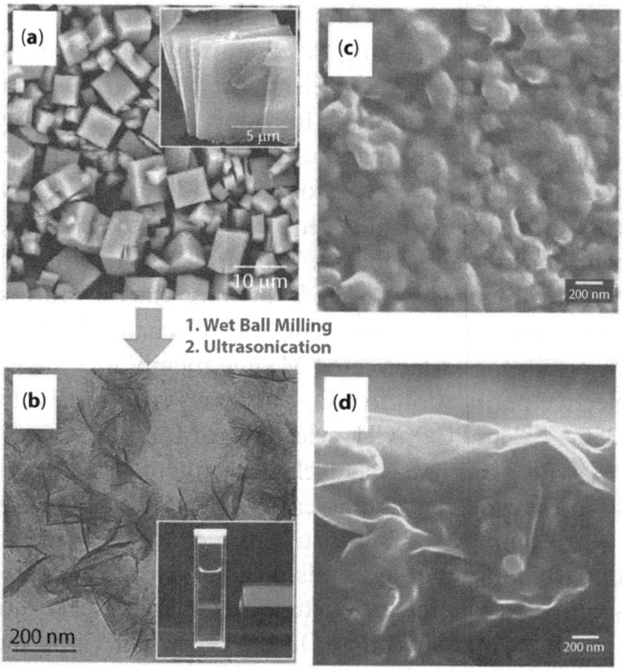

Figure 10.6 (a) Scanning electron microscopy (SEM) image of as-synthesized $Zn_2(bim)_4$ crystals. The inset image shows the typical flake-like morphology of $Zn_2(bim)_4$ crystals. (b) Transmission electron microscopy (TEM) image of $Zn_2(bim)_4$ MSNs. The inset shows the Tyndall effect of a colloidal suspension. (c) SEM top view and d) cross-sectional view of a $Zn_2(bim)_4$ nanosheet layer on α-Al_2O_3 support.

the g-C_3N_4 nanosheets produced from dicyandiamide and undergoing 4 h of thermal etching (DCN-4 nanosheets) showed the optimal CO_2 sorption and sieving properties. The membrane with 0.25 wt% DCN-4 nanosheets exhibited simultaneous enhancement in CO_2 permeance and CO_2/N_2 selectivity compared with pure PEBA membrane. Zhou *et al.* [41] have successfully fabricated g-C_3N_4 nanosheet membranes with high hydrogen permselectivity using a facile GO-modified strategy. Through the covalent interactions between the terminal -NH_2/-NH in g-C_3N_4 and -COOH in GO, the non-selective defects in g-C_3N_4 membranes were well remedied, contributing to excellent H_2/CO_2 selectivity.

10.4 Applications for Gas Separation

10.4.1 Hydrogen Purification

Hydrogen is an important chemical raw material and one of the main energy sources in the future. With the rapid development of organic synthesis, semiconductors, petrochemicals, fuel cells, etc., the demand for hydrogen in all fields is also increasing. The separation and purification technology of hydrogen has also attracted more and more attention. 2D-material membranes with precisely tuned gas transport channels have exhibited huge potential in H_2 purification.

Celebi *et al.* [42] developed an optimized CVD growth method followed by a focused ion beam approach to prepare bilayer graphene with narrowly distributed pore sizes ranging from < 10 nm to 1 μm and high-density nanopores (~10^3 to 10^6 per membrane). As shown in Figure 10.7a, the H_2 permeance of the porous graphene membrane was as high as ~10^2 mol m^{-2}

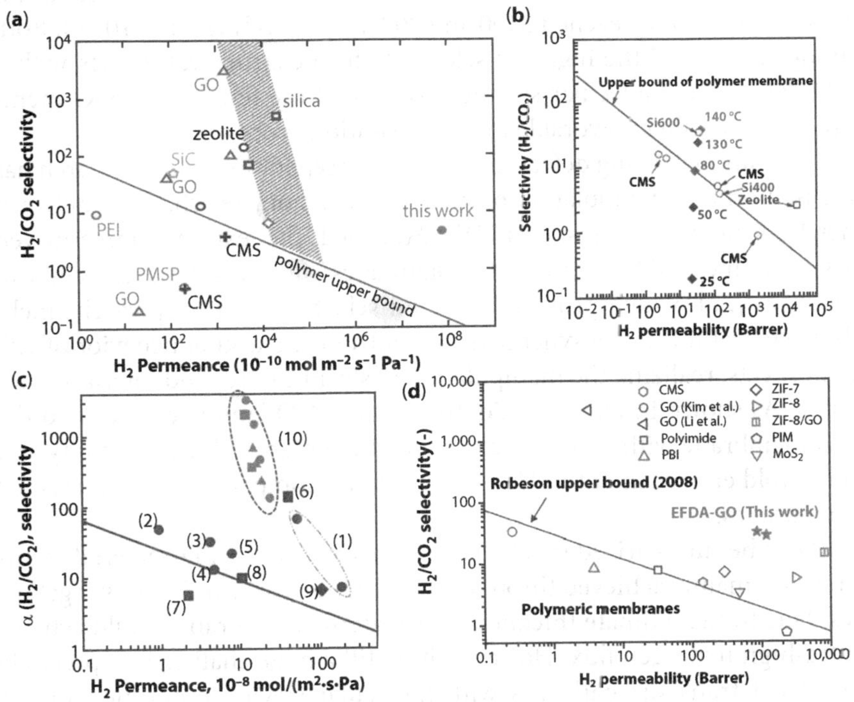

Figure 10.7 Comparison of H_2/CO_2 gas-separation performances of (a) the porous graphene membranes (7.6 nm pore diameter, with 4.0% porosity), (b) and (c) GO membranes, (d) EFDA-GO membranes and other membranes.

s^{-1} Pa^{-1}, which was three orders of magnitude higher than those of inorganic, polymeric, GO, and composite membranes. The selectivity is comparable to some conventional polymer or CMS membranes. For certain applications, permeance becomes more crucial than selectivity, in which large amounts of gases with comparable molar ratios are separated.

Kim *et al.* [24] found that GO membranes with higher permeance can be prepared by simple heat treatment to create pores on the basal plane. Heat treatment at 140 °C will cause GO to decompose and release primarily water and CO_2, and CO and CO_2 are released from oxygen-containing functional groups on basal planes or at edges, resulting in irreversible defects. The H_2 and CO_2 permeance of the membrane was tested under a pressure of 1 bar. Below 130 °C, H_2 permeance increased gradually while CO_2 permeance decreased steadily. At ~130 °C to 140 °C, H_2 permeance increased abruptly and CO_2 permeance increased slightly. The H_2/CO_2 permselectivity at 140 °C was as high as 40, as shown in Figure 10.7b, which is one of the highest values reported for other comparable membranes. Yu *et al.* [26] used vacuum filtration to prepare ultra-thin GO membranes with a thickness of approximately 1.8 nm. The separation selectivity of these membranes for H_2/CO_2 and H_2/N_2 reached 3400 and 900, respectively (Figure 10.7c). Both groups attributed the high gas selectivity to the structural defects within GO flakes. The molecular sieving property of the few-layered GO membranes makes them favorable for H_2 separation processes.

In addition to using defects on GO for gas separation, more research has focused on the manipulation of the 2D sub-nanometer channels between neighboring GO nanosheets. We proposed the external force-driven assembly method to fabricate GO laminates with an interlayer height of ~0.4 nm [27], which provided fast and selective gas transport channels. The external forces are synergistic to control the stacking behavior of GO nanosheets, realizing the manipulation of slit-like pores and interlayer galleries. As shown in Figure 10.7d, the ordered GO laminates endowed the GO membrane with 2–3 orders of magnitude higher H_2 permeability and three-fold enhancement in H_2/CO_2 selectivity compared with commercial membranes.

It can be summarized that hydrogen purification via 2D-material membranes is mainly achieved through the in-plane defects or interlayer galleries. Due to the ultimate thickness, 2D-material membranes could achieve ultra-high hydrogen flux. However, it is still a huge challenge to precisely construct transport pathways with molecular sieving properties for gas separation.

10.4.2 CO_2 Capture

The sustainable development of the environment encourages people to pay attention to energy conservation and emission reduction, especially carbon emissions in various industrial processes. Therefore, CO_2 capture in pre-combustion processes (CO_2/H_2) or separating CO_2 from flue gases (CO_2/N_2) is very important. In addition, CO_2 separation is important in natural gas processing and biogas upgrading (CO_2/CH_4) [43, 44].

Ammonia absorption is currently the most mature CO_2 separation and capture technology, but the regeneration of absorbents requires huge costs [44]. Among the many alternative technologies, membrane separation has obvious advantages in terms of floor space, energy consumption, and cost, making it the most promising. However, the performance of the current membranes still needs to be further improved particularly to meet the targeted performance for industrial application and commercialization. For large applications of CO_2 removal from natural gas, CO_2 permeance is targeted at over 100 GPU with CO_2/CH_4 selectivity in the range of 20–30. For CO_2 capture from flue gas, the membrane is expected to achieve CO_2 permeance over 1000–5000 GPU and CO_2/N_2 selectivity of 30–50 [45].

Kim *et al.* [24] evaluated the gas transport behavior of GO membranes (prepared by method one mentioned above) by changing the relative humidity of the feed. They found that the permeation rate of nearly all gases decreased with increasing the feed humidity, but CO_2 permeance increased by nearly 50 times in the presence of water, thereby increasing the selectivity of CO_2/gas. GO membranes prepared by method two formed highly interlocked structures, leading to lower gas permeance but higher selectivity compared with those prepared by method one. The CO_2 intrinsic permeation rate of the thin GO membrane is ~120 GPU, and the CO_2/N_2 selectivity is ~20. As shown in Figure 10.8a, the gas permeances decreased gradually as the humidity increased, but the CO_2 permeance remained constant, therefore, the selectivity of CO_2/gas is gradually increased.

Our group proposed to employ favorable molecular interactions between GO and polymers to stimulate the stacking of GO nanosheets into well-defined laminates [31], and the interlayer space can serve as molecular sieving channels. Moreover, the oxygen-containing groups on GO nanosheets can interact with CO_2 to increase the affinity of the membrane to CO_2. In addition to straight diffusion pathways, strong CO_2 affinity and molecular sieving properties contribute to highly permeable and selective membranes. As shown in Figure 10.8b, membranes with GO concentrations of 0.05, 0.075, and 0.1 wt% were fabricated and denoted as GO-05,

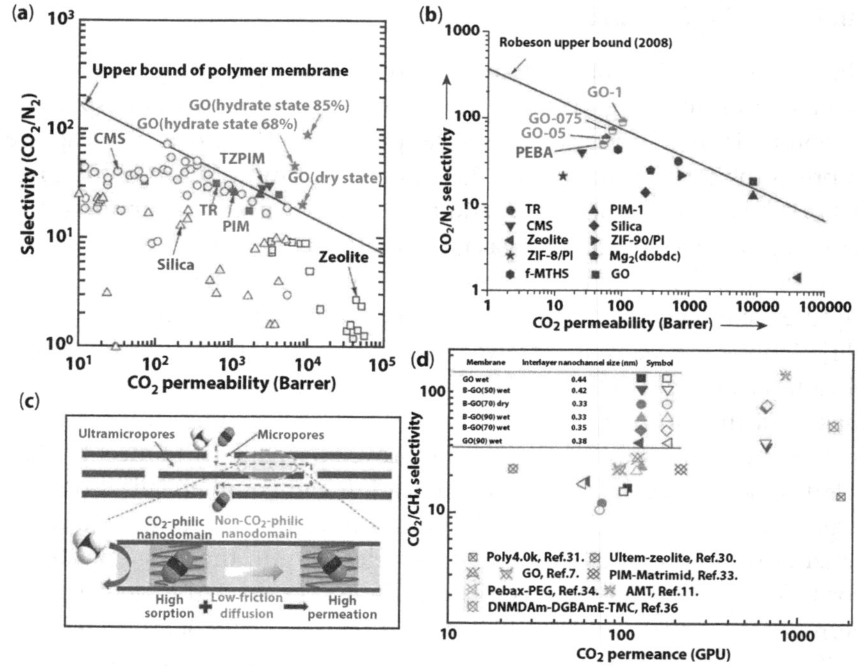

Figure 10.8 (a) Relation between CO_2 permeability and CO_2/N_2 selectivity of method two GO membranes under dry and humidified conditions. (b) CO_2/N_2 separation performance of PEBA, GO-05, GO-075, and GO-1 membranes and representative membranes reported in the literature. (c) Representation of gas molecules transport in the GO stacking membranes channels containing both CO_2-philic and non-CO_2-philic nanodomains. (d) CO_2/CH_4 separation performance of the B-GO membranes.

GO-075, and GO-1, respectively. With the increase of GO content, CO_2 permeability and CO_2/N_2 selectivity were simultaneously enhanced. The CO_2/N_2 separation performance of optimized GO membranes transcends the upper bound for polymeric membranes.

In addition to using GO as a filler for mixed matrix membranes, cross-linking agents with affinity to CO_2 can also be introduced into GO layers. Jiang *et al.* [46] incorporated poly(ethylene glycol) diamines (PEGDA) into GO layers to prepare MMM for gas separation. As shown in Figure 10.8c, PEGDA reacts with the epoxy groups on the GO surface to produce a CO_2-philic area, which can increase CO_2 adsorption, while the unreacted area serves as a non-CO_2-philic fast transport channel. Owing to the orderly stacking of nanochannels through cross-linking and the heterogeneous nanodomains with moderate CO_2 affinity, a GO-PEGDA500 membrane exhibits a high CO_2 permeance of 175.5 GPU and a CO_2/CH_4

selectivity of 69.5. Yu *et al.* [29] grafted molecular brushes of piperazine onto the GO layer to prepare a 20-nm GO hollow fiber membrane by a facile coating process. The membrane exhibits excellent separation performance under simulated flue gas conditions with CO_2 permeance of 1020 GPU and CO_2/N_2 selectivity as high as 680.

Inspired by aquaporin, Jiang *et al.* [28] prepared an ultra-thin GO membrane cross-linked with boric acid. By fine-tuning the interlayer nanochannels of GO through borate cross-linking and associated water of hydration, nanochannels with favorable physical microenvironments and distinct size sieving properties are constructed, which allow the facilitated transport of small CO_2 molecules while blocking the passage of slightly larger molecules such as CH_4 and N_2. A favorable water environment is also sustained by humidified feed gas and the hydrophilic nature of the modified GO nanochannels. As shown in Figure 10.8d, the as-prepared membrane exhibited a high CO_2 permeance of 650 GPU, and a CO_2/CH_4 selectivity of 75, which is currently the best performance reported for GO-based membranes.

Although CO_2 is a nonpolar gas, the polarity of the individual C=O bonds in the molecule allows for interaction with polar groups in the membrane. Thus, CO_2 can act as a Lewis acid or a Lewis base and can participate in hydrogen bonding [24, 47]. This allows us to have a variety of means to regulate the transport behavior of CO_2. Different from hydrogen purification, in addition to constructing sub-nanometer channels with molecular sieving properties, 2D-material membranes for CO_2 capture use the affinity to CO_2 to enhance CO_2 adsorption, realizing its preferential permeation. Humidified feed is often utilized to promote the reversible reaction of CO_2 with the groups on the membrane, leading to enhanced separation performance [47]. Although there has been a lot of research on 2D material membranes for CO_2 separation, only a few of them possess the separation performance required for industrial application.

10.4.3 More Challenging Gas Mixtures

In addition to the above gas pairs, 2D-material membranes are also applied in more challenging gas pairs, such as O_2/N_2 separation and olefin/paraffin separation. These gas pairs have a very small difference in molecular dynamic diameter (O_2: 0.346 nm, N_2: 0.364 nm; C_2H_4: 0.416 nm, C_2H_6: 0.444 nm), so it is difficult to efficiently separate them.

The few-layer graphene membrane prepared by Kim *et al.* [24] on a porous Poly(1-methylsilyl-1-propyne) (PTMSP) substrate can effectively separate O_2 and N_2. They found that as the number of graphene layers increases, the permeation of O_2 decreases, and the selectivity of O_2/N_2

increases. The O_2/N_2 selectivity increased from 1.5 (for PTMSP) to 6 (for G5/PTMSP), whereas the O_2 permeability decreased from 730 to 29 Barrer, surpassing the O_2/N_2 upper bound of polymeric membranes with a selectivity similar to that of carbon molecular sieve (CMS) membranes. The increase in selectivity of the G#/PTMSP films, with increasing the number of graphene layers deposited, indicates the formation of slit-like interlayer spacing with sizes similar to those in CMS membranes.

Qi *et al.* [48] prepared GO membranes on porous stainless steel hollow fibers (PSSHF) by electrophoresis deposition (ED) method. They reduced GO *in situ* during the membrane formation process. They obtained a reduced interlayer spacing of approximately 0.36 nm, which allowed molecules (H_2, CO_2, N_2, CH_4, C_2H_4, and C_2H_6) with a kinetic diameter of less than 0.39 nm to pass, but provided a molecular sieving cut-off between the permeation of C_2 and C_3 hydrocarbons due to the larger kinematic diameter of propane. As shown in Figures 10.9a and 10.9b, the ideal selectivities for C_2/C_3 hydrocarbons were 551 (C_2H_4/C_3H_8), 319 (C_2H_4/C_3H_6), 332 (C_2H_6/C_3H_8), and 192 (C_2H_6/C_3H_6).

Inspired by the hierarchical structure and facilitated gas transport of biological membranes, Jiang *et al.* [49] prepared a highly selective ethylene/ethane separation membrane. They introduced the silver carrier and ionic liquid (IL) into GO laminates. By adjusting the amount of impregnated IL and GO-IL/carrier interactions, the in-plane nano-wrinkles can be generated and further tailored for enhancing the ethylene/ethane sieving property of GO membranes; meanwhile, the carriers fixed within 2D nanochannels exhibit high activity, thus greatly facilitating ethylene transport. As shown in Figures 10.9c and 10.9d, by virtue of synergistic effects of molecular sieving and facilitated transport, an unprecedented combination of high ethylene permeance (72.5 GPU) and superhigh ethylene/ethane selectivity (215) is achieved, overwhelming currently reported advanced membranes. Moreover, a continuous permeation test was carried out for 200 h, the membrane exhibited stable gas permeance and selectivity. Similarly, they also reported on boron nitride/ionic liquid membranes, which exhibit an unprecedented separation performance with superhigh C_2H_4 permeance of 138 GPU and C_2H_4/C_2H_6 selectivity of 128 [50].

Two-dimensional materials, especially graphene oxide, can form 2D sub-nanometer channels via the ordered stacking of nanosheets. Therefore, by precisely manipulating the size of interlayer channels, ethylene/ethane can be sieved via "slit-like" pores similar to those in rigid CMS [51]. However, it is still a huge challenge to precisely tune the interlayer channels of graphene oxide membranes at the sub-angstrom scale. It is a relatively

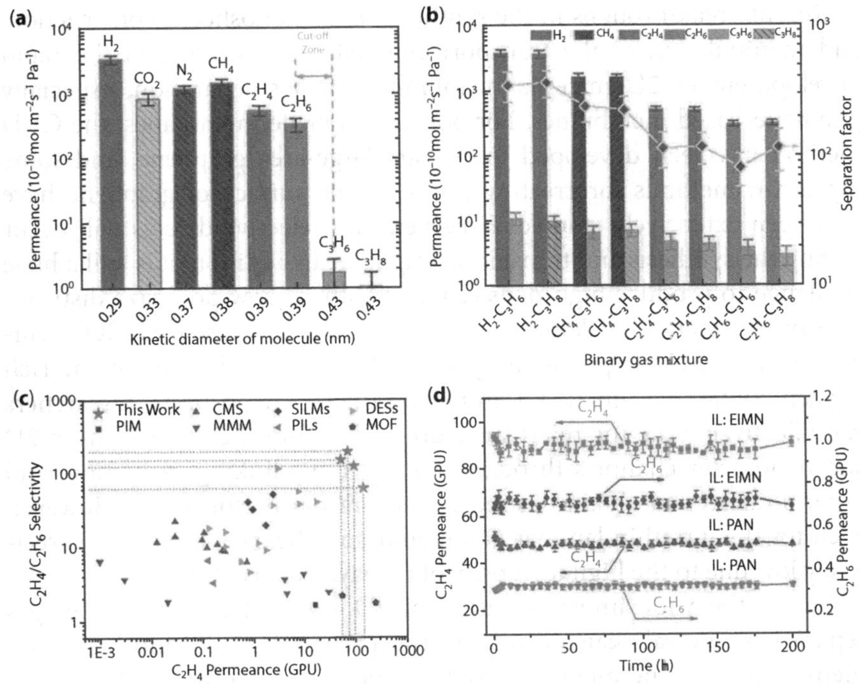

Figure 10.9 Gas permeation performance of ED-GO@PSSHF composite membrane. (a) Single-gas measurements of small gas molecules and light hydrocarbon permeation through ED-GO@PSSHF membrane. (b) Separation performance of ED-GO@PSSHF membrane for binary gas mixtures. (c) Comparison of C_2H_4/C_2H_6 separation performance with different types of membranes reported during the period 2010-2019. (d) Long-term stability for C_2H_4/C_2H_6 separation of Ag/IL-GO membranes with IL content of 33 wt%.

common method to introduce carriers that promote the permeation of target molecules, such as silver ions [52].

10.5 Conclusions and Perspectives

In summary, compared with the upper bound of state-of-the-art polymeric membranes, 2D-material membranes exhibit excellent separation performance and have been considered as one of the most promising membranes for gas separation. The ultrathin thickness minimizes gas transport resistance in the membrane, and the in-plane pores or interlayer galleries can provide transport channels with molecular sieving properties, contributing to ultra-high gas permeance and separation selectivity at the same time.

Recent breakthroughs in the synthesis of 2D nanosheets, construction, and manipulation of the transport channels have stimulated the rapid development of 2D material membranes for gas separation, especially graphene-based membranes. For porous graphene membranes, the CVD method has been developed to prepare large-area graphene, and some top-down methods for creating pores on the surface of graphene have also been extensively studied. However, non-selective defects will occur during the synthesis and transfer of graphene. In addition, it is still a huge challenge to produce high-density pores with narrow pore size distribution on graphene. For laminar membranes, centimeter-sized GO membranes can be easily prepared by a variety of methods. Due to the rich oxygen-containing functional groups on the GO nanosheets, researchers can effectively tune the microstructure and chemical environment of 2D sub-nanometer channels through grafting, cross-linking, reduction, and other strategies to obtain desired gas separation performance. However, even for the ultrathin laminar GO membrane, the gas permeance is relatively low due to the high tortuosity of the layered structure.

Before the wide implementation of 2D material membranes in gas separation, several issues involving the preparation of nanosheets and membranes and the membrane stability need to be solved: i) More economical, efficient, and universal strategies for synthesizing 2D nanosheets with well-defined structures need to be explored. ii) Methods for precisely constructing and manipulating 2D sub-nanometer channels with molecular sieving properties need to be investigated. Moreover, the methods for fabricating m^2-level membranes urgently need to be proposed to promote the practical application of 2D-material membranes. iii) The performance stability under harsh operation conditions (e.g., high temperature, high pressure, and high humidity) must be considered as well. iv) The transport behaviors of gas molecules in membranes are mainly described by the solution-diffusion mechanism. Considering the unique 2D sub-nanometer channels within the membranes, the model of confined mass transfer should be updated to describe the particularly confined transport behavior through 2D-material membranes accurately based on advanced characterization and simulation technology.

References

1. Bakonyi, P., Nemestothy, N., Belafi-Bako, K., Biohydrogen purification by membranes: An overview on the operational conditions affecting the performance of non-porous, polymeric and ionic liquid based gas separation membranes. *Int. J. Hydrogen. Energ.*, 38, 23, 9673–9687, 2013.
2. Baker, R.W., Future directions of membrane gas separation technology. *Ind. Eng. Chem. Res.*, 41, 6, 1393–1411, 2002.
3. Goh, P.S., Ismail, A.F., Sanip, S.M., Ng, B.C., Aziz, M., Recent advances of inorganic fillers in mixed matrix membrane for gas separation. *Sep. Purif. Technol.*, 81, 3, 243–264, 2011.
4. Bernardo, P., Drioli, E., Golemme, G., Membrane Gas Separation: A Review/State of the Art. *Ind. Eng. Chem. Res.*, 48, 10, 4638–4663, 2009.
5. Robeson, L.M., The upper bound revisited. *J. Membr. Sci.*, 320, 1-2, 390–400, 2008.
6. Geim, A.K. and Novoselov, K.S., The rise of graphene. *Nat. Mater.*, 6, 3, 183–191, 2007.
7. Li, D. and Kaner, R.B., Materials science - Graphene-based materials. *Science*, 320, 5880, 1170–1171, 2008.
8. Geim, A.K. and Grigorieva, I.V., Van der Waals heterostructures. *Nature*, 499, 7459, 419–425, 2013.
9. Dubertret, B., Heine, T., Terrones, M., The Rise of Two-Dimensional Materials. *Acc. Chem. Res.*, 48, 1, 1–2, 2015.
10. Xu, M.S., Liang, T., Shi, M.M., Chen, H.Z., Graphene-Like Two-Dimensional Materials. *Chem. Rev.*, 113, 5, 3766–3798, 2013.
11. Liu, G.P., Jin, W.Q., Xu, N.P., Two-Dimensional-Material Membranes: A New Family of High-Performance Separation Membranes. *Angew Chem. Int. Edit.*, 55, 43, 13384–13397, 2016.
12. Liu, G.P., Jin, W.Q., Xu, N.P., Graphene-based membranes. *Chem. Soc. Rev.*, 44, 15, 5016–5030, 2015.
13. Peng, Y., Li, Y.S., Ban, Y.J., Jin, H., Jiao, W.M., Liu, X.L., Yang, W.S., Metal-organic framework nanosheets as building blocks for molecular sieving membranes. *Science*, 346, 6215, 1356–1359, 2014.
14. Liu, Y., Wang, N.Y., Cao, Z.W., Caro, J., Molecular sieving through interlayer galleries. *J. Mater. Chem. A*, 2, 5, 1235–1238, 2014.
15. Achari, A., Sahana, S., Eswaramoorthy, M., High performance MoS2 membranes: effects of thermally driven phase transition on CO2 separation efficiency. *Energ. Environ. Sci.*, 9, 4, 1224–1228, 2016.
16. Ding, L., Wei, Y.Y., Li, L.B., Zhang, T., Wang, H.H., Xue, J., Ding, L.X., Wang, S.Q., Caro, J., Gogotsi, Y., MXene molecular sieving membranes for highly efficient gas separation. *Nat. Commun.*, 9, 155, 2018.

17. Shen, J., Liu, G.Z., Ji, Y.F., Liu, Q., Cheng, L., Guan, K.C., Zhang, M.C., Liu, G.P., Xiong, J., Yang, J., Jin, W.Q., 2D MXene Nanofilms with Tunable Gas Transport Channels. *Adv. Funct. Mater.*, 28, 31, 1801511, 2018.
18. Berry, V., Impermeability of graphene and its applications. *Carbon*, 62, 1–10, 2013.
19. Bunch, J.S., Verbridge, S.S., Alden, J.S., van der Zande, A.M., Parpia, J.M., Craighead, H.G., McEuen, P.L., Impermeable atomic membranes from graphene sheets. *Nano Lett.*, 8, 8, 2458–2462, 2008.
20. Jiang, D.E., Cooper, V.R., Dai, S., Porous Graphene as the Ultimate Membrane for Gas Separation. *Nano Lett.*, 9, 12, 4019–4024, 2009.
21. Yoo, B.M., Shin, J.E., Lee, H.D., Park, H.B., Graphene and graphene oxide membranes for gas separation applications. *Curr. Opin. Chem. Eng.*, 16, 39–47, 2017.
22. Koenig, S.P., Wang, L.D., Pellegrino, J., Bunch, J.S., Selective molecular sieving through porous graphene. *Nat. Nanotechnol.*, 7, 11, 728–732, 2012.
23. Zhao, J., He, G., Huang, S., Villalobos, L.F., Dakhchoune, M., Bassas, H., Agrawal, K.V., Etching gas-sieving nanopores in single-layer graphene with an angstrom precision for high-performance gas mixture separation. *Sci. Adv.*, 5, 1, eaav1851, 2019.
24. Kim, H.W., Yoon, H.W., Yoon, S.M., Yoo, B.M., Ahn, B.K., Cho, Y.H., Shin, H.J., Yang, H., Paik, U., Kwon, S., Choi, J.Y., Park, H.B., Selective Gas Transport Through Few-Layered Graphene and Graphene Oxide Membranes. *Science*, 342, 6154, 91–95, 2013.
25. Dreyer, D.R., Park, S., Bielawski, C.W., Ruoff, R.S., The chemistry of graphene oxide. *Chem. Soc. Rev.*, 39, 1, 228–240, 2010.
26. Li, H., Song, Z.N., Zhang, X.J., Huang, Y., Li, S.G., Mao, Y.T., Ploehn, H.J., Bao, Y., Yu, M., Ultrathin, Molecular-Sieving Graphene Oxide Membranes for Selective Hydrogen Separation. *Sci.*, 342, 6154, 95–98, 2013.
27. Shen, J., Liu, G.P., Huang, K., Chu, Z.Y., Jin, W.Q., Xu, N.P., Subnanometer Two-Dimensional Graphene Oxide Channels for Ultrafast Gas Sieving. *ACS Nano*, 10, 3, 3398–3409, 2016.
28. Wang, S.F., Wu, Y.Z., Zhang, N., He, G.W., Xin, Q.P., Wu, X.Y., Wu, H., Cao, X.Z., Guiver, M.D., Jiang, Z.Y., A highly permeable graphene oxide membrane with fast and selective transport nanochannels for efficient carbon capture. *Energ. Environ. Sci.*, 9, 10, 3107–3112, 2016.
29. Zhou, F.L., Tien, H.N., Xu, W.W.L., Chen, J.T., Liu, Q.L., Hicks, E., Fathizadeh, M., Li, S.G., Yu, M., Ultrathin graphene oxide-based hollow fiber membranes with brush-like CO2-philic agent for highly efficient CO2 capture. *Nat. Commun.*, 8, 2107, 2017.
30. Cheng, L., Guan, K.C., Liu, G.P., Jin, W.Q., Cysteamine-crosslinked graphene oxide membrane with enhanced hydrogen separation property. *J. Membr. Sci.*, 595, 117568, 2020.

31. Shen, J., Liu, G.P., Huang, K., Jin, W.Q., Lee, K.R., Xu, N.P., Membranes with Fast and Selective Gas-Transport Channels of Laminar Graphene Oxide for Efficient CO2 Capture. *Angew Chem. Int. Edit.*, 54, 2, 578–582, 2015.
32. Shen, J., Zhang, M.C., Liu, G.P., Guan, K.C., Jin, W.Q., Size Effects of Graphene Oxide on Mixed Matrix Membranes for CO2 Separation. *Aiche J.*, 62, 8, 2843–2852, 2016.
33. Li, X.Q., Cheng, Y.D., Zhang, H.Y., Wang, S.F., Jiang, Z.Y., Guo, R.L., Wu, H., Efficient CO2 Capture by Functionalized Graphene Oxide Nanosheets as Fillers To Fabricate Multi-Permselective Mixed Matrix Membranes. *ACS Appl. Mater. Inter.*, 7, 9, 5528–5537, 2015.
34. Dong, G., Hou, J., Wang, J., Zhang, Y., Chen, V., Liu, J., Enhanced CO2/N2 separation by porous reduced graphene oxide/Pebax mixed matrix membranes. *J. Membr. Sci.*, 520, 860–868, 2016.
35. Ha, H., Park, J., Ando, S., Kim, C.B., Nagai, K., Freeman, B.D., Ellison, C.J., Gas permeation and selectivity of poly(dimethylsiloxane)/graphene oxide composite elastomer membranes. *J. Membr. Sci.*, 518, 131–140, 2016.
36. Peng, Y., Li, Y.S., Ban, Y.J., Yang, W.S., Two-Dimensional Metal-Organic Framework Nanosheets for Membrane-Based Gas Separation. *Angew Chem. Int. Edit.*, 56, 33, 9757–9761, 2017.
37. Rodenas, T., Luz, I., Prieto, G., Seoane, B., Miro, H., Corma, A., Kapteijn, F., Xamena, F.X.L.I., Gascon, J., Metal-organic framework nanosheets in polymer composite materials for gas separation. *Nat. Mater.*, 14, 1, 48–55, 2015.
38. Liu, Y.T., Wu, H., Min, L.F., Song, S.Q., Yang, L.X., Ren, Y.X., Wu, Y.Z., Zhao, R., Wang, H.J., Jiang, Z.Y., 2D layered double hydroxide membranes with intrinsic breathing effect toward CO2 for efficient carbon capture. *J. Membr. Sci.*, 598, 117663, 2020.
39. Xu, X.Z., Wang, J.J., Zhou, A.W., Dong, S.Y., Shi, K.Q., Li, B.A., Han, J.B., O'Hare, D., High-efficiency CO2 separation using hybrid LDH-polymer membranes. *Nat. Commun.*, 12, 1, 3069, 2021.
40. Cheng, L., Song, Y.Y., Chen, H.M., Liu, G.Z., Liu, G.P., Jin, W.Q., g-C3N4 nanosheets with tunable affinity and sieving effect endowing polymeric membranes with enhanced CO2 capture property. *Sep. Purif. Technol.*, 250, 117200, 2020.
41. Zhou, Y.S., Zhang, Y., Xue, J., Wang, R., Yin, Z.J., Ding, L., Wang, H.H., Graphene oxide-modified g-C3N4 nanosheet membranes for efficient hydrogen purification. *Chem. Eng. J.*, 420, 129574, 2021.
42. Celebi, K., Buchheim, J., Wyss, R.M., Droudian, A., Gasser, P., Shorubalko, I., Kye, J.I., Lee, C., Park, H.G., Ultimate Permeation Across Atomically Thin Porous Graphene. *Science*, 344, 6181, 289–292, 2014.
43. D'Alessandro, D.M., Smit, B., Long, J.R., Carbon Dioxide Capture: Prospects for New Materials. *Angew Chem. Int. Edit.*, 49, 35, 6058–6082, 2010.
44. Prasetya, N., Himma, N.F., Sutrisna, P.D., Wenten, I.G., Ladewig, B.P., A review on emerging organic-containing microporous material membranes for carbon capture and separation. *Chem. Eng. J.*, 391, 123575, 2020.

45. Galizia, M., Chi, W.S., Smith, Z.P., Merkel, T.C., Baker, R.W., Freeman, B.D., 50th Anniversary Perspective: Polymers and Mixed Matrix Membranes for Gas and Vapor Separation: A Review and Prospective Opportunities. *Macromolecules*, 50, 20, 7809–7843, 2017.
46. Wang, S.F., Xie, Y., He, G.W., Xin, Q.P., Zhang, J.H., Yang, L.X., Li, Y.F., Wu, H., Zhang, Y.Z., Guiver, M.D., Jiang, Z.Y., Graphene Oxide Membranes with Heterogeneous Nanodomains for Efficient CO2 Separations. *Angew Chem. Int. Edit.*, 56, 45, 14246–14251, 2017.
47. Park, S.C., Chae, I.S., Moon, G.H., Kim, B.S., Jang, J., Wessling, M., Kang, Y.S., Lewis acidic water as a new carrier for facilitating CO2 transport. *J. Mater. Chem. A*, 7, 10, 5190–5194, 2019.
48. Qi, B.Y., He, X.F., Zeng, G.F., Pan, Y.C., Li, G.H., Liu, G.J., Zhang, Y.F., Chen, W., Sun, Y.H., Strict molecular sieving over electrodeposited 2D-interspacing-narrowed graphene oxide membranes. *Nat. Commun.*, 8, 825, 2017.
49. Dou, H.Z., Xu, M., Jiang, B., Wen, G.B., Zhao, L., Wang, B.Y., Yu, A.P., Bai, Z.Y., Sun, Y.L., Zhang, L.H., Chen, Z.W., Jiang, Z.Y., Bioinspired Graphene Oxide Membranes with Dual Transport Mechanisms for Precise Molecular Separation. *Adv. Funct. Mater.*, 29, 50, 1905229, 2019.
50. Dou, H.Z., Jiang, B., Xu, M., Zhang, Z., Wen, G.B., Peng, F.F., Yu, A.P., Bai, Z.Y., Sun, Y.L., Zhang, L.H., Jiang, Z.Y., Chen, Z.W., Boron Nitride Membranes with a Distinct Nanoconfinement Effect for Efficient Ethylene/Ethane Separation. *Angew Chem. Int. Edit.*, 58, 39, 13969–13975, 2019.
51. Rungta, M., Zhang, C., Koros, W.J., Xu, L.R., Membrane-based ethylene/ethane separation: The upper bound and beyond. *Aiche J.*, 59, 9, 3475–3489, 2013.
52. Campos, A.C.C., dos Reis, R.A., Ortiz, A., Gorri, D., Ortiz, I., A Perspective of Solutions for Membrane Instabilities in Olefin/Paraffin Separations: A Review. *Ind. Eng. Chem. Res.*, 57, 31, 10071–10085, 2018.

Index

Printed in the USA/Agawam, MA
September 18, 2026

910661.116